Mechanics of the Cell

David Boal

Simon Fraser University, British Columbia

CAMBRIDGE
UNIVERSITY PRESS

PUBLISHED BY THE PRESS SYNDICATE OF THE UNIVERSITY OF CAMBRIDGE
The Pitt Building, Trumpington Street, Cambridge, United Kingdom

CAMBRIDGE UNIVERSITY PRESS
The Edinburgh Building, Cambridge CB2 2RU, UK
40 West 20th Street, New York, NY 10011–4211, USA
477 Williamstown Road, Port Melbourne, VIC 3207, Australia
Ruiz de Alarcón 13, 28014 Madrid, Spain
Dock House, The Waterfront, Cape Town 8001, South Africa

http://www.cambridge.org

First published 2002
Reprinted 2003

Printed in the United Kingdom at the University Press, Cambridge

Typeface Times New Roman MT 9.5/13 pt.   *System* QuarkXPress™   [SE]

A catalogue record for this book is available from the British Library

Library of Congress Cataloguing in Publication data

Boal, David H.
  Mechanics of the cell / David H. Boal.
    p.   cm.
  Includes bibliographical references and index.
  ISBN 0 521 79258 4 (hardback) – ISBN 0 521 79681 4 (pbk.)
  1. Cells – Mechanical properties.   2. Physical biochemistry.   I. Title.
 QH645.5.B63   2001
 571.6′34–dc21   2001028709

ISBN 0 521 79258 4 hardback
ISBN 0 521 79681 4 paperback

Contents

Preface

The cells of our bodies represent a very large class of systems whose structural components often are both complex and soft. A system may be *complex* in the sense that it may comprise several components having quite different mechanical characteristics, with the result that the behavior of the system as a whole reflects an interplay between the characteristics of the components in isolation. Further, the individual structural units of biological systems tend to be *soft*: for example, the compression resistance of a protein network may be more than an order of magnitude lower than that of the air we breathe. While some of the mechanics relevant to such soft biomaterials has been established for more than a century, there are other aspects, for example the thermal undulations of fluid and polymerized sheets, that have been investigated only in the past few decades.

The general strategy of this text is first to identify common structural features of the cell, then to investigate its mechanical components in isolation, and lastly to assemble these components into simple cells. The first chapter introduces metaphors for the cell, describes its architecture and develops some intuition about the properties of soft materials. The remaining nine chapters are grouped into three sections. Parts I and II are devoted to biopolymers and membranes, respectively, taking a conventional reductionist approach, while acknowledging that the soft materials of the cell are anything but conventional. Part III applies our understanding of soft materials from Parts I and II to the problems of cell motion and the mechanics of complete, albeit simple, cells. We focus on equilibrium properties of cells, and only limited aspects of cell dynamics are considered.

Each chapter begins with an experimental view of the phenomena to be addressed, and a statement of the principal concepts to be developed. For the more mathematical topics, the reader may then skip to the last section of the chapter, without going through the traditional derivations, to see the results applied in a biological context. The approach is to keep the mathematical detail manageable without losing rigor. Care has been taken to choose topics that can be approached within a common theoretical framework; this has forced the omission of some phenomena where a disproportionate effort would be required to assemble the related theoretical machinery. Some of the longer proofs and other support material may be found at http://www.cambridge.org and http://www.sfu.ca/~boal.

Reflecting the multidisciplinary nature of biophysics, four appendices provide quick reviews of undergraduate material related to the text. For the physicist, Appendices A and B focus on animal cell types and the molecular composition of the cell's mechanical components, respectively. For the biologist, Appendices C and D introduce some core results from statistical mechanics and elasticity theory. The end-of-chapter problems are grouped according to (i) applications to biological systems, and (ii) formal extensions and supplementary derivations. The text is suitable for a one-semester course delivered to senior undergraduate and beginning graduate students with an interest in biophysics. Some formal results from the problem sets may provide additional lecture material for a more mathematical course aimed at graduate students in physics.

Financial support for my research comes from the Natural Sciences and Engineering Research Council of Canada. Thanks go to my colleagues Michael Plischke and Michael Wortis at Simon Fraser University and Myer Bloom and Evan Evans at the University of British Columbia for their continuing conversations and insight; in addition, the program on soft surfaces and interfaces of the Canadian Institute for Advanced Research has provided an excellent forum for the exchange of ideas. Gerald Lim deserves credit for thoroughly reading the manuscript and checking the problem sets. This project was started while I enjoyed the hospitality of Rashmi Desai and the Physics Department at the University of Toronto in 1997–98 and its completion has spanned a three-year period. I thank my wife Heather, along with our children Adrie and Alex, for their support and understanding when the task of writing this text inevitably spilled over into evenings and weekends.

The staff at Cambridge University Press, particularly Rufus Neal, have helped immensely in bringing focus to the book and placing its content in the larger context. However, the lingering errors, omissions and obfuscations are my responsibility, and I am always appreciative of suggestions for ways to improve the quality of the text.

Dave Boal
boal@sfu.ca

List of symbols

A, A_v	area, area per vertex of a network
$\mathbf{a}$	area element of a surface
a, a_o	interface area per molecule (a_o at equilibrium)
$\mathbf{b}_i, b$	monomer bond vector and length
B_{eff}	effective bond length
$b_{\alpha\beta}$	second fundamental form
C, C_o	curvature, spontaneous curvature
C_1, C_2	principal curvatures to a surface
C_m, C_p	curvature along, or perpendicular, to a meridian
C_{ijkl}	elastic moduli
C_{vdw}	van der Waals interaction parameter
$C(i,j)$	binomial coefficient
c	concentration, units of [*moles/volume*] or [*mass/volume*]
D	diffusion coefficient
D_f	filament diameter for rods or chains
D_s	distance between two membranes or plates
d	dimensionality of a system
d_{bl}, d_p, d_{sh}	thickness of a bilayer, plate or shell
E	energy
$\mathbf{E}$	electric field
E_{bind}	energy to separate an amphiphile from an aggregate
E_{sphere}, E_{disk}	energy of a spherical shell or flat disk
e	elementary unit of charge
$\mathbf{F}$	force
F_{buckle}	buckling force of a rod
$F, \mathcal{F}$	free energy, free energy density
F_{sol}	free energy of a solution phase
G	Gibb's free energy
$G(t)$	time-dependent relaxation modulus
$G'(\omega), G''(\omega)$	shear storage and loss moduli
$g, g_{\alpha\beta}$	metric, metric tensor

H, H_v	enthalpy, enthalpy per vertex
h	Planck's constant
$h(x, y)$	height of a surface in Monge representation
h_x, h_{xx}	first and second derivatives of $h(x, y)$ in the x-direction
$\mathcal{I}$	moment of inertia of cross section
K_A, K_V	compression modulus for area, volume
k_B	Boltzmann's constant
k_{on}, k_{off}	capture and release rates in polymerization
k_{sp}	spring constant
L	length of a rod
$\mathcal{L}(y)$	Langevin function
L_c	contour length of a filament
$\mathcal{L}_K$	Kuhn length of a polymer
ℓ_B	Bjerrum length
ℓ_{cc}	C–C distance projected on the axis of hydrocarbon chain
ℓ_D	Debye length
ℓ_{hc}	length of hydrocarbon chain along its axis
$\mathcal{M}$	bending moment
$[M]$	monomer concentration
$[M]_c, [M]_{ss}$	minimum $[M]$ for filament growth, $[M]$ at treadmilling
m	molecular mass
N, n	number of monomers in a polymer chain
N_K	number of Kuhn lengths in a polymer
n_c	number of carbon atoms in a hydrocarbon chain
$\mathbf{n}$	unit normal vector to a curve or surface
$\mathbf{n}_x, \mathbf{n}_y$	derivative of $\mathbf{n}$ in the x or y direction
P	pressure
$\mathcal{P}, \mathcal{P}(x), \mathcal{P}_r$	probabilities and probability densities
$\mathcal{P}_L, \mathcal{P}_R, \mathcal{P}_{net}$	probabilities of motion for thermal ratchets
$\mathbf{p}$	momentum
p	bond or site occupation probability on a lattice
p_C, p_R	connectivity and rigidity percolation thresholds
Q, q	electric charge
R, R_p	radius; pipette radius
R_{hc}	effective radius of a hydrocarbon chain
$R_v, R_v{}^*$	vesicle radius, minimum vesicle radius
R_1, R_2	principal radii of curvature
$\mathbf{r}$	position vector with Cartesian components (x, y, z)
S	entropy
S_{gas}	entropy of an ideal gas
S_{ijkl}	elastic constants
s	arc length

s, s_o	spring length (s_o at equilibrium)
T	temperature
$\mathbf{t}$	unit tangent vector to a curve
U^α	eigenvalues of strain tensor
$\mathbf{u}$	displacement vector in a deformation
u, u_{ij}	strain tensor
V	volume
$V(x)$	potential energy function
V_{mol}, V_{slab}	van der Waals potentials
V_o	volume of undeformed object
v_{ex}	excluded volume parameter of a polymer
v_{hc}	volume of a hydrocarbon chain
v_{red}	reduced volume parameter
W_{ad}	adhesion energy per unit area
w	width parameter of triangular probability distribution
Y	Young's modulus
Z	partition function
β	inverse temperature $(k_B T)^{-1}$
γ	surface tension
$\varepsilon, \varepsilon_o$	permittivity, permittivity of vacuum
η	viscosity
κ_f	flexural rigidity of polymer
κ_b, κ_G	bending rigidity (Gaussian bending rigidity) of a membrane
κ_{nl}	non-local bending resistance
λ	edge tension of bilayer
λ^*	minimum edge tension for membrane stability
λ_p	mass per unit length of polymer
$\Lambda, \Lambda_x, \Lambda_y, \Lambda_z$	deformation scaling parameter
μ_p, μ_s	pure, simple shear moduli
ν, ν_{Fl}	scaling exponent; Flory exponent for self-avoiding polymers
ξ_p	persistence length
Π_{exp}, Π_{coll}	stress at expansion and collapse transitions in 2D
ρ	number density (for chains or molecules)
ρ^*	transition density between dilute and semi-dilute solutions
ρ^{**}	transition density between semi-dilute and concentrated solutions
ρ_{agg}	critical aggregation density
ρ_{ch}	charge per unit volume
ρ_m	mass per unit volume
ρ_L	contour length of polymer per unit volume

ρ_N	transition density between isotropic and nematic phases
ρ_+, ρ_-, ρ_s	number densities of ions in solution
$\sigma_{ij}, \sigma_\theta, \sigma_z$	stress tensor; hoop and axial stress of a cylinder
σ_p	Poisson ratio
σ_s	charge per unit area
$\tau, \tau_\theta, \tau_z$	two-dimensional tension; hoop and axial tension of a cylinder
Ψ, ψ	electric potential

Chapter 1

Introduction to the cell

The number of cells in the human body is literally astronomical, about three orders of magnitude more than the number of stars in the Milky Way. Yet, for their immense number, the variety of cells is much smaller: only about 200 different cell types are represented in the collection of about 10^{14} cells that make up our bodies. These cells have diverse capabilities and, superficially, have remarkably different shapes, as illustrated in Fig. 1.1. Some cells, like certain varieties of bacteria, are not much more than inflated bags, shaped like the hot-air or gas balloons invented more than two centuries ago. Others, such as nerve cells, may have branched structures at each end connected by an arm that is more than a thousand times long as it is wide. The basic structural elements of most cells, however, are the same: fluid sheets enclose the cell and its compartments, while networks of filaments maintain the cell's shape and help organize its contents. Further, the chemical composition of these structural elements bears a strong family resemblance from one

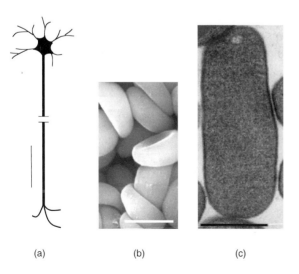

(a)   (b)   (c)

Fig. 1.1. Examples of cell shapes. (a) A neuron is a highly elongated cell usually with extensive branching where it receives sensory input or dispatches signals; indicated by the scale bar, its length may be hundreds of millimeters. (b) Mammalian red blood cells adopt a biconcave shape once they lose their nucleus and enter the circulatory system (bar is 4 μm; courtesy of Dr Elaine Humphrey, University of British Columbia). (c) Cylindrically shaped, the bacterium *Escherichia coli* has a complex boundary but little internal structure (bar is 0.9 μm; courtesy of Dr Terry Beveridge, University of Guelph). The image scale changes by two orders of magnitude from (a) to (c).

cell to another, perhaps reflecting the evolution of cells from a common ancestor; for example, the protein *actin*, which forms one of the cell's principal filaments, is found in organisms ranging from yeasts to humans.

The many chemical and structural similarities of cells tempt us to search for systematics in their architecture and components. We find that the structural elements of the cell are *soft*, in contrast to the hard concrete and steel of buildings and bridges. This is not a trivial observation: the mechanical properties of soft materials may be quite different from their hard, conventional counterparts and may reflect different microscopic origins. For instance, the fact that soft rubber becomes more resistant to stretching when heated, compared with the tendency of most materials to become more compliant, reflects the genesis of rubber elasticity in the variety of a polymer's molecular configurations. The theoretical framework for understanding soft materials, particularly flexible networks and membranes, has been assembled only in the past few decades, even though our experimental knowledge of soft materials goes back two centuries to the investigation of natural rubber by John Gough in 1805.

The functions performed by a cell can be looked upon from a variety of perspectives. Some functions are chemical, such as the manufacture of proteins, while others could be regarded as information processing, such as how a cell recognizes another cell as friend or foe. In this text, we concentrate on the *physical* attributes of cells, addressing questions such as the following.

- How does a cell maintain or change its shape? Some cells, such as the red blood cell, must be flexible enough to permit very large deformations, while others, such as plant cells, act cooperatively to produce a mildly stiff multicellular structure. What are the properties of the cell's components that are responsible for its strength and elasticity?
- How do cells move? Most cells are more than just inert bags, and some can actively change shape, permitting them to jostle past other cells in a tissue or locomote on their own. What internal structures of a cell are responsible for its movement?
- How do cells transport material internally? For most cells, especially meter-long nerve cells, diffusion is a slow and inefficient means of transporting proteins from their production site to their working site. What mechanisms, generating what forces, does a cell use for efficient transportation?
- How do cells stick together, as a multicellular organism such as ourselves, or how do they avoid adhering when it is unwanted? Do the thermal fluctuations of the cell's flexible membranes affect adhesion, or is it strictly a chemical process?

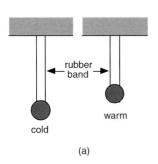

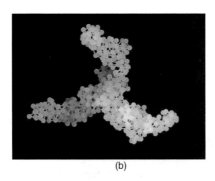

Fig. 1.2. (a) When natural rubber is heated, it becomes more resistant to stretching, an effect which can be demonstrated by hanging an object from an elastic band. When heated, the elastic band contracts, causing the object to rise. (b) A fluid membrane extends arms and fingers at large length scales, although it is smooth at short length scales (simulation from Boal and Rao, 1992a).

• What are the stability limits of the cell's components? A biological filament may buckle, or a membrane may tear, if subjected to strong enough forces. Are there upper and lower limits to the sizes of functioning cells?

To appreciate the mechanical operation of a whole cell, we must understand how its components behave both in isolation and as a composite structure. In the first two sections of this text, we treat the components individually, describing filaments and networks in Part I, followed by fluid and polymerized membranes in Part II. These two sections provide an experimental picture of the cell's structural elements and develop theoretical techniques to interpret and predict their mechanical characteristics. We demonstrate that many properties of soft materials are novel to the point of being counter-intuitive; as illustrated in Fig. 1.2, some networks may shrink or stiffen when heated, while fluid sheets form erratic arms and fingers over long distances. Both of these effects are driven by entropy, as we will establish below.

Although it is important to understand the individual behavior of the cell's components, it is equally important to assemble the components and observe how the cell functions as a whole. Often, a given structural element plays more than one role in a cell, and may act cooperatively with other elements to produce a desired result. In Part III, we examine several aspects of multicomponent systems, including cell mobility, adhesion and deformation. We close the text with a review of the forces at work in a cell, and some speculations on the limits of cell sizes imposed by the properties of their molecular constituents.

1.1 Designs for a cell

Although his own plans for Chicago skyscrapers were not devoid of decoration, architect Louis Sullivan (1856–1924) argued that functionally unnecessary embellishments detracted from a building's appeal. His celebrated dictum, "Form follows function", has found application in

Fig. 1.3. (a) An early fifteenth century merchant ship displays the efficient use of materials in the design of the reinforced hull and rigging (original illustration by fifteenth century engraver Israel von Mekenem; redrawn by Gordon Grant in Culver, 1992; ©1994 by Dover Publications). (b) Cross-linked filaments inside a nerve cell from a frog. Neurofilaments (running vertically) are 11 nm in diameter, compared with the cross-links with diameters of 4–6 nm. Bar is 0.1 μm (reprinted with permission from Hirokawa, 1982; ©1982 by the Rockefeller University Press).

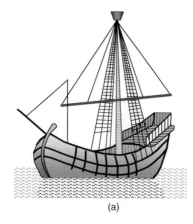

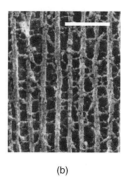

(a)   (b)

many areas beyond architecture and engineering, and is particularly obvious in the designs that evolution has selected for the cell.

Let's begin our discussion of the cell, then, by reviewing some strategies used in our own architectural endeavors, particularly in situations where functionality is demanded of minimal materials. Viewing the cell as a self-contained system, we look to the construction of boats, balloons and old cities for common design themes, although we recognize that none of these products of human engineering mimics a complete cell.

Thin membranes for isolating a cell's contents

Sailing ships, particularly older ships built when materials were scarce and landfalls for provisioning infrequent, face many of the same design challenges as cells. For example, both require that the internal workings of the system, including the crew and cargo in the case of a boat, be isolated in a controlled way from the system's environment. As illustrated by the merchant ship of Fig. 1.3(a), the naval architects of the fifteenth century opted for complex, multicomponent structures in their designs. The boundary of the boat is provided by a wooden "membrane" which need not be especially thick to be largely impermeable. However, thin hulls have little structural strength to maintain the boat's shape or integrity. Rather than make the planks uniformly thicker to increase the strength of the hull, naval architects developed a more efficient solution by reinforcing the hull at regular intervals. The reinforcing elements in Fig. 1.3(a) are linear, linked together to form a tension-resistant scaffolding around the hull. In the design of all but the simplest cells, evolution has similarly selected a cytoskeleton or cell wall composed of molecular filaments to reinforce the thin plasma membrane of the cell boundary.

Networks for tensile strength

The rigging of the sailing ship of Fig. 1.3(a) illustrates another design adopted by the cell. Rather than use stout poles placed on either side of the mast to *push* it into position, a boat uses ropes on either side of the mast to *pull* it into position. By pulling, rather than pushing, the structural elements need only have good tension resistance, rather than the more demanding compression or buckling resistance of poles. Employing strings and ropes with good tensile strength but little resistance to buckling, rigging provides the required functionality with minimal materials. Because the tensile strength of a rope is needed just along one direction, between the top of the mast and the attachment point on the hull, rigging uses only weak lateral links between the strong ropes connected to the mast. This is a design observed in the cross-linking of filaments in nerve cells, as seen in Fig. 1.3(b), and in the cell walls of cylindrical bacteria, which are composed of stiff filaments oriented in the direction bearing the largest stress, linked together transversely by floppy molecular chains.

The relationship between the mast and rigging of a boat exhibits an intriguing balance of tension and compression: the mast has a strong resistance to compression and bending but is held in place by rigging with little resistance to bending. The cytoskeleton of the cell also contains a mix of filaments with strong and weak bending resistance, although these filaments span a more modest range of stiffness than ropes and masts. It has been suggested that diverse filaments may be linked together in a cell to form tension/compression couplets which could have a role in maintaining cell shape or aiding cell movement (see Ingber, 1997).

Composite structures for materials efficiency

The forces on a boat are not quite the same as the forces experienced by a cell. An important difference is that the external pressure on the hull from the surrounding water is greater than the interior pressure, so that the internal structure of the boat must contain bracing with good compression resistance to prevent the hull from collapsing. In contrast, the interior pressure of some cells, such as many varieties of bacteria, may be much higher than their surroundings. Thus, the engineering problem facing a bacterium is one of explosion rather than collapse, and such cells have a mechanical structure which more closely resembles the hot-air balloon illustrated in Fig. 1.4(a). Balloons have a thin, impermeable membrane to confine the low-density gas that gives the balloon its buoyancy. Outside of the balloon is a network to provide extra mechanical

Fig. 1.4. (a) In a hot-air balloon or gas balloon, a thin membrane confines the gas within the balloon, and an external network provides mechanical attachment points and may aid in maintaining the balloon's shape. (b) A two-dimensional network of the protein spectrin is attached to the inside of the red blood cell membrane to provide shear resistance. Shown partially expanded in this image, the separation between the six-fold junctions of the networks reaches 200 nm when fully stretched (courtesy of A. McGough and R. Josephs, University of Chicago; see McGough and Josephs, 1990).

(a)

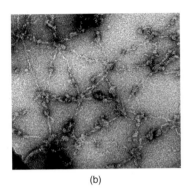

(b)

strength to the membrane and to provide attachment sites for structures such as the passenger gondola. By placing the network on the outside of the membrane and allowing the interior pressure to force physical contact between the network and membrane, there is no need to further reinforce the membrane to prevent tearing at the interior attachment sites of the network. Again, the network is under tension, so its mechanical strength can be obtained from light-weight ropes rather than heavy poles. Plant cells and most bacteria make use of external walls to reinforce their boundary membrane and balance the pressure difference across it. In a red blood cell, the two-dimensional network illustrated in Fig. 1.4(b) is attached to the membrane's *interior* surface to help the cell recover its rest shape after deformation in the circulatory system.

Internal organization for efficient operation

Advanced cells have a complex internal structure wherein specialized tasks, such as energy production or protein synthesis and sorting, are carried out by specific compartments collectively referred to as organelles. An equivalent system of human design might be a city, in which conflicting activities tend to be geographically isolated. Residential areas might be localized in one part of the city, food distribution in another, manufacturing in yet a third. How can these activities be best organized for the efficient transport of people and material within the city? Consider the plan of the walled city illustrated in Fig. 1.5(a). First note that this city has a boundary defined by the town wall, designed less to confine the inhabitants of the city than to keep hostile forces from entering it. The walls of old cities also reflect the optimal deployment of limited resources, such as the stones used in their construction and the skilled labor needed to assemble them. The minimal town wall needed to enclose a given land area is a circle, just as the minimal cell boundary to enclose a given protein-rich volume is a spherical shell. Of course,

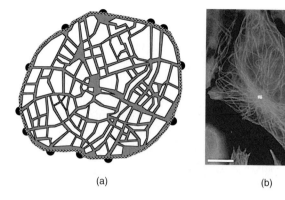

(a)   (b)

Fig. 1.5. (a) Town plan of Malines (now Mechelen), Belgium. By laying out the town wall in the form of a circle, the largest amount of land can be enclosed by the least amount of wall. Streets within the walls form an irregular web, with entry points indicated by black disks (after Girouard, 1985; © 1985 by Yale University Press). (b) The array of microtubules in a cultured fibroblast helps organize the cell's organelles and provides transportation corridors. Bar is 10 μm. (Reprinted with permission from Rodionov *et al.*, 1999; © 1999 by the National Academy of Sciences (USA.))

other factors, such as their function or mechanisms for growth and division, also influence the design and shape of cells. In a city, an efficient transportation system to direct the flow of people and materials is mandatory; for instance, it would be chaotic if visitors arriving at the gates to the city were forced to randomly diffuse through a jumble of houses in search of their destination. Such diffusive processes are very slow: the displacement from the start of a path, as the proverbial crow flies, increases only as the square root of the total path length walked by the visitor. To overcome this problem, cities use dedicated rights-of-way, including roads and railways, to provide high-speed transportation between specific locations. Depending on its layout, the most efficient street pattern in a city may be an irregular web, rather than a grid, although the latter has become commonplace in modern cities because it simplifies the layout of lots for building construction. In the design of cells, stiff filaments may provide pathways along which specialized molecules can carry their cargo; for instance, microtubules crowd the transportation corridor of the long section of the nerve cell of Fig. 1.1(a) and their layout also can be seen in the fibroblast of Fig. 1.5(b).

Materials to match the expected usage

Lastly, what about the choice of construction materials? Buildings and bridges are subject to a variety of forces that degrade a structure over time and may ultimately cause it to fail. Thus, the engineering specifications for structural materials will depend not only on their cost and availability, but also upon the building's environment and the nature of the forces to which it is subjected. As far as mechanical failure is concerned, each material has its own Achilles' heel, which may limit its applicability to certain structures. For instance, steel provides the flexibility needed to accommodate vibrations from the traffic on a suspension bridge, but has a lifetime imposed by its resistance to corrosion and fatigue. Further, the longevity we expect for our buildings and bridges is influenced by anticipated usage, public taste, and technological change, to name a few criteria. It is senseless, therefore, to overdesign a building that is likely to be torn down long before the mechanical strength of its components is in doubt. Similarly, the choice of materials for the construction of a cell is influenced by many competing requirements or limitations. For instance, some molecules that are candidates for use in a cell wall may produce a wall that is strong, but not easily repairable or amenable to the process of cell division. Further, the availability of materials and their ease of manufacture by the cell are also important considerations in selecting molecular building blocks and in designing the cellular structure. Lastly, all of these conflicting interests must be resolved so that the organism is sufficiently robust and long-lived to compete in its environment.

What we have done in this section is search for common architectural themes in the designs of boats and balloons, and the plans of towns and cities. We find that designs making effective use of available materials often employ specialized structural elements that must act cooperatively in order to function: thin membranes for boundaries, ropes for tensile strength, and walls to balance internal pressure. The choice of construction materials for a given structural element is determined by many factors, such as availability or ease of assembly and repair, with the overall aim of producing a structure with an acceptable lifetime. As we will see in the following sections, evolution has selected many of the same effective design principles as human engineering to produce cells that are adaptable, repairable and functional in a wide range of environments. As we better understand Nature's building code, we will discover subtle features that may have application beyond the cellular world.

1.2 Cell shapes, sizes and structures

Despite their immense variety of shapes and sizes, cells display common architectural themes reflecting the similarity of their basic functions. For instance, all cells have a semi-permeable boundary that selectively segregates the cell's contents from its environment. Frequently, cells adopt similar strategies to cope with mechanical stress, such as the reinforced membrane strategy of boats and balloons discussed in Section 1.1. Further, the chemical similarities among the structural elements of different cells are remarkably strong. In this section, we first review some of the basic mechanical necessities of all cells and then provide a general overview of the construction of several representative cells, namely simple cells such as bacteria, as well as complex plant or animal cells. A longer introduction to cell structure can be found in Appendix A or textbooks such as Alberts *et al.* (1994) or Prescott *et al.* (1996).

As described in Section 1.1, the outer boundaries of boats and balloons are fairly thin compared with the linear size of the vessels themselves. Modern skyscrapers also display this design: the weight of the building is carried by an interior steel skeleton, and the exterior wall is often just glass cladding. A cell follows this strategy as well, by using for its boundary a thin membrane whose tensile strength is less important than its impermeability to water and its capability for self-assembly and repair. By using thin, flexible membranes, the cell can easily adjust its shape as it responds to its changing environment or reproduces through division.

Whether this membrane needs reinforcement depends upon the stresses it must bear. Some proteins embedded in the membrane function as mechanical pumps, allowing the cell to accumulate ions and molecules in its interior. If the ion concentrations differ across its membrane, the cell may operate at an elevated osmotic pressure, which may be an order of magnitude larger than atmospheric pressure in some bacteria (bicycle tires are commonly inflated to about double atmospheric pressure). The cell may accommodate such pressure by reinforcing its membrane with a network of strings and ropes or by building a rigid wall. Even if the membrane bears little tension, networks may be present to help maintain a cell's shape.

What other mechanical attributes does a cell have? Some cells can locomote or actively change shape, permitting them to pursue foes. For example, our bodies have specialized cells that can remove dead cells or force their way through tissues to attack foreign invaders. One way for a cell to change its shape is to possess a network of stiff internal poles that push the cell's surface in the desired manner. Consequently, several types of structural filament, each with differing stiffness, are present in the cell: some filaments are part of reinforcing networks while others are associated with locomotion or internal transportation. Of course, conservation of momentum tells us that there is more to cell motion than pushing on its boundary: to generate relative motion, the cell must adhere to a substratum or otherwise take advantage of the inertial properties of its environment.

The operative length scale for cells is the micron or μm, a millionth of a meter. The smallest cells are a third of a micron in diameter, while the largest ones may be more than a hundred microns across. Nerve cells have particularly long sections called axons running up to a meter from end to end, although the diameter of an axon is in the micron range. Structural elements of a cell, such as its filaments and sheets, generally have a transverse dimension within a factor of two of 10^{-2} μm, which

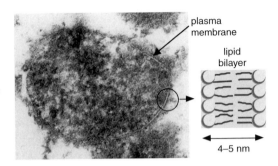

Fig. 1.6. Thin section of *Mycoplasma hominis*, illustrating the plasma membrane isolating the contents of the cell from its surroundings. As shown in the cartoon enlargement, the plasma membrane contains a bilayer of dual-chain lipid molecules, about 4–5 nm thick in its pure form, and somewhat thicker in the presence of other membrane components. Proteins are found in both the plasma membrane and the cell's interior (the fuzzy patches are protein-manufacturing sites called ribosomes). Typical diameter of a mycoplasma is a third of a micron, although the one shown here is 0.6 μm across. Adhering to the external surface of this cell is a precipitate from the animal blood serum in which the cell was grown (courtesy of Dr Terry Beveridge, University of Guelph).

is equal to 10 nm (a nanometer is 10^{-9} m); that is, they are very thin in at least one direction. For comparison, a human hair has a diameter of order 10^2 μm.

Let's now examine a few representative cells to see how membranes, networks and filaments appear in their construction. The two principal categories of cells are procaryotes (without a nucleus) such as mycoplasmas and bacteria, and eucaryotes (with a nucleus) such as plant and animal cells. Having few internal mechanical elements, some of today's procaryotic cells are structural cousins of the earliest cells, which emerged more than 3.5 billion years ago. Later in the Earth's history, eucaryotes adopted internal membranes to further segregate their contents and provide additional active surface area within the cell. We begin by discussing generic designs of procaryotic cells.

Mycoplasmas and bacteria

Mycoplasmas are among the smallest known cells and have diameters of perhaps a third of a micron. As displayed in Fig. 1.6, the cell is bounded by a plasma membrane, which is a two-dimensional fluid sheet composed primarily of lipid molecules. Described further in Appendix B, the principal lipids of the membrane have a polar or charged head group, to which are attached two hydrocarbon chains. The head groups are said to be hydrophilic, reflecting their affinity for polar molecules such as water, whereas the non-polar hydrocarbon chains are hydrophobic, and tend to shun contact with water. In an aqueous environment, some types of lipids can self-assemble into a fluid sheet consisting of two layers, referred to as a lipid bilayer, with a combined thickness of 4–5 nm. Like slices of bread in a sandwich, the polar head groups of the lipids form the two surfaces of the bilayer, while the hydrocarbon chains are tucked inside. The bilayer is a two-dimensional fluid and does not have the same elastic properties as a piece of cloth or paper. For instance, if you wrap an apple with a flat sheet of paper, the paper develops folds to accommodate the shape of the apple; however, if you dip the apple in thick syrup, the syrup flows and forms a smooth, two-dimensional fluid coating on the apple's surface.

The interior of a mycoplasma contains, among other things, the cell's genetic blueprint, DNA, as well as large numbers of proteins, which also may be embedded in the plasma membrane itself. Although the beautiful world of biochemistry is not the focus of this text, we pause briefly to mention the size and structure of proteins and DNA (more details are provided in Appendix B). Ranging in mass upwards of 10^5 daltons (one dalton, or Da, is one-twelfth the mass of a carbon-12 atom), proteins are linear polymers of amino acids containing an amino group ($-NH_3^+$) and an organic acid group ($-COO^-$). Many proteins fold up into compact structures with diameters ranging from a few to tens of nanometers, depending upon the mass of the protein, and some varieties of these globular proteins, in turn, are monomers in still larger structures such as cytoskeletal filaments with diameters ranging up to 25 nm. Like proteins, DNA (deoxyribonucleic acid) and RNA (ribonucleic acid) are polymers with a backbone consisting of a sugar/phosphate repeat unit, to each of which is attached one member of a small set of organic bases, generating the linear pattern of the genetic code. In the form of a double helix, DNA is 2.0 nm in diameter.

Most bacteria both are larger and have a more complex boundary than the simple plasma membrane of the mycoplasma. For instance, the bacterium *Escherichia coli*, an inhabitant of the intestinal tract, has the shape of a cylinder about 1 μm in diameter, and several microns in length, capped by hemispheres at each end. The interior of a bacterium may be under considerable pressure but the presence of a cell wall prevents rupture of its plasma membrane. Except in archaebacteria, the cell wall is composed of layers of a sugar/amino-acid network called peptidoglycan: Gram-positive bacteria have a thick layer of peptidoglycan encapsulating a single plasma membrane (Fig. 1.7(b)), whereas Gram-negative bacteria have just a thin layer of peptidoglycan sandwiched

Fig. 1.7. Boundary structure of bacteria. (a) A Gram-negative bacterium, such as *Aermonas salmonicida*, has a very thin layer of peptidoglycan sandwiched between two membranes. Thin strings visible in the cell's cytoplasm are DNA. (b) In Gram-positive bacteria, such as *Bacillus stearothermophilus*, only one bilayer is present, and the peptidoglycan blanket is much thicker. These bacteria have an additional layer of proteinaceous subunits on their surface (S-layer), which is not always present in other bacteria (courtesy of Dr Terry Beveridge, University of Guelph). (c) The molecular structure of peptidoglycan displays stiff chains of sugar rings, oriented around the girth of the bacterium, cross-linked by floppy strings of amino acids oriented along its axis.

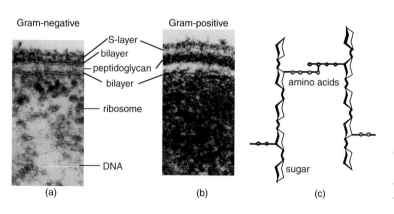

Gram-negative

Gram-positive

S-layer
bilayer
peptidoglycan
bilayer

ribosome

DNA

amino acids

sugar

(a)　　　　(b)　　　　(c)

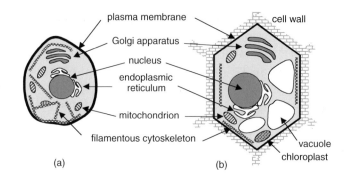

(a)　　　　(b)

Fig. 1.8. Schematic sections through generic animal (a) and plant (b) cells showing the layout of the organelles and other structural components (after Alberts *et al.*, 1994; reproduced with permission; © 1994 by Routledge, Inc., part of The Taylor and Francis Group). Having a more complex internal structure than bacteria, these cells are correspondingly larger, typically 10–100 μm for plants and 10–30 μm for animals.

between two lipid bilayers (Fig. 1.7(a)), the inner one of which is the plasma membrane. Parenthetically, this classification reflects the ability of a bacterium to retain Gram's stain. As shown in Fig. 1.7(c), the peptidoglycan network is anisotropic (i.e. does not have the same structure in all directions), with its sugar chains oriented around the girth of the bacterium, the direction which bears the greatest surface stress. The sugar chains are linked transversely with loose strings of amino acids, the latter bearing only half the surface stress that the sugar chains bear. In analogy with a ship's rigging, the sugar chains are like the strong ropes attached to the mast, while the amino-acid strings are weaker cords that link the ropes into a network.

Plant and animal cells

The general layout of a plant cell is displayed in Fig. 1.8(b). Exterior to their plasma membrane, plant cells are bounded by a cell wall, permitting them to withstand higher internal pressures than a wall-less animal cell can support. However, both the thickness and the chemical composition of the wall of a plant cell are different from those of a bacterium. The thickness of the plant cell wall is in the range of 0.1 to 10 μm, which may be larger than the total length of some bacteria, and the plant cell wall is composed of cellulose, rather than the peptidoglycan of bacteria. Also in contrast to bacteria, both plant and animal cells contain many internal membrane-bounded compartments called organelles. Plant cells share most organelles with animal cells, although plant cells alone contain chloroplasts; liquid-filled vacuoles are particularly large in plant cells and may occupy a large fraction of the cell volume.

As illustrated in Fig. 1.8(a), animal cells share a number of common features with plant cells, but their lack of space-filling vacuoles means that animal cells tend to be smaller in linear dimension. Neither do they have a cell wall, relying instead upon the cytoskeleton for much of their mechanical rigidity, although the cytoskeleton does not provide the strength that would allow an animal cell to support a significant internal pressure. Some components of the cytoskeleton meet at the centriole, a cylindrical organelle about 0.4 μm long. The most important organelles of plant and animal cells are the following.

- The nucleus, with a diameter in the range 3–10 μm, contains almost all of the cell's DNA and is bounded by a pair of membranes, attached at pores to allow the passage of material across the nuclear envelope.
- The endoplasmic reticulum surrounds the nucleus and is continuous with its outer membrane. As a series of folded sheets, the rough endoplasmic reticulum has a small volume compared to its surface, on which proteins are synthesized by strings of ribosomes, like beads on a necklace.
- The membrane-bounded Golgi apparatus is the site of protein sorting, and appears like layers of flattened disks with diameters of a few microns. Small vesicles, with diameters in the range 0.2–0.5 μm, pinch off from the Golgi and transport proteins and other material to various regions of the cell.
- Mitochondria and chloroplasts, the latter present only in plant cells, produce the cell's energy currency, ATP (adenosine triphosphate). Shaped roughly like a cylinder with rounded ends (often 0.5 μm in diameter), a mitochondrion is bounded by a double membrane. Also bounded by a double membrane, but containing internal compartments as well, chloroplasts are about 5 μm long and are the site of photosynthesis.

All of the material within the cell, with the exclusion of its nucleus, is defined as the cytoplasm, which contains organelles as well as the largely aqueous cytosol. Permeating the cytosol in some cells, or simply attached to the plasma membrane in others, is the cytoskeleton, a network of filaments of varying size and rigidity. The various filament types of the cytoskeleton may be organized into separate networks with different mechanical properties. In the schematic epithelial cell shown in Fig. 1.9, slender filaments of actin are associated with the plasma membrane, while thicker intermediate filaments are connected to cell attachments sites, and stiff microtubules radiate from the microtubule organizing center. In addition to providing the cell with mechanical strength, elements of the cytoskeleton may be used as pathways for the transportation of material, much like the roads of a city.

The above survey of cell structure demonstrates several things. First, membranes are ubiquitous components of the cell, providing boundaries for the cell itself and for the cell's organelles and other compartments. Membranes contain both proteins and dual-chain lipids, the

Fig. 1.9. Schematic organization of cytoskeletal filaments – actin (a), intermediate filaments (b) and microtubules (c) – in an epithelial cell. Intermediate filaments also form a lamina around the nucleus in part (b) (see Chapter 3); microtubules radiate from a microtubule organizing center in part (c). Further details on the connectivity of the cytoskeleton can be found in Appendix A (after Alberts *et al.*, 1994; reproduced with permission; © 1994 by Routledge, Inc., part of The Taylor and Francis Group).

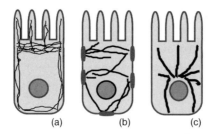

(a)   (b)   (c)

latter capable of self-assembly into bilayers under the appropriate conditions, as we will establish in Chapter 5. Being only 4–5 nm thick, bilayers are very flexible and can adapt to the changing shape of the cell as needed. Second, filaments are present in the cell in a variety of forms, sometimes as isolated molecules like DNA or RNA, sometimes as part of a network like the cytoskeleton or cell wall. The filaments range up to 25 nm in diameter; the thickest filaments appear stiff on the length scale of the cell, while the thinnest filaments appear to be highly convoluted. Most individual filaments, networks and membranes of cells are then *soft*, in that they may be easily deformed by forces commonly present in a cell. We now discuss in more detail what is meant by "soft", and describe the thermal fluctuations in the size and shape of filaments and membranes.

1.3 Soft strings and sheets

In a cell, most filaments are not straight like the beams of a skyscraper nor are the membranes flat like the steel sheets in the hull of a boat. For instance, Fig. 1.10(a) is an image of DNA immobilized on a substrate, looking much like pieces of thread thrown casually onto a table. Similarly, Fig. 1.10(b) demonstrates how membranes, in this case isolated membranes from *Escherichia coli*, can sustain regions of very high curvature. These images hint that the strings and sheets in cells are soft, a hypothesis that is confirmed by observing the ease with which they are deformed by ambient forces within the cell and by those applied externally.

Words like *hard* and *soft* are used in at least two contexts when describing the elastic characteristics of a material. Sometimes the terms are used in a comparative sense to reflect the property of one material relative to another; for example, Moh's scale is a well-known logarithmic measure of hardness (talc has a hardness of 1 on this scale, while diamond has a hardness of 10). In another usage, softness indicates the response of an object to forces routinely present in its environment. In thermal equilibrium, an object can acquire energy from its surroundings, permitting or forcing it to change shape even if energy is needed to

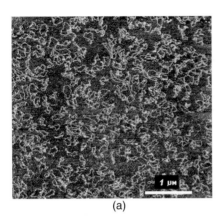

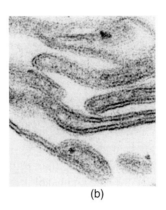

(a)   (b)

Fig. 1.10. (a) Atomic force microscope image of DNA on a mica substrate. Bar is 1 μm. (Reprinted with permission from Shlyakhtenko *et al.*, 1999; © 1999 by the Biophysical Society.) (b) Thin section of outer membranes extracted from *E. coli* after staining with osmium tetroxide and uranyl acetate. The bilayer, clearly visible as parallel black lines, is 7.5 nm thick (courtesy of Dr Terry Beveridge, University of Guelph).

do so. The amplitude of these thermal fluctuations in shape depends, of course, on the softness of the material. While rigid objects may not even modestly change shape in response to thermal fluctuations in their energy, flexible filaments may bend from side to side and membranes may undulate at non-zero temperature. We now describe the rigidity and thermally driven shape changes of biological filaments and sheets on cellular length scales; the elastic behavior of macroscopic objects like leaves and feathers can be found in Vogel's very readable book on the subject (Vogel, 1998).

Soft filaments

The resistance that any filament offers to bending depends upon its size and material composition: thick ropes are stiffer than thin strings and steel is more rigid than cooked pasta. To qualitatively understand their bending behavior, consider what happens when a force is applied to the free end of a thin filament, while the other end is fixed, as indicated in Fig. 1.11. The energy required to gently bend a rod into the shape of an arc depends on three factors:

- the size of the bending angle, θ in Fig. 1.11; the deformation energy increases like the square of the angle, just as the potential energy of a simple spring increases like the square of its displacement from equilibrium;
- the material composition of the rod; for instance, it takes about one hundred times more energy to bend a metallic filament (like copper) than an otherwise identical biofilament, through the same bending angle;
- the diameter (and cross sectional shape) of the filament; for example, the rigidity of thin, cylindrical rods increases like the fourth power of their diameter.

Fig. 1.11. The energy required to gently bend a straight rod (a) is proportional to the square of the bending angle θ, defined in (b).

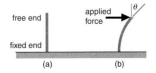

(a)   (b)

This last factor means that thick cellular filaments such as microtubules, with diameters of 25 nm, have a bending resistance about two orders of magnitude larger than thin actin filaments, of diameter 8 nm: the fourth power of their diameters has the ratio $(25/8)^4 = 95$. Thus, the cell has at its disposal a selection of filaments spanning a large range of bending rigidity, from floppy threads to stiff molecular ropes, and these filaments bend much more easily than if they were made of metal or similar materials.

What about the thermal motion of a filament as it waves back and forth, exchanging energy with its environment? Very stiff filaments hardly move from their equilibium positions; for them, the value of θ in Fig. 1.11 is usually close to zero. In contrast, highly flexible filaments may sample a bewildering variety of shapes. For instance, at room temperature, an otherwise straight microtubule 10 μm long would be displaced, on average, by about a tenth of a radian (or 6 degrees of arc) because of thermal motion. Were the microtubule made of steel, its mean angular displacement would be less than one degree of arc. In comparison, a 10 μm length of spectrin, which is much thinner and more flexible, would look like a ball of thread.

Soft sheets

In analogy to flexible filaments, membranes display a bending resistance that depends on their geometry and material composition. For the membranes of the cell, this resistance covers a much smaller range – perhaps only a factor of ten – than the many orders of magnitude spanned by the bending resistance of cellular filaments. The reason for modest range in bending rigidity is mainly geometrical: reflecting their rather generic lipid composition, viable cell membranes tend to have a thickness of about 4–5 nm, insufficient to take advantage of the power-law dependence of the bending rigidity on membrane thickness. However, these biological membranes do exhibit about two orders of magnitude less resistance to bending than would an otherwise identical membrane made from copper or steel.

The energy required to bend an initially flat bilayer into a closed spherical shape like a cell is neither trivial nor insurmountable, as is demonstrated in Chapter 6. However, gentle undulations of membranes, such as those illustrated in Fig. 1.12(a), can be generated by thermal fluctuations alone. The figure displays images of a pure lipid vesicle, which has a mechanical structure like a water-filled balloon, except that the boundary is a fluid membrane. The images in Fig. 1.12(a) are separated by an elapsed time of 1 s, and have been superimposed to show the amplitude of the undulations. The types of motion executed by the surface are shown schematically in Fig. 1.12(b), where the grey

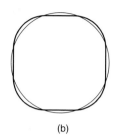

(a)   (b)

Fig. 1.12. (a) Pure lipid vesicles, whose surface area is slightly more than the minimum needed to contain its aqueous contents, display thermal undulations like those shown in these superimposed video images. (b) Schematic representation of some surface oscillations of a vesicle; a circle representing the mean position of the surface is drawn in grey for comparison (reprinted with permission from Yeung and Evans, 1995, courtesy of Dr Evan Evans, University of British Columbia, © 1995 by Les Editions de Physique).

circle indicates the mean position of the membrane. Clearly, the membrane is sufficiently stiff that the wavelength of the undulations is of the order of the size of the vesicle and not dramatically smaller.

Soft vs. hard: entropy vs. energy

What is the origin of the elastic properties of soft materials? For guidance, let's look to the changing appearance of spaghetti as it is cooked in a boiling pot of water. In its initial dry state, the spaghetti is rod-like, and resists bending because of the *energy* needed to deform the pasta from its molded shape. As the pasta absorbs water, it softens and adopts ever more sinuous configurations, its playful wiggling driven by the turbulence of boiling water. This results in a different mode of deformation resistance than the bending of rods – now, a force must be supplied to straighten out the dancing filament as it is pushed and pulled by its turbulent environment. Very soft filaments in the cell also wave and wiggle, although not because the cytosol is boiling! Rather, the energy of the filament fluctuates because it is in thermal contact with its surroundings, permitting the filament to tumble through a collection of shapes, even if they contain some energy-consuming kinks and bends.

Forcing a filament to straighten out restricts its configurations and reduces its entropy, a process that does not happen spontaneously according to thermodynamics. To see how this works, let's consider a chain with four rigid segments that are forced to lie horizontally. Our unphysical chain is permitted to double back on itself, and has 16 distinguishable configurations, assuming that we can tell one end of the chain from the other, as displayed in Fig. 1.13. Reviewed in Appendix C, the entropy of a system depends on the number of accessible configurations; the entropy of the chain in the figure is proportional to $\ln(16) = 2.8$, where "ln" refers to the natural logarithm. The configurations in the figure are unique, and are grouped according to the displacement between the ends of the chain (the displacement is the end-to-end distance, without regard to the path in between); for instance, there are six

Fig. 1.13. Configurations of a four-segment chain with inequivalent ends in one dimension. The displacement between ends of the chain is 4, 2 and 0 segment lengths in groups (a), (b) and (c), respectively. Switchbacks have been offset for clarity.

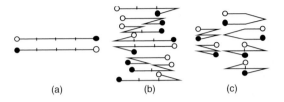

(a)   (b)   (c)

configurations having zero displacement. Suppose that we apply a tension to the ends of the chains to pull them apart, such that the configurations with zero displacement are no longer available. The result of this operation is to reduce the number of accessible configurations to ten; the entropy is now proportional to $\ln(10) = 2.3$. In other words, the stretching force reduces the entropy of the chain, which is not a process favored thermodynamically.

Elasticity arising from the resistance to entropy loss, as we have just described, is well-established in polymer physics (Flory, 1953). Another example of such elasticity is an ideal gas, whose resistance to compression reflects the loss of entropy associated with herding its molecular constituents into a smaller volume. J.J. Waterston, a pioneer in the theory of gases, invoked a cloud of flying insects as a metaphor for a gas and we can imagine the effort needed to force a swarm of wasps into a tiny volume (see Section 2-1 of Kauzmann, 1966). From these examples, we see that while the deformation resistance of stiff materials is dominated by energy considerations, that of soft materials is also influenced by entropy.

The whole cell

How does the behavior of the whole cell express, and take advantage of, the softness of its components? The human red blood cell provides a good example of a mechanically simple cell: its plasma membrane easily deforms as the cell passes through narrow capillaries, and its cytoskeleton helps restore the cell to its rest shape once passage is complete. The magnitude of the deformations that the red cell can sustain is demonstrated in Fig. 1.14(a), which displays the cell's shape as it is drawn up a micropipette by suction. Part (b) is an image of the density of the membrane-associated cytoskeleton, to which fluorescent molecules have been attached, showing how the cytoskeleton becomes ever more dilute, and hence less visible in the image, as it stretches up the pipette. Part (c) is a computer simulation of a similar deformation, based on a model in which the elasticity of the cytoskeleton arises from the entropy of its thin spectrin filaments. The images indicate the magnitude of the deformation to which the materials of a cell can be subjected without failure: the fluid membrane permits the cell to squeeze into a narrow

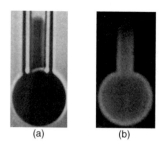

(a)   (b)   (c)

Fig. 1.14. Deformation of a human red blood cell as it is drawn up a pipette approximately 1 μm in diameter. Part (a) is a bright-field microscope image of the cell and the pipette, while (b) is a fluorescence image showing the density of the cytoskeleton. Part (c) is a computer simulation of the experiment, based on the entropic contribution of the cytoskeleton ((a) and (b) reprinted with permission from Discher et al. (1994), © 1994 by the American Association for the Advancement of Science; (c) from Discher et al. (1998)).

pipette, and the density of the cytoskeleton itself evidently drops by a factor of two along the length of the aspirated segment.

1.4 Summary

Many of the design principles that have been developed for structures such as buildings and bridges are equally applicable to the architecture of the cell. For example, the efficient usage of available materials may be most readily achieved through composite systems, in which the required functionality is obtained only from the combined properties of the individual structural elements. We see this strategy in the design of the cell, where the plasma membrane provides the necessary barrier for confining the cell's contents and is reinforced as necessary by networks and walls to withstand the stresses of the membrane's environment. We observe that at least one linear dimension (i.e., width or thickness) of most structural elements of the cell is very small, say 4–5 nm for the thickness of a membrane or 8–25 nm for the diameter of many biofilaments, so that most mechanical components of the cell are soft in some respects because of their small dimensions. Further, the resistance of soft biological materials to deformation may be a hundred times less than conventional hard materials such as metals. In thermal equilibrium, soft filaments and sheets oscillate and undulate because the energy required for modest changes in shape is available in the cell. The presence of soft materials is required not only for the everyday tasks of the cell but also for its growth and division, necessitating the development of a much broader building code than what is applicable to human engineering.

In the next several chapters, we investigate the generic characteristics of soft filaments and sheets. The cytoskeleton is one of the primary topics of this text, and Part I is entirely devoted to the behavior of flexible filaments, both in isolation (Chapter 2), and as they are welded into networks (Chapters 3 and 4). Part II approaches our second principal topic, membranes, in a similar vein: the molecular structure and self-assembly of bilayers are treated in Chapter 5, while membrane undulations and their consequences are the subject of Chapter 6. Throughout

Parts I and II, many prominent features of soft structures are shown to be rooted in their entropy, and are most easily understood by means of statistical mechanics. Appendices C and D provide quick tours of some concepts from statistical mechanics and elasticity that appear in the mathematical sections of Parts I and II.

We assemble these isolated ropes and sheets into some simple, but complete, cells in Part III. The shapes of biological structures such as vesicles and mammalian red blood cells are interpreted with the aid of mechanical models in Chapter 7. Next, Chapter 8 reintroduces membrane fluctuations and considers their role in the interaction and adhesion of cells. The motion of a cell, including its molecular basis, is the subject of Chapter 9. We then devote the last chapter of the text to the study of the forces in cells, and to the speculative question of the mechanical challenges faced by the largest and smallest cells.

Part I
Rods and ropes

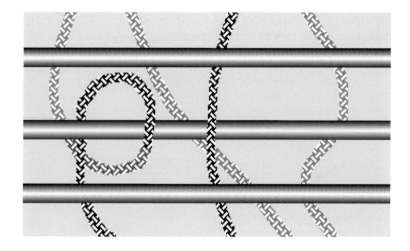

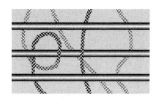

Chapter 2

Polymers

The structural elements of the cell can be broadly classified as filaments or sheets, where by the term *filament*, we mean a string-like object whose length is much greater than its width. Some filaments, such as DNA, function as independent units, but most structural filaments in the cell are linked to form two- or three-dimensional networks. As seen on the cellular length scale of a micron, individual filaments may be relatively straight or highly convoluted, reflecting, in part, their resistance to bending. This opening chapter to Part I concentrates on the mechanical properties of individual filaments, such as their bending or stretching resistance; the two chapters making up the remainder of Part I consider how filaments are knitted together to form networks, perhaps closely associated with a membrane as a two-dimensional web (Chapter 3) or perhaps extending though the three-dimensional volume of the cell (Chapter 4).

2.1 Filaments in the cell

Our discussion of cellular ropes and rods begins with a look at their molecular composition and linear dimensions. All of the ropes are linear polymers, in the sense that they are constructed from individual monomeric units to form an unbranched chain. The monomers need not be identical, and may themselves be constructed of more elementary chemical units. For example, the monomeric unit of DNA and RNA is a troika of phosphate, sugar and organic base, with the phosphate and sugar units alternating along the backbone of the polymer (see Appendix B). However, the monomers are not completely identical because the base may vary from one monomer to the next. The double helix of DNA contains two sugar–base–phosphate strands, with a length along the helix of 0.34 nm per pair of organic bases, and a corresponding molecular mass per unit length of about 1900 Da/nm (Saenger, 1984).

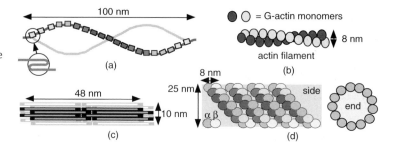

Fig. 2.1. (a) Two spectrin chains intertwine in a filament, where the boxes represent regions in which the protein chain has folded back on itself (as in the inset). The two strings are stretched and separated for clarity.
(b) Monomers of G-actin associate to form a filament of F-actin, which superficially appears like two intertwined strands. (c) A hollow intermediate filament is composed of eight protofilaments, each of which is itself four monomers (see Fig. 2.2). (d) Microtubules usually contain 13 protofilaments, whose elementary unit is an 8 nm long dimer of the proteins α and β tubulin. Both side and end views of the cylinder are shown.

The principal components of the cytoskeleton – actin, intermediate filaments and microtubules – are themselves composite structures made from protein subunits, each of which is a linear chain hundreds of amino acids long (Hesketh and Prym, 1996). In addition, some cells contain fine strings of the protein spectrin, whose structure we discuss first before considering the thicker filaments of the cytoskeleton. As organized in the human erythrocyte, two pairs of chains, each pair containing two intertwined and inequivalent strings of spectrin (called α and β), are joined end-to-end to form a filament about 200 nm in contour length. The α and β chains have molecular masses of 230000 and 220000 Da, respectively, giving a mass per unit length along the tetramer of 4500 Da/nm. An individual chain folds back on itself repeatedly like a letter Z, so that each monomer is a series of 19 or 20 relatively rigid barrels 106 amino acid residues long, as illustrated in Fig. 2.1(a) (see Rief *et al.*, 1999, for spectrin folding).

Forming somewhat thicker filaments than spectrin, the protein actin is present in many different cell types and plays a variety of roles in the cytoskeleton. The elementary actin building block is the protein G-actin ("G" for globular), a single chain of approximately 375 amino acids having a molecular mass of 42000 Da. G-actin units can assemble into a long string called F-actin ("F" for filamentous), which, as illustrated in Fig. 2.1(b), has the superficial appearance of two strands forming a coil, although the strands are not, in fact, independently stable. The filament has a width of about 8 nm and a mass per unit length of 16000 Da/nm. Typical actin monomer concentrations in the cell are 1–5 mg/ml; as a benchmark, a concentration of 1 mg/ml is 24 μM for a molecular mass of 42 kDa.

Intermediate filaments are slightly wider than F-actin and have a more complex hierarchical structure, as illustrated by the model in Fig. 2.1(c) and Fig. 2.2. The basic building blocks of the filament are two protein chains intertwined as a helix. Pairs of helices lie side-by-side to form a linear protofilament some 2–3 nm in width. The intermediate filament itself is a bundle of eight protofilaments in a roughly cylindrical shape

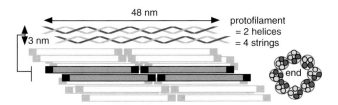

about 10 nm in diameter. Many intermediate-filament monomers have masses in the 40 000–70 000 Da range and lengths of the order 50 nm, such that the mass per unit length of a four-strand protofilament is about 4 500 Da/nm, and that of a 32-strand filament is about 35 000 Da/nm, with some variation.

The thickest individual filaments are microtubules of the protein tubulin, present as a heterodimer of α-tubulin and β-tubulin, each with a molecular mass of about 50 000 Da. Pairs of α- and β-tubulin form a unit 8 nm in length, and these units can assemble α to β successively into a hollow microtubule consisting of 13 linear protofilaments (in almost all cells), as shown in Fig. 2.1(d). The overall molecular mass per unit length of a microtubule is about 160 000 Da/nm, ten times that of actin. Tubulin is present at concentrations of a few milligrams per milliliter in a common cell; given a molecular mass of 100 kDa for a tubulin dimer, a concentration of 1 mg/ml corresponds to 10 μM.

Biological filaments can also be found outside of the cell. Frequently present in connective tissue such as tendon, the many types of collagen are fibrous proteins that are organized into hierarchical structures such as ropes and mats. The string of amino acids in type I collagen, for example, is arranged in a helix with three amino-acid residues per turn, and three such strands associate to form a larger molecule called tropocollagen. Displayed in Fig. 2.3, tropocollagen is about 300 nm long and 1.5 nm in diameter, with a mass per unit length of about 1000 Da/nm. Many threads of tropocollagen are organized into a collagen fibril, which may have a diameter of 10–300 nm. The collagen fibrils themselves can assemble in parallel formation into a collagen fiber.

As a last example of cellular filaments, we mention cellulose, which, as the principal tension-bearing component of the plant cell, is a linear

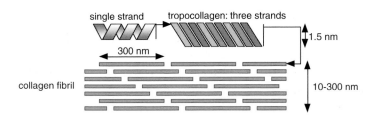

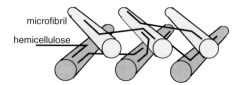

Fig. 2.4. In the plant cell wall, 60–70 individual cellulose polymers are bundled together to form a microfibril. Hemicellulose links the microfibrils at 20–40 nm intervals to form oriented sheets. The pectin network of the plant cell wall is not displayed. All elements have been separated in the drawing for clarity (after Alberts *et al.*, 1994; reproduced with permission; © 1994 by Routledge, Inc., part of The Taylor and Francis Group).

polysaccharide made from the sugar glucose (see Appendix B for more details). A bundle of 60–70 molecular strands of cellulose lie parallel to each other to form a linear microfibril that typically has a diameter of about 10 μm. In the plant cell wall, microfibrils are themselves linked by hemicellulose molecules to form a sheet-like structure in which the fibrils are approximately aligned with each other. The cell wall then consists of multiple sheets of microfibrils, with successive sheets having different orientations of the fibrils, as in Fig. 2.4.

The design of cellular filaments has been presented in some detail in order to illustrate both their similarities and differences. Most of the filaments possess a hierarchical organization of threads wound into strings, which then may be wound into ropes. The filaments within a cell are, to an order of magnitude, about 10 nm across, which is less than 1% of the diameters of the cells themselves. As one might expect, the visual appearance of the cytoskeletal filaments on cellular length scales varies with their thickness. The thickest filaments, microtubules, are stiff on the length scale of a micron, such that isolated filaments are only gently curved. In contrast, intertwined strings of spectrin are relatively flexible: at ambient temperatures, a 200 nm filament of spectrin adopts such convoluted shapes that the distance between its end-points is only 75 nm on average (for spectrin filaments that are part of a network).

The biological rods and ropes of a cell may undergo a variety of deformations, depending upon the nature of the applied forces and the mechanical properties of the filament. Analogous to the tension and compression experienced by the rigging and masts of a sailing ship, some forces lie along the length of the filament, causing it to stretch, shorten or perhaps buckle. In other cases, the forces are transverse to the filament, causing it to bend or twist. Whatever the deformation mode, energy is required to distort the filament from its "natural" shape, by which we mean its shape at zero temperature and zero stress. Consider, for example, a uniform straight rod of length L bent into an arc of a circle of radius R, as illustrated in Fig. 2.5(a). Within a simple model for the bending of rods introduced in Section 2.2, the energy E_{arc} required to perform this deformation is given by

$$E_{arc} = \kappa_f L/2R^2 = Y \mathcal{I} L/2R^2, \qquad (2.1)$$

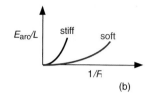

(a)   (b)

Fig. 2.5. Bending a rod of length L into the shape of an arc of radius R (in part (a)) requires an input of energy E_{arc} whose magnitude depends upon the severity of the deformation and the stiffness of the rod (b).

where κ_f is called the flexural rigidity of the rod: large κ_f corresponds to stiff rods. The flexural rigidity of uniform rods can be written as $\kappa_f = Y\mathcal{I}$, where Y and $\mathcal{I}$ are terms reflecting the composition and geometry of the rod, respectively, as will be explained momentarily. Fig. 2.5(b) displays how the bending energy behaves according to Eq. (2.1): a straight rod has $R=\infty$, and hence $E_{arc}=0$, while a strongly curved rod might have L/R near unity, and hence $E_{arc} \gg 0$, depending on the magnitude of κ_f.

Young's modulus Y is a measure of the bulk elasticity of the material from which the rod is fabricated. Stiff materials, such as steel, have $Y \sim 2 \times 10^{11}$ J/m^3, while softer materials, such as plastics, have $Y \sim 10^9$ J/m^3. The moment of inertia of the cross section, $\mathcal{I}$ (not to be confused with the moment of inertia of the mass, familiar from rotational motion), depends upon the geometry of the rod; for instance, a cylindrical rod of constant density has $\mathcal{I} = \pi R^4/4$. Owing to its power law dependence on filament radius, the flexural rigidity of polymers in the cell spans nearly five orders of magnitude.

Now, the energy of an object in thermal equilibrium is not constant, but fluctuates with time. The reader is familiar with this phenomenon in the context of the kinetic theory of gases, wherein the kinetic energy of an individual gas molecule changes through collisions, such that the average kinetic energy per molecule is $3/2\,k_B T$, where T is the absolute temperature and k_B is Boltzmann's constant (1.38×10^{-23} J/K). Thus, the thermal energy scale is set by $k_B T$, equal to 4×10^{-21} J at room temperature. At finite temperature, then, an otherwise straight rod bends as it exchanges energy with its environment, as suggested by Fig. 2.6. One way of quantifying the amplitude of the shape fluctuations at finite temperature is finding the typical distance along the rod over which it undergoes a significant change in direction: flexible rods change direction over shorter distances than stiff rods. This length scale must be directly

Fig. 2.6. Sample configurations of a very flexible rod at non-zero temperature as it exchanges energy with its surroundings. The base of the filament in the diagram is fixed. The configuration at the far left is an arc of a circle subtending an angle of L/R radians.

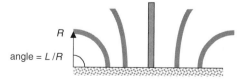

R

angle = L/R

Fig. 2.7. Two samples from the set of configurations available to a highly flexible filament. The end-to-end displacement vector r_{ee} is indicated by the arrow in part (a). The number of configurations available at a given end-to-end distance is reduced as a force F is applied to the ends of the filament in (a) to stretch it out like that in (b).

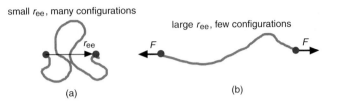

small r_{ee}, many configurations

large r_{ee}, few configurations

(a)   (b)

proportional to the flexural rigidity κ_f (stiffer filaments are straighter) and inversely proportional to temperature $k_B T$ (colder filaments are straighter). In fact, the combination $\kappa_f/k_B T$ has the units of length, and is defined as the persistence length ξ_p of the filament:

$$\xi_p = \kappa_f/k_B T = Y\mathcal{I}/k_B T, \tag{2.2}$$

an expression arising naturally in the treatment of shape fluctuations in Section 2.2. Some intuition about the meaning of the persistence length can be found by considering a filament with a length $L = \xi_p$. From Eq. (2.1), the energy E_{arc} rises as a filament bends, reaching the thermal energy scale $k_B T$ when $R = \xi_p/\sqrt{2}$ for $L = \xi_p$. This corresponds to a highly curved configuration, where the arc subtends an angle $L/R = \sqrt{2}$ radians $= 81°$, according to Fig. 2.6. In other words, a filament displays significant bending due to thermal fluctuations at length scales of the order of ξ_p.

So long as its persistence length is large compared to its contour length, i.e. $\xi_p \gg L$, a filament appears relatively straight, and recognizable as a rod. However, if $\xi_p \ll L$, the filament adopts more convoluted shapes, such as that in Fig. 2.7(a). What is the likelihood that a flexible filament will be observed in a convoluted shape, as opposed to a straight one? Using the end-to-end displacement r_{ee} as an observable of a configuration, there are many convoluted shapes with r_{ee} close to zero, but very few extended configurations with $r_{ee} \sim L$, as shown in the example in Fig. 1.13. Recall that entropy is proportional to the logarithm of the number of configurations (see Appendix C), so that r_{ee}/L near zero corresponds to large entropy and r_{ee}/L near unity corresponds to low entropy. Thus, entropy favors convoluted configurations, and consequently, r_{ee}/L is more likely to be near 0 than 1. Further, it will be shown in Section 2.3 that the average value of r_{ee} does not grow linearly with the contour length, but rather as a fractional power like the square root.

Lastly, what happens as we stretch a flexible filament by pulling on its ends, as in Fig. 2.7(b)? Stretching the filament reduces the number of configurations available to it, thus lowering its entropy; thermodynamics tells us that this is not a desirable situation – systems do not spontaneously lower their entropy, all other things being equal. Because of this, a force must be applied to the ends of the filament to pull it straight and the filament is elastic by virtue of its entropy, as explained in Section 1.3. For small extensions, this force is proportional to the change in r_{ee} from its equilibrium value, just like Hooke's law for springs. In fact, the elastic behavior of convoluted filaments can be represented by an effective spring constant k_{sp} given by

$$k_{sp} = 3k_B T / 2L\xi_p,$$
(2.3)

which is valid for a chain in three dimensions near equilibrium (see Section 2.4).

Thus, we see that filaments exhibit elastic behavior with differing microscopic origins. At low temperatures, a filament may resist stretching and bending for purely energetic reasons associated with displacing atoms from their most energetically favored positions. On the other hand, at high temperatures, the shape of a very flexible filament may fluctuate strongly, and entropy discourages such filaments from straightening out. In Sections 2.2–2.4, we investigate these situations using the formalism of statistical mechanics, before returning in Section 2.5 to apply the formal results to biological polymers.

2.2 Flexible rods

The various polymers and filaments in the cell display bending resistances whose numerical values span six orders of magnitude, from highly flexible alkanes through somewhat stiffer protein polymers such as F-actin, to moderately rigid microtubules. Viewed on micron length scales, these filaments may appear to be erratic, rambunctious chains or gently curved rods, and their elastic properties may be dominated by entropic or energetic effects. In selecting a formalism for interpreting the characteristics of cellular filaments, one can choose among several simple pictures of linear polymers, each picture emphasizing different aspects of the polymer. In this section, we view the filament as a smoothly curving rod, while in Section 2.3, our picture is that of a wiggly segmented chain. These two pictures of linear polymers overlap, of course, and there are links between their parametrizations.

Arc length and curvature

We first describe a rod as a continuous curve with no kinks or discontinuities, and ignore, for the time being, its cross-sectional shape and material composition. As displayed in Fig. 2.8(a), each point on the curve corresponds to a position vector $\mathbf{r}$, represented by the familiar Cartesian triplet (x, y, z). It is often convenient to write $\mathbf{r}$ and other characteristics of the curve in a parametric representation as a function of the arc length s, say

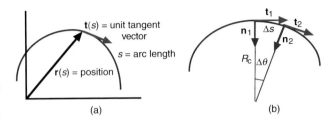

Fig. 2.8. (a) A point on the curve at arc length s is described by a position vector $\mathbf{r}(s)$ and a unit tangent vector $\mathbf{t}(s) = \partial\mathbf{r}/\partial s$. (b) Two locations are separated by an arc length Δs subtending an angle $\Delta\theta$ at a vertex formed by extensions of the unit normals $\mathbf{n}_1$ and $\mathbf{n}_2$. Extensions of $\mathbf{n}_1$ and $\mathbf{n}_2$ intersect at a distance R_c from the curve.

$\mathbf{r}(s)$ or $[x(s), y(s), z(s)]$, where s follows along the contour of the curve, running from 0 at one end to the full contour length L_c at the other. As an illustration, the equation $x^2 + y^2 = R^2$ of a circle of radius R lying in the xy plane is represented in the parametric approach by $x(s) = R\cos(s/R)$ and $y(s) = R\sin(s/R)$, where the arc length s is zero at $(x,y) = (R,0)$.

The unit tangent vector $\mathbf{t}$ characterizes the direction of the curve as it winds its way through space, as shown in Fig. 2.8. For example, in two dimensions, $\mathbf{t}$ has (x,y) components $(\cos\theta, \sin\theta)$, where θ is the angle between $\mathbf{t}$ and the x-axis. For a short section of arc Δs, over which the curve appears straight, the pair $(\cos\theta, \sin\theta)$ can be replaced by $(\Delta r_x/\Delta s, \Delta r_y/\Delta s)$, which becomes $(\partial r_x/\partial s, \partial r_y/\partial s)$ in the infinitesimal limit, or

$$\mathbf{t}(s) = \partial\mathbf{r}/\partial s, \tag{2.4}$$

a result valid in any number of dimensions. How does $\mathbf{t}$ change along the curve? Consider two nearby positions, which we shall call 1 and 2 on the curve illustrated in Fig. 2.8(b). If the curve were straight, the unit tangent vectors $\mathbf{t}_1$ and $\mathbf{t}_2$ at points 1 and 2 would be parallel; in other words, unit tangent vectors to a straight line are independent of position. However, such is not the case with curved lines, and the rate of change of $\mathbf{t}$ with s provides a measure of the curvature at any given position. As we recall from introductory mechanics, the vector $\Delta\mathbf{t} = \mathbf{t}_2 - \mathbf{t}_1$ is perpendicular to the curve in the limit where positions 1 and 2 are infinitesimally close. Thus, the rate of change of $\mathbf{t}$ with s is proportional to the unit normal to the curve $\mathbf{n}$, and we define the proportionality constant to be the curvature C

$$\partial\mathbf{t}/\partial s = C\mathbf{n}, \tag{2.5}$$

where C has units of inverse length. We can substitute Eq. (2.4) into (2.5) to obtain

$$C\mathbf{n} = \partial^2\mathbf{r}/\partial s^2. \tag{2.6}$$

The reciprocal of C is the local radius of curvature of the arc, which is displayed in Fig. 2.8(b) by extrapolating the unit normals $\mathbf{n}_1$ and $\mathbf{n}_2$ to their point of intersection. If positions 1 and 2 are close by on the contour, then the arc is approximately a segment of a circle with radius R_c and defines an angle $\Delta\theta = \Delta s / R_c$, where Δs is the segment length. However, $\Delta\theta$ is also the angle between $\mathbf{t}_1$ and $\mathbf{t}_2$; that is, $\Delta\theta = |\Delta\mathbf{t}| / t = |\Delta\mathbf{t}|$, the second equality following from $|\mathbf{t}| = t = 1$. Equating these two expressions for $\Delta\theta$ yields $|\Delta\mathbf{t}| / \Delta s = 1/R_c$, which can be compared with Eq. (2.5) to give

$$C = 1/R_c. \tag{2.7}$$

Lastly, the unit normal vector $\mathbf{n}$, which is $\Delta\mathbf{t} / |\Delta\mathbf{t}|$, can be rewritten by using $|\Delta\mathbf{t}| = \Delta\theta$

$$\mathbf{n} = \partial\mathbf{t}/\partial\theta. \tag{2.8}$$

Bending energy of a thin rod

Suppose that we take a straight rod of length L_c with uniform density and cross section, and bend it into an arc with radius R_c, as in Fig. 2.6. The problem of finding the energy E_{arc} associated with this deformation is solved in many texts on continuum mechanics, and has the form (Landau and Lifshitz, 1986)

$$E_{arc}/L_c = \kappa_f/2R_c^2 = Y\mathscr{I}/2R_c^2, \tag{2.9}$$

where κ_f is called the flexural rigidity (units of $[energy]\cdot[length]$), Y is Young's modulus of the rod, and $\mathscr{I}$ is the moment of inertia of the cross section (see Fig. 2.9). Young's modulus appears in expressions of the form $[stress] = Y[strain]$, and has the same units as stress, since strain is dimensionless (see Appendix D for a review of elasticity theory). For three-dimensional materials, Y has units of energy density, and typically ranges from 10^9 J/m^3 for plastics to 10^{11} J/m^3 for metals.

The moment of inertia of the cross section is defined somewhat similarly to the moment of inertia of the mass: it is an area-weighted integral of the squared distance from an axis

$$\mathscr{I}_y = \int x^2 \, dA, \tag{2.10}$$

where the xy plane defined by the integration axes is perpendicular to the length of the rod, and dA is an element of surface area in that plane. For example, if the rod is a cylinder of radius R, the cross section has the shape of a solid disk with an area element dA at position x given by $dA = 2(R^2 - x^2)^{1/2}dx$, as shown in Fig. 2.9. Hence,

$$\mathscr{I}_y = 4 \int_0^R x^2 \, (R^2 - x^2)^{1/2} \, dx = \pi R^2/4 \text{ (solid cylinder).} \tag{2.11}$$

Should the rod have a hollow core of radius R_i, like a microtubule, then the moment of inertia in Eq. (2.11) would be reduced by the moment of inertia $\pi R_i^2/4$ of the core:

$$\mathscr{I}_y = \pi(R^4 - R_i^4)/4 \text{ (hollow cylinder).} \tag{2.12}$$

Fig. 2.9. Section through a cylindrical rod showing the xy-axes used to evaluate the moment of inertia of the cross section $\mathscr{I}_y$ in Eq. (2.10).

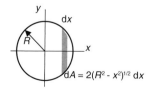

Other rods of varying cross sectional shape are treated in the problem set.

The deformation energy per unit length of the arc in Eq. (2.9) is inversely proportional to the square of the radius of curvature, or, equivalently, is proportional to the square of the curvature C from Eq. (2.7). In fact, one would expect on general grounds that the leading order contribution to the energy per unit length must be C^2, just as the potential energy of an ideal spring is proportional to the square of the displacement from equilibrium. Alternatively, then, the energy per unit length could be written as $E_{arc}/L = \kappa_f (\partial t/\partial s)^2/2$ by using Eq. (2.5). Further, there is no need for the curvature to be constant along the length of the filament, and the general expression for the total energy of deformation E_{bend} is, to lowest order,

$$E_{bend} = (\kappa_f/2) \int_0^{L_c} (\partial t/\partial s)^2 \, ds, \tag{2.13}$$

where the integral runs along the length of the filament. This form for E_{bend} is called the Kratky–Porod model; it can be trivially modified to represent a rod with an unstressed configuration that is intrinsically curved.

Fluctuations and persistence length

At zero temperature, a filament adopts a shape that minimizes its energy, which corresponds to a straight rod if the energy is governed by Eq. (2.13). At non-zero temperature, the filament exchanges energy with its environment, permitting the shape to fluctuate, as illustrated in Fig. 2.10(a). According to Eq. (2.13), the bending energy of a filament rises as its shape becomes more contorted and the local curvature along the filament grows; hence, the bending energy of the configurations increases from left to right in Fig. 2.10(a). Now, the probability $\mathcal{P}(E)$ of the filament being found in a specific configuration with energy E is proportional to the Boltzmann factor $\exp(-\beta E)$, where β is the inverse temperature $\beta = 1/k_B T$ (see Appendix C for a review). The Boltzmann factor tells us that the larger the energy required to deform the filament into a specific shape, the lower the probability that the filament will have that shape, all other things being equal. Thus, a filament will adopt

Fig. 2.10. (a) Sample of configurations available to a filament; for a given κ_f the bending energy of the filament rises as its shape becomes more contorted. (b) If the filament is a section of a circle, the angle subtended by the arc length s is the same as the change in the direction of the unit tangent vector **t** along the arc.

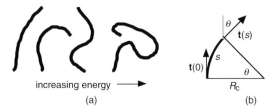

increasing energy →

(a)

(b)

configurations with small average curvature if their flexural rigidity is high or the temperature is low; their shape will resemble sections of circles, becoming contorted only at high temperatures.

Let us assume that our filament can sustain only gentle curves and has constant curvature; neither is our filament so long that it closes upon itself. The shape can then be uniquely parametrized by the angle θ between the unit tangent vectors $\mathbf{t}(0)$ and $\mathbf{t}(s)$ at the two ends of the filament (see Fig. 2.10(b)) and has a specific energy E_{bend} given by Eq. (2.9). For an arc of a circle with radius R_c, the angle θ is the same as that subtended by the arc of length s (that is, $\theta = s/R_c$) such that Eq. (2.9) can be written as

$$E_{arc} = \kappa_f s/2R_c^{\,2} = \kappa_f \theta^2/2s. \tag{2.14}$$

For now, we use s to denote the length of filament, rather than L_c, since the result that we are about to obtain also applies to short segments of more sinuous filaments, just as long as the segment has constant curvature.

The angle θ changes as the filament waves back and forth: at higher temperatures, the oscillations have a larger amplitude and the filament samples larger values of θ than at lower temperatures. To characterize the magnitude of the oscillations, we evaluate the mean value of θ^2, denoted by the conventional $\langle \theta^2 \rangle$. If the filament has a constant length, $\langle \theta^2 \rangle$ involves a weighted average of the three-dimensional position sampled by the end of the filament. That is, with one end of the filament defining the direction of a coordinate axis (say the z-axis), the other end is described by the polar angle θ and the azimuthal angle ϕ. Assuming that the shapes in the ensemble are arcs of circles, the probability of each configuration is equal to $\mathcal{P}(E_{arc})$, so that

$$\langle \theta^2 \rangle = \int \theta^2 \, \mathcal{P}(E_{arc}) \mathrm{d}\Omega / \int \mathcal{P}(E_{arc}) \mathrm{d}\Omega, \tag{2.15}$$

where the integral must be performed over the solid angle $\mathrm{d}\Omega = \sin\theta \, \mathrm{d}\theta \, \mathrm{d}\phi$. The bending energy E_{arc} is independent of ϕ, allowing the azimuthal angle to be integrated out, leaving

$$\langle \theta^2 \rangle = \int \theta^2 \exp(-\beta E_{arc}) \sin\theta \, \mathrm{d}\theta / \int \exp(-\beta E_{arc}) \sin\theta \, \mathrm{d}\theta, \tag{2.16}$$

where the minimum value of θ in the integrals is 0, and $\mathcal{P}(E_{arc})$ has been replaced by the Boltzmann factor.

By assumption, our filaments are sufficiently stiff that E_{arc} increases rapidly with θ; that is, we consider only small oscillations in the filament shape. As a consequence, the Boltzmann factor decays rapidly with θ, meaning that $\sin\theta$ is sampled only at small θ, and can be replaced by the small angle approximation $\sin\theta \sim \theta$. Hence, Eq. (2.16) becomes

$$\langle \theta^2 \rangle = (2s/\beta\kappa_f) \int x^3 \exp(-x^2) \, \mathrm{d}x / \int x \exp(-x^2) \, \mathrm{d}x, \tag{2.17}$$

after substituting Eq. (2.14) for E_{arc} and changing variables to $x = (\beta \kappa_f / 2s)^{1/2} \theta$. In the small oscillation approximation, the upper limits of the integrals in Eq. (2.17) can be extended to infinity with little error, whence both integrals are equal to 1/2 and cancel.

Thus, the expression for the mean square value of θ is

$$\langle \theta^2 \rangle \cong 2s / \beta \kappa_f \text{ (small oscillations).} \tag{2.18}$$

The combination $\beta \kappa_f$ has the units of length, and is defined as the persistence length ξ_p of the filament:

$$\xi_p \equiv \beta \kappa_f. \tag{2.19}$$

Note that the persistence length decreases with increasing temperature.

The variation in the direction of the tangent vectors provides a geometrical interpretation of the persistence length. Consider the scalar product of the unit tangent vectors $\mathbf{t}(0) \cdot \mathbf{t}(s)$, which has its maximum value of unity only if the tangent vectors are parallel at the ends of the filament. At non-zero temperature, the filament samples a variety of orientations such that the ensemble average $\langle \mathbf{t}(0) \cdot \mathbf{t}(s) \rangle = \langle \cos \theta \rangle$ has a maximum absolute value of unity. The quantity $\langle \mathbf{t}(0) \cdot \mathbf{t}(s) \rangle$ is referred to as the correlation function of the tangent vector: it describes the correlation between the direction of the tangent vectors at different positions along the curve. At low temperatures where θ is usually small, $\cos \theta$ can be approximated by $\cos \theta \sim 1 - \theta^2 / 2$, permitting the correlation function to be written as

$$\langle \mathbf{t}(0) \cdot \mathbf{t}(s) \rangle \sim 1 - \langle \theta^2 \rangle / 2. \tag{2.20}$$

The dispersion in θ in this small oscillation limit is given by Eq. (2.18), so that

$$\langle \mathbf{t}(0) \cdot \mathbf{t}(s) \rangle \sim 1 - s / \xi_p \qquad (s / \xi_p \ll 1), \tag{2.21}$$

which can be used to obtain the mean squared difference in the tangent vectors

$$\langle [\mathbf{t}(s) - \mathbf{t}(0)]^2 \rangle = 2 - 2 \langle \mathbf{t}(0) \cdot \mathbf{t}(s) \rangle \sim 2s / \xi_p \qquad (s / \xi_p \ll 1). \tag{2.22}$$

Thus, the persistence length measures the distance along the filament over which the orientation of the curve becomes decorrelated.

For a rigid rod, meaning a rod whose contour length L_c is short compared with ξ_p, Eq. (2.21) correctly predicts that $\langle \mathbf{t}(0) \cdot \mathbf{t}(L_c) \rangle \sim 1$, and also correctly predicts that the correlations initially die off linearly as contour length grows. However, if $L_c \gg \xi_p$, the filament appears floppy and $\langle \mathbf{t}(0) \cdot \mathbf{t}(L_c) \rangle$ should vanish as the tangent vectors at the extreme ends of the filament become uncorrelated, a behavior not seen in Eq. (2.21) because it was derived in the limit of small oscillations. Rather, the correct expression for the tangent correlation function applicable at short and long distances is

$$\langle \mathbf{t}(0) \cdot \mathbf{t}(s) \rangle = \exp(-s / \xi_p), \tag{2.23}$$

from which we see that Eq. (2.21) is the leading-order approximation via $\exp(-x) \sim 1 - x$ at small x. Intuitively, one would expect to obtain an expression like Eq. (2.23) by applying Eq. (2.21) repeatedly to successive sections of the filament; a more detailed derivation can be found in Doi and Edwards (1986).

2.3 Sizes of polymer chains

A function of both temperature and bending resistance, the persistence length of a filament sets the scale of its thermal undulations. As will be seen in Section 2.5, the persistence lengths of the polymers and filaments in the cell span an immense range, from a thousand times smaller to a thousand times larger than the cellular length scale of a micron. If the contour length of the filament is much smaller than its persistence length, the filament can be viewed as a relatively stiff rod undergoing only limited excursions from its equilibrium shape. In contrast, a filament with a short persistence length compared to its contour length may appear highly convoluted and display a large array of configurations.

The curve drawn in Fig. 2.11 might represent a filament with a contour length L_c many times its persistence length ξ_p. The configuration is not excessively convoluted, but would become more so as L_c grows much larger than ξ_p. The direction of the curve, as characterized by the unit tangent vector $\mathbf{t}(s)$ at arc length s, changes constantly, such that its correlation function $\langle \mathbf{t}(s) \cdot \mathbf{t}(0) \rangle$ decays exponentially with s as $\exp(-s/\xi_p)$. Over small distances $s \ll \xi_p$ (positions 1 and 2 on Fig. 2.11) the tangent vector undergoes just a modest change in direction, while over large distances $s \gg \xi_p$ (positions 1 and 3), the directions are uncorrelated. Clearly, very flexible polymers sample an extensive collection of contorted shapes with erratically changing directions. Do the configurations in this collection have any large scale characteristics, or are they just an unruly mob of rapidly changing tangents and curvatures? If the ensemble of configurations do have common or universal features, upon what properties of the filaments do they depend? We discuss several polymer families, characterized by their connectivity and interactions, to answer these questions.

Fig. 2.11. Representative configuration of a filament whose contour length is many times its persistence length ξ_p. Positions 1 and 2 are separated by an arc length s with $s < \xi_p$, while positions 1 and 3 have $s \gg \xi_p$. The displacement vector between the ends of the filament is denoted by $\mathbf{r}_{ee}$.

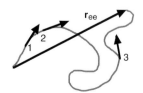

Ideal chains and filaments

Our first investigation of polymer geometry is based upon the continuous filament model introduced in Section 2.2. From among the several different observables that characterize the size of a polymer configuration, we choose to evaluate the end-to-end displacement vector $\mathbf{r}_{ee} = \mathbf{r}(L_c) - \mathbf{r}(0)$, where $\mathbf{r}(s)$ denotes the position of the filament at arc length s (see Fig. 2.11). The mean square value of $\mathbf{r}_{ee}$ is then

$$\langle \mathbf{r}_{ee}^2 \rangle = \langle [\mathbf{r}(L_c) - \mathbf{r}(0)]^2 \rangle. \tag{2.24}$$

Integrating the unit tangent vector $\mathbf{t}(s)$ of Eq. (2.4)

$$\mathbf{r}(s) = \mathbf{r}(0) + \int_0^s du\, \mathbf{t}(u), \tag{2.25}$$

permits Eq. (2.24) to be recast as

$$\langle \mathbf{r}_{ee}^2 \rangle = \int_0^{L_c} du \int_0^{L_c} dv\, \langle \mathbf{t}(u) \cdot \mathbf{t}(v) \rangle. \tag{2.26}$$

According to Eq. (2.23), the correlation function $\langle \mathbf{t}(s) \cdot \mathbf{t}(0) \rangle$ decays exponentially as $\exp(-s/\xi_p)$, leaving us to evaluate

$$\langle \mathbf{r}_{ee}^2 \rangle = \int_0^{L_c} du \int_0^{L_c} dv\, \exp(-|u-v|/\xi_p). \tag{2.27}$$

The condition that the argument of the exponential must be negative can be enforced by breaking the integral into two identical pieces where one integration variable is kept less than the other:

$$\langle \mathbf{r}_{ee}^2 \rangle = 2 \int_0^{L_c} du \int_0^u dv\, \exp(-[u-v]/\xi_p). \tag{2.28}$$

It is straightforward to solve this integral using a few changes of variables

$$2 \int_0^{L_c} \exp(-u/\xi_p)\, du \int_0^u dv\, \exp(v/\xi_p) = 2 \int_0^{L_c} du\, \exp(-u/\xi_p) \cdot$$

$$\xi_p \cdot [\exp(u/\xi_p) - 1] = 2\xi_p^2 \int_0^{L_c/\xi_p} dw\, [1 - \exp(-w)]. \tag{2.29}$$

Evaluating the last integral gives

$$\langle \mathbf{r}_{ee}^2 \rangle = 2\xi_p L_c - 2\xi_p^2 [1 - \exp(-L_c/\xi_p)]. \tag{2.30}$$

This equation simplifies in two limits. If $\xi_p \gg L_c$, Eq. (2.30) reduces to $\langle \mathbf{r}_{ee}^2 \rangle^{1/2} = L_c$ using the approximation $\exp(-x) \sim 1 - x + x^2/2 \ldots$ valid at small x; the filament appears rather rod-like with an end-to-end displacement close to its contour length. At the other extreme where $\xi_p \ll L_c$, Eq. (2.30) is approximately

$$\langle \mathbf{r}_{ee}^2 \rangle \cong 2\xi_p L_c \text{ (if } L_c \gg \xi_p), \tag{2.31}$$

implying that, over long distances compared to the persistence length, $\langle r_{ee}^2 \rangle^{1/2}$ grows like the square root of the contour length, not as the contour length itself. In other words, long polymers appear convoluted, and their average linear dimension increases much more slowly than their contour length.

Fig. 2.12. End-to-end displacement vector r_{ee} resulting from a chain with four segments.

Let's now repeat this calulation in a discrete representation to relate the geometry of a continuous polymer to its monomeric constituents. The filament is replaced by a chain of N segments each of which is described by a vector $\mathbf{b}_i$ having the same length and orientation as the segment. From all N vectors along the chain, one can construct the end-to-end displacement $\mathbf{r}_{ee}$ from

$$\mathbf{r}_{ee} = \sum_{i=1,N} \mathbf{b}_i, \tag{2.32}$$

as illustrated graphically in Fig. 2.12. Taking the ensemble average over all chains with the same number of links N, the mean squared end-to-end displacement $\langle \mathbf{r}_{ee}^2 \rangle$ is

$$\langle \mathbf{r}_{ee}^2 \rangle = \sum_i \sum_j \langle \mathbf{b}_i \cdot \mathbf{b}_j \rangle. \tag{2.33}$$

Now we specialize to the case where all segments of the chain have the same length b_i although their orientations differ. In a random chain, bond vector $\mathbf{b}_i$ can assume any orientation independent of $\mathbf{b}_j$, with the result that $\langle \mathbf{b}_i \cdot \mathbf{b}_j \rangle$ vanishes. Thus, the only terms surviving in the double sum of Eq. (2.31) are the diagonal elements $i=j$, each of which equals b^2, so that a chain with random orientations obeys

$$\langle \mathbf{r}_{ee}^2 \rangle = Nb^2 \text{ (random chain)}. \tag{2.34}$$

An alternative expression makes use of the contour length, $L_c = Nb$, of the chain

$$\langle \mathbf{r}_{ee}^2 \rangle = L_c b. \tag{2.35}$$

This expression is the same as that found in the continuum description in the limit where $L_c \gg \xi_p$, namely $\langle \mathbf{r}_{ee}^2 \rangle = 2\xi_p L_c$, except that the persistence length ξ_p has been replaced by $b/2$. In other words, both descriptions show that the linear dimension of very flexible filaments increases as the square root of the contour length. The scaling behavior $\langle \mathbf{r}_{ee}^2 \rangle^{1/2} \sim N^{1/2}$ or $L_c^{1/2}$ in Eqs. (2.31) and (2.34) is referred to as *ideal* scaling. Note that our determination of the ideal scaling exponent does not depend on the dimension of space in which the chain resides: random chains in two dimensions (i.e., confined to a plane) or three dimensions both exhibit the same scaling behavior.

Even if the bond geometry is somewhat restricted, ideal scaling still applies if the chains are permitted to intersect themselves. In the freely rotating chain model, successive chains elements $\mathbf{b}_i$ and $\mathbf{b}_{i+1}$ are forced to have the same polar angle α, although the bonds may swivel around each other. This model is solved in the problem set (see also Flory (1953), p. 414), and obeys

$$\langle \mathbf{r}_{ee}^2 \rangle = Nb^2 (1 - \cos\alpha)/(1 + \cos\alpha), \tag{2.36}$$

in the large N limit. Now, Eq. (2.36) has the same scaling exponent for $\langle \mathbf{r}_{ee}^2 \rangle$ as a function of N as does Eq. (2.34), namely $N^{1/2}$, demonstrating that self-intersecting freely rotating chains are ideal. Further, Eq. (2.36) reduces to (2.34) when the chain is measured on a length scale of $b \, [(1 - \cos\alpha)/(1 + \cos\alpha)]^{1/2}$, suggesting that an *effective bond length* B_{eff} can be defined for freely rotating chains via

$$B_{eff} = b \, [(1 - \cos\alpha)/(1 + \cos\alpha)]^{1/2}. \tag{2.37}$$

The parametrizations of $\langle \mathbf{r}_{ee}^2 \rangle$ employed most commonly for ideal chains with N segments are

$$\langle \mathbf{r}_{ee}^2 \rangle = \begin{cases} NB_{eff}^2 \\ L_c \mathscr{L}_K \\ 2L_c \xi_p \end{cases} \tag{2.38}$$

where the Kuhn length, $\mathscr{L}_K$, is defined in analogy with the monomer length: $\langle \mathbf{r}_{ee}^2 \rangle = N_K \mathscr{L}_K^2$ and $L_c = N_K \mathscr{L}_K$, with N_K the number of Kuhn lengths in the contour length.

Self-avoiding linear chains

Our treatment of random chains places no restriction on the interaction between chain segments: nothing in the mathematical representation of the chains prevents the displacement vectors from crossing one another. However, physical systems have an excluded volume that enforces self-avoidance of the chain, as illustrated in Fig. 2.13 for two-dimensional chains. This steric interaction among the chain elements is important for chains in one-, two- and three-dimensional systems. As an illustration, consider the simple situation in which a chain lies along the x-axis. Self-avoidance forbids the chain from reversing on itself from one step to the next, so that the end-to-end distance must be just the contour length Nb: i.e., $\langle \mathbf{r}_{ee}^2 \rangle^{1/2} \sim N^1$ for a straight chain in one dimension. But Eq. (2.34) shows that $\langle \mathbf{r}_{ee}^2 \rangle^{1/2}$ for ideal chains scales like $N^{1/2}$, *independent of embedding dimension*. Thus, we conclude that in one dimension, self-avoidance of a chain dramatically affects its scaling properties: N^1 for self-avoiding chains and $N^{1/2}$ for ideal chains. Similar conclusions can be drawn for chains in two and three dimensions, although the scaling exponents are different. As shown by Flory, rather general arguments lead to the

Fig. 2.13. Self-avoidance changes the scaling properties of chains in one-, two- and three-dimensional systems. In the two-dimensional configurations displayed here, (a) is a random chain and (b) is a self-avoiding chain.

(a)

(b)

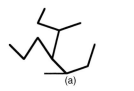
(a)

(b)

Fig. 2.14. Sample
configurations of a branched
polymer (a) and a dense chain
(b) in two dimensions. To aid
the argument in the text, the
chain in (b) consists of linked
squares, which, when packed
tightly together, cover an area
$\sim r^2$ in two dimensions.

prediction that the scaling exponents of self-avoiding linear chains should obey (see Section 6.4)

$$\nu_{FL} = 3/(2+d), \tag{2.39}$$

where d is the embedding dimension. Eq. (2.39) gives $\nu_{FL} = 1$, 3/4, 3/5 and 1/2, in one to four dimensions, respectively, predictions which have been shown to be exact or nearly so. As the ideal scaling exponent cannot be less than 1/2, Eq. (2.39) is not valid in more than four dimensions; hence, the effects of self-avoidance are irrelevant in four or more dimensions and the scaling is always ideal.

Branched polymers

The polymers discussed in most of this text are linear chains; however, there are many examples of polymers with extensive side branches. The scaling behavior of such *branched polymers* should not be the same as single chains, since branching adds more monomers along the chain length, as illustrated in Fig. 2.14(a). Because a branched polymer has more than two ends, the end-to-end displacement has to be replaced by a different measure of the polymer size, such as the radius of gyration, R_g (see end-of-chapter problems). The radius of gyration for branched polymers is found to have a scaling form

$$\langle R_g^2 \rangle^{1/2} \sim N^\nu, \tag{2.40}$$

where N is the number of polymer segments and $\nu = 0.64$ and 0.5 in two and three dimensions, respectively (see Section 6.4). In comparison, self-avoiding linear chains have scaling exponents of 3/4 and 0.59, respectively (see Eq. (2.39)), meaning that the spatial region occupied by branched polymers grows more slowly with N than does that of linear chains; i.e., linear chains are less dense than branched polymers. Fluid membranes also behave like branched polymers at large length scales (see Section 6.4).

Collapsed chains

None of the chain configurations described so far in this section is as compact as it could be. Consider a system of identical objects, say

Table 2.1. *Exponents for the scaling law $\langle R_g^2 \rangle^{1/2} \sim N^\nu$ for ideal (or random) chains, self-avoiding chains and branched polymers, as a function of embedding dimension d. Collapsed chains have the highest density and obey $\langle R_g^2 \rangle^{1/2} \sim N^{1/d}$*

Configuration	$d=2$	$d=3$	$d=4$
Ideal chains	1/2	1/2	1/2
Self-avoiding chains	3/4	0.59	1/2
Branched polymers	0.64	1/2	—
Collapsed chains	1/2	1/3	1/4

squares in two dimensions or cubes in three dimensions, having a length b to the side such that each object has a "volume" of b^d in d dimensions, and N of these objects have a volume Nb^d. The configuration of the N objects with the smallest surface area is the most compact or the most *dense* configuration, as illustrated in Fig. 2.14(b), and we denote by r the linear dimension of this configuration. Ignoring factors of π and the like, the total volume Nb^d of the most compact configuration is proportional to r^d, so that r itself scales like

$$r \sim N^{1/d} \text{ (dense)}. \qquad (2.41)$$

Polymers can be made to collapse into their most dense configurations by a variety of experimental means, including changes in the solvent, and it is observed that the collapse of the chains occurs at a well-defined phase transition.

The scaling exponents of all the systems that we have considered in this section are summarized in Table 2.1. If the chains are self-avoiding, $1/d$ represents the lower bound on the possible scaling exponents, and the straight rod scaling of $\langle R_g^2 \rangle^{1/2} \sim N^1$ represents the upper bound. One can see from the table that random or self-avoiding chains, as well as branched polymers, exhibit scaling behavior that lies between these extremes.

2.4 Chain configurations and elasticity

Averaged quantities such as the tangent correlation function or $\langle r_{ee}^2 \rangle$ provide only limited information about the behavior of chains or filaments. In this section, we develop a more complete picture of chain geometry by determining the probability distribution for the length of their end-to-end displacements. These distributions confirm that it is highly unlikely for a random chain to be found in a fully stretched con-

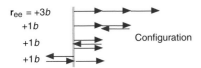

$r_{ee} = +3b$

$+1b$

$+1b$ Configuration

$+1b$

Fig. 2.15. Configurations for a one-dimensional chain with three segments of equal length b. Only half of the allowed configurations are shown, namely those with $r_{ee} > 0$.

figuration: the most likely value of r_{ee}^2 for a freely jointed chain is not far from its mean value of Nb^2. That the chain has an average size much smaller than its contour length is a reflection of its entropy: there are far more configurations available for $r_{ee} \sim N^{1/2}b$ than there are for the fully extended situation $r_{ee} = Nb$. Because of this, as the chain is made to straighten out by an external force, its entropy decreases and work must be done on the chain to stretch it: in other words, the chain behaves elastically because of its entropy.

Random chain in one dimension

Consider the set of one-dimensional random chains with three segments, as shown in Fig. 2.15: each chain starts off at the origin, and a link in the chain can point to the right or the left. As each link has two possible orientations, there are a total of $2^3 = 8$ possible configurations for the chain as a whole. Using $C(\mathbf{r}_{ee})$ to denote the number of configurations with a particular end-to-end displacement $\mathbf{r}_{ee}$, the eight configurations are distributed according to:

$$C(+3b) = 1 \quad C(+1b) = 3 \quad C(-1b) = 3 \quad C(-3b) = 1. \tag{2.42}$$

The reader will recognize that these values of $C(\mathbf{r}_{ee})$ are equal to the binomial coefficients in the expansion of $(p+q)^3$; i.e., the values are the same as the coefficients $N\,!/i\,!j\,!$ in the expansion

$$(p+q)^N = \sum_{i=0,N} \{N\,!/i\,!j\,!\}\, p^i q^j, \tag{2.43}$$

where $j = N - i$. Is this fortuitous? Not at all; the different configurations in Fig. 2.15 just reflect the number of ways that the left- and right-pointing vectors can be arranged. So, if there are i vectors pointing left, and j pointing right, such that $N = i + j$, then the total number of ways in which they can be arranged is just the binomial coefficient

$$C(i, j) = N!/i!\,j!. \tag{2.44}$$

One can think of the configurations in Fig. 2.15 as random walks in which each step (or link) along the walk occurs with probability 1/2. Thus, the probability $\mathcal{P}(i, j)$ for there to be a configuration with (i, j) steps to the (left, right) is equal to the product of the total number of

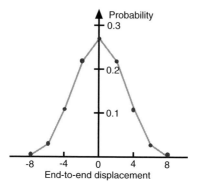

Fig. 2.16. Probability distribution from Eq. (2.45) for a one-dimensional chain with six segments.

configurations ($C(i, j)$ from Eq. (2.44)) with the probability of an individual configuration, which is $(1/2)^i (1/2)^j$:

$$\mathscr{P}(i, j) = \{N!/i!\,j!\}\,(1/2)^i\,(1/2)^j. \tag{2.45}$$

Note that the probability in Eq. (2.45) is appropriately normalized to unity, as can be seen by setting $p = q = 1/2$ in Eq. (2.43)

$$\Sigma_{i=0,N}\,\mathscr{P}(i, j) = \Sigma_{i=0,N}\,\{N!/i!\,j!\}\,(1/2)^i\,(1/2)^j = (1/2 + 1/2)^N = 1. \tag{2.46}$$

What happens to the probability distribution as the number of steps becomes larger and the distribution consequently appears more continuous? The probability distribution for a one-dimensional chain with $N = 6$ is shown in Fig. 2.16, where we note that the end-to-end displacement $\mathbf{r}_{ee} = (j - i) = (2j - N)$ changes by 2 for every unit change in i or j. The distribution is peaked at $\mathbf{r}_{ee} = 0$, as one would expect, and then falls off towards zero at large values of $|\mathbf{r}_{ee}|$ where $i = 0$ or N. As becomes ever more obvious for large N, the shape of the curve in Fig. 2.16 resembles a Gaussian distribution, which has the form

$$\mathscr{P}(x) = (2\pi\sigma^2)^{-1/2}\,\exp[-(x - \mu)^2/2\sigma^2]. \tag{2.47}$$

Normalized to unity, this expression is a probability density (i.e., a probability per unit value of x) such that the probability of finding a state between x and $x + \mathrm{d}x$ is $\mathscr{P}(x)\mathrm{d}x$. The mean value μ of the distribution can be obtained from

$$\mu = \langle x \rangle = \int_{-\infty}^{\infty} x\,\mathscr{P}(x)\,\mathrm{d}x, \tag{2.48}$$

and its variance σ^2 is

$$\sigma^2 = \langle (x - \mu)^2 \rangle = \langle x^2 \rangle - \mu^2, \tag{2.49}$$

as expected. Eq. (2.47) is the general form of the Gaussian distribution; the values of μ and σ are specific to the system of interest. For one Cartesian component of $\mathbf{r}_{ee}$ in a random chain, $\mu = 0$ because the vectors $\mathbf{r}_{ee}$ are isotropically distributed; taken together, $\mu = 0$ and $\langle x^2 \rangle = Nb^2$ imply

$$\sigma^2 = Nb^2 \text{ (one dimension)}, \tag{2.50}$$

for random chains. Proofs of the equivalence of the Gaussian and binomial distributions at large N can be found in most statistics textbooks. However, the Gaussian distribution provides a surprisingly accurate approximation to the binomial distribution even for modest values of N, as can be seen from Fig. 2.16.

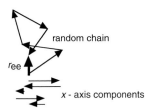

Fig. 2.17. Projection of the segments of a two-dimensional chain onto the x-axis.

Random chain in three dimensions

By projecting their configurations onto a set of Cartesian axes, as illustrated in Fig. 2.17, three-dimensional random chains can be treated as three separate one-dimensional systems. For example, the x-component of the end-to-end displacement, $r_{ee,x}$, is just the sum of the individual monomer vectors as projected onto the x-axis:

$$r_{ee,x} = \sum_i b_{i,x} \tag{2.51}$$

where $b_{i,x}$ is the x-projection of the monomer vector $\mathbf{b}_i$. For freely jointed chains, the component $b_{i,x}$ is independent of the component $b_{i+1,x}$, so the projections form a random walk in one dimension, although the x-axis projections are of variable lengths even if all monomers have the same b. If the number of segments is large, the probability distribution with variable segment length has the same form as the distribution with uniform segment length (Chapter 1 of Reif, 1965),

$$\mathcal{P}(x) = (2\pi\sigma^2)^{-1/2} \exp(-r_{ee,x}^2/2\sigma^2), \tag{2.52}$$

with a variance given by $\sigma^2 = N\langle b_x^2 \rangle$.

Now $\langle b_x^2 \rangle$ refers to the expectation of the projection of the individual segments on the x-axis. Because of symmetry, we anticipate that the mean projections are independent of direction, so

$$\langle b_x^2 \rangle = \langle b_y^2 \rangle = \langle b_z^2 \rangle = b^2/3, \tag{2.53}$$

the last equality arising from $\langle b_x^2 \rangle + \langle b_y^2 \rangle + \langle b_z^2 \rangle = \langle b^2 \rangle = b^2$ for segments of constant length. Hence, the variance in Eq. (2.52) is

$$\sigma^2 = Nb^2/3 \text{ (for three dimensions).} \tag{2.54}$$

Returning to the properties of three-dimensional chains, the probability of finding the end-to-end displacement in a volume $dx\,dy\,dz$ centered on the position (x,y,z) is $\mathcal{P}(x,y,z)dx\,dy\,dz$, where $\mathcal{P}(x,y,z)$ is

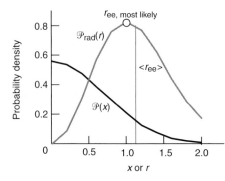

Fig. 2.18. Probability distributions for random chains in three dimensions. Two cases are shown: the three-dimensional distribution, Eq. (2.57), as a function of $r=r_{ee}$, and the x-axis projection, Eq. (2.52), as a function of $x=r_{ee,x}$ ($\sigma^2=1/2$ in both distributions).

the product of the probability distributions in each of the Cartesian directions

$$\mathcal{P}(x,y,z)=\mathcal{P}(x)\,\mathcal{P}(y)\,\mathcal{P}(z)=(2\pi\sigma^2)^{-3/2}$$
$$\exp[-(x^2+y^2+z^2)/2\sigma^2], \tag{2.55}$$

where σ^2 is still given by Eq. (2.54), and where $x\equiv r_{ee,x}$, etc. Eq. (2.55) says that, of all possible chain configurations, the most likely set of coordinates for the tip of the chain is (0,0,0), which is the origin or tail of the chain; it does *not* say that the most likely value of r_{ee} is zero. Indeed, the distribution of the magnitude of $\mathbf{r}_{ee}$ must reflect the fact that many different coordinate positions have the same r. The probability for the chain having a radial end-to-end distance between r and $r+dr$ is $\mathcal{P}_{rad}(r)dr$, where $\mathcal{P}_{rad}(r)$ is the probability per unit length obtained from

$$\mathcal{P}(x,y,z)dx\,dy\,dz=\mathcal{P}_{rad}(r)dr, \tag{2.56}$$

so that

$$\mathcal{P}_{rad}(r)=4\pi r^2\,(2\pi\sigma^2)^{-3/2}\,\exp(-r^2/2\sigma^2). \tag{2.57}$$

It's the extra factor of r^2 outside of the exponential that shifts the most likely value of r_{ee} away from zero. Eq. (2.57), as well as the projection of the chain on the x-axis, is plotted in Fig. 2.18. We can equate to zero the derivative of $\mathcal{P}_{rad}(r)$ with respect to r to find the most likely value of r_{ee}. A summary of the results for ideal chains in three dimensions is:

$$r_{ee,\,most\,likely}=(2/3)^{1/2}\,N^{1/2}b, \tag{2.58}$$

$$\langle r_{ee}\rangle=(8/3\pi)^{1/2}\,N^{1/2}b, \tag{2.59}$$

and, of course,

$$\langle \mathbf{r}_{ee}^2\rangle=Nb^2. \tag{2.60}$$

Note that r_{ee} in Eqs. (2.58) and (2.59) is the scalar radius $r_{ee}=(\mathbf{r}_{ee}^2)^{1/2}$.

Entropic elasticity

The probability distribution functions, as illustrated in Fig. 2.18, confirm our intuition that far more chain configurations have end-to-end displacements close to the mean value of r_{ee} than to the chain contour length L_c. Being proportional to the logarithm of the number of configurations, the entropy S of a chain must decrease as the chain is stretched from its equilibrium length. Now the free energy of an ensemble of chains is $F = E - TS$, which is simply $F = -TS$ for freely jointed chains, since their configurations all have vanishing energy E. Thus, S decreases and F increases as the chain is stretched at non-zero temperature; in other words, work must be done to stretch the chain, and the chain is elastic by virtue of its entropy.

Viewed as a spring obeying Hooke's law, the effective force constant of the chain can be extracted by comparing the distributions for the end-to-end displacement of the chain with that of a spring, whose fluctuations can be calculated using statistical mechanics. Now, a Hookean spring has a potential energy $V(x)$ equal to $k_{sp}x^2/2$, where x is the displacement from equilibrium and k_{sp} is the force constant of the spring. Aside from an overall normalization factor, the probability distribution $\mathcal{P}(x)$ for the spring displacement x is proportional to the usual Boltzmann factor $\exp(-E/k_B T)$, which becomes, for the Hooke's Law potential

$$\mathcal{P}(x) \sim \exp(-k_{sp}x^2/2k_B T). \tag{2.61}$$

The probability distribution for the displacement of an ideal chain from Eq. (2.52) is $\mathcal{P}(x) \sim \exp(-x^2/2\sigma^2)$, again aside from an overall normalization factor. Comparing the functional form of the two distributions at large x gives $k_{sp} = k_B T/\sigma^2$, where $\sigma^2 = Nb^2/d$ for ideal chains embedded in d dimensions (the dimensionality can be seen from Eqs. (2.50) and (2.54)). Hence, in three dimensions, we expect

$$k_{sp} = 3k_B T/Nb^2 = 3k_B T/2\xi_p L_c \text{ (three dimensions)}, \tag{2.62}$$

using $L_c = Nb$ and $\xi_p = b/2$ for an ideal chain. Observe that k_{sp} increases with temperature, which is readily demonstrated experimentally by hanging a weight from an elastic band, and then using a device (like a hair dryer) to heat the elastic. The weight will be seen to rise as the elastic heats up, since k_{sp} rises simultaneously and provides greater resistance to the stretching of the elastic by the weight.

Highly stretched chains

The Gaussian probability distribution, Eq. (2.52), gives a good description of chain behavior at small displacements from equilibrium. It predicts, from Eq. (2.62), that the force f required to produce an extension x in the end-to-end displacement is $f = (3k_{\mathrm{B}}T/2\xi_{\mathrm{p}}L_{\mathrm{c}})x$, which can be rewritten as

$$x/L_{\mathrm{c}} = (2\xi_{\mathrm{p}}/3k_{\mathrm{B}}T)\, f. \tag{2.63}$$

If the chain segments are individually inextensible, the force required to extend the chain should diverge as the chain approaches its maximal extension, $x/L_{\mathrm{c}} \to 1$. Such a divergence is not present in Eq. (2.63), indicating that the Gaussian distribution must be increasingly inaccurate and ultimately invalid as an inextensible chain is stretched towards its contour length.

Of course, the Gaussian distribution is only an approximate representation of freely jointed chains; fortunately, the force–extension relation of rigid, freely jointed rods can be obtained analytically. For those familiar with the example, the problem is analogous to the alignment of spin vectors in an external field, where the spin vectors represent the projection of the polymer segments along the direction of the applied field. It is straightforward to show (Kuhn and Grün, 1942; James and Guth, 1943; see also Flory, 1953, p. 427) that the solution has the form

$$x/L_{\mathrm{c}} = \mathcal{L}(2\xi_{\mathrm{p}}f/k_{\mathrm{B}}T), \tag{2.64}$$

where $\mathcal{L}(y)$ is the Langevin function

$$\mathcal{L}(y) = \coth(y) - 1/y. \tag{2.65}$$

Note that x in Eq. (2.64) is the projection of the end-to-end displacement along the direction of the applied force. For small values of f, Eq. (2.64) reduces to the Gaussian expression Eq. (2.63); for very large values of f, the Langevin function tends to 1 so that x asymptotically approaches L_{c} in Eq. (2.64), as desired.

The force–extension relation of freely jointed rods provides a reasonably accurate description of biopolymers. Its weakness lies in viewing the polymer as a chain of rigid segments: thick filaments such as microtubules and DNA surely look more like continuously flexible ropes than chains of rigid rods. A more appropriate representation of flexible filaments can be derived from the Kratky–Porod energy expression, Eq. (2.13), and is referred to as the worm-like chain (WLC). Although the general form of the WLC force–extension relationship is numerical, an accurate interpolation formula has been obtained by Marko and Siggia (1995):

$$\xi_{\mathrm{p}}f/k_{\mathrm{B}}T = (1/4)(1 - x/L_{\mathrm{c}})^{-2} - 1/4 + x/L_{\mathrm{c}}. \tag{2.66}$$

Again, the force diverges in this expression as $x/L_{\mathrm{c}} \to 1$, as desired. Eq. (2.66) and the freely jointed chain display the same behavior at both

large and small forces, although their force–extension curves may dis-
agree by as much as 15% for intermediate forces.

2.5 Elasticity of cellular filaments

The bending deformation energy of a filament can be characterized by
its flexural rigidity κ_f. Having units of [*energy · length*], the flexural rigidity of uniform rods can be written as a product of the Young's modulus
Y (units of [*energy/length^3*]) and the moment of inertia of the cross
section $\mathcal{I}$ (units of [*length^4*]): $\kappa_f = Y\mathcal{I}$. At finite temperature T, the rod's
shape fluctuates, with the local orientation of the rod changing strongly
over length scales characterized by the persistence length $\xi_p = \kappa_f/k_B T$,
where k_B is Boltzmann's constant. We now review the experimental
measurements of κ_f or ξ_p for a number of filaments in the cell, and then
interpret them using results from Sections 2.2–2.4.

Measurements of persistence length

Mechanical properties of the principal structural filaments of the cytoskeleton – spectrin, actin, intermediate filaments and microtubules – have
been obtained through a variety of methods. In first determining the persistence length of spectrin, Stokke *et al.* (1985a) related the intrinsic viscosity of a spectrin dimer to its root-mean-square radius, from which the
persistence length could be extracted via a relationship like Eq. (2.30).
The resulting values of ξ_p covered a range of 15–25 nm, depending upon
temperature. A more recent measurement (Svoboda *et al.*, 1992) employs
optical tweezers to hold a complete erythrocyte cytoskeleton in a flow
chamber while the appearance of the cytoskeleton is observed as a function of the salt concentration of the medium. It is found that a persistence
length of 10 nm is consistent with the mean squared end-to-end displace-
ment $\langle r_{ee}^2 \rangle$ of the spectrin tetramer and with the dependence of the skeleton's diameter on salt concentration. Both measurements comfortably
exceed the lower bound of 2.5 nm placed on the persistence length of a
spectrin *monomer* by viewing it as a freely jointed chain of segment length
$b = 5$ nm and invoking $\xi_p = b/2$ from Eqs. (2.31) and (2.35) (5 nm is the
approximate length of each of approximately 20 barrel-like subunits in a
spectrin monomer of contour length 100 nm; see Fig. 2.1(a)).

The persistence length of F-actin has been extracted from the analysis of perhaps a dozen experiments, although we cite here only a few
recent works as an introduction to the literature. The measurements
involve both native and fluorescently labelled actin filaments, which may
account for some of the variation in the reported values of ξ_p. The principal techniques include:

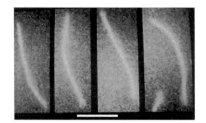

Fig. 2.19. Thermal fluctuations of a rhodamine-labelled actin filament observed by fluorescence microscopy at intervals of 6s. The bar is 5 μm in length (reprinted with permission from Isambert *et al.*, 1995; © 1995 by the American Society for Biochemistry and Molecular Biology).

(i)   dynamic light scattering, which has given a rather broad range of results, the most recent converging on $\xi_p \sim 16$ μm (Janmey *et al.*, 1994),

(ii) direct microscopic observation of the thermal fluctuations of fluorescently labelled actin filaments, as illustrated in Fig. 2.19. Actin filaments stabilized by phalloidin are observed to have $\xi_p = 18 \pm 1$ μm (Gittes *et al.*, 1993; Isambert *et al.*, 1995), while unstabilized actin filaments are more flexible, at $\xi_p = 9 \pm 0.5$ μm (Isambert *et al.*, 1995),

(iii) direct microscopic observation of the driven oscillation of labelled actin filaments, giving $\xi_p = 7.4 \pm 0.2$ μm (Riveline *et al.*, 1997).

Taken together, these experiments and others indicate that the persistence length of F-actin lies in the 10–20 μm range, about a thousand times larger than spectrin dimers.

Microtubules have been measured with several of the same techniques as employed for extracting the persistence length of actin filaments. Again, both pure and treated (in this case, taxol-stabilized) microtubules have been examined by means of:

(i)   direct microscopic observation of the bending of microtubules as they move within a fluid medium, yielding ξ_p in the range of 1–8 mm (Venier *et al.*, 1994; Kurz and Williams, 1995; Felgner *et al.*, 1996),

(ii) direct microscopic observation of the thermal fluctuations of microtubules, giving a range of 1–6 mm (Gittes *et al.*, 1993; Venier *et al.*, 1994; Kurz and Williams, 1995),

(iii) direct microscopic observation of the buckling of a single, long microtubule confined within a vesicle under controlled conditions, leading to $\xi_p = 6.3$ mm (Elbaum *et al.*, 1996).

Thus, the persistence length of microtubules is more than an order of magnitude larger than a typical cell diameter, with several measurements yielding results near the high end of the 1–6 mm range.

The filaments of the cytoskeleton are not the only polymers whose mechanical properties are important to the operation of the cell. For example, the packing of DNA into the restricted volume of the cell is a

Table 2.2. *Linear density* λ_p *(mass per unit length) and persistence length* ξ_p *of some biologically important polymers*

Polymer	Configuration	λ_p (D/nm)	ξ_p (nm)
Long alkanes	linear polymer	~110	~0.5
Spectrin	two-strand filament	4500	10–20
DNA	double helix	1900	53 ± 2
F-actin	filament	16000	10–20×10^3
Intermediate filaments	32-strand filament	~35000	—
Tobacco mosaic virus		~140000	$\sim 1 \times 10^6$
Microtubules	13 protofilaments	160000	1–6×10^6

significant challenge, given both the contour length and persistence length of a DNA molecule. Measurements of the DNA persistence length date back at least a decade to the work of Taylor and Hagerman (1990), who found $\xi_p = 45 \pm 1.5$ nm by observing the rate at which a linear strand of DNA closes into a circle. A more recent experiment directly manipulates a single DNA molecule by attaching a magnetic bead to one end of the filament (while the other is held fixed) and applying a force by means of an external magnetic field. The resulting force–extension relation for DNA from bacteriophage lambda (a virus that attacks bacteria such as *E. coli*) is found to be well-described by the worm-like chain model, Eq. (2.66), which involves only two parameters – the contour length and the persistence length (Bustamante *et al.*, 1994). As well as yielding a fitted contour length in agreement with the crystallographic value, the procedure gives a fitted persistence length of 53 ± 2 nm, in the same range as found earlier by Taylor and Hagerman. Yet another approach records the motion of a fluorescently labelled DNA molecule in a fluid (Perkins *et al.*, 1995), the analysis of which gives $\xi_p \sim 68$ nm (Stigter and Bustamante, 1998), a slightly higher value than that of unlabelled DNA. The torsion resistance of DNA has also been measured (Strick *et al.*, 1996).

The above measurements are summarized in Table 2.2, which also displays the mass per unit length of the filament (from Section 2.1). For comparison, the table includes alkanes with a very short persistence length (from Flory, 1969) and the tobacco mosaic virus, a hollow rod-like structure with a linear density similar to that of a microtubule and a persistence length to match. Experimentally, the flexural rigidity of the virus on a substrate is obtained by observing the response of the virus when probed by the tip of an atomic force microscope (Falvo *et al.*, 1997). In this case and several others, the persistence length quoted in Table 2.2 is found from the flexural rigidity via Eq. (2.19).

ξ_p and Young's modulus

The measured persistence lengths in Table 2.2 span more than six orders of magnitude, a much larger range than the linear density, which covers about three orders of magnitude. We can understand this behavior by viewing the polymers as flexible rods, whose flexural rigidity from Eq. (2.9) is

$$\kappa_f = Y\mathcal{I}, \tag{2.67}$$

and corresponding persistence length, according to Eq. (2.19), is

$$\xi_p = Y\mathcal{I}/k_B T, \tag{2.68}$$

where the moment of inertia of the cross section for hollow rods of inner radius R_i and outer radius R is (from Eq. (2.12))

$$\mathcal{I} = \pi(R^4 - R_i^4)/4.$$

For some hollow biofilaments like the tobacco mosaic virus, for which $R/R_i \sim 4.5$, only a small error is introduced by neglecting R_i^4 in the expression for $\mathcal{I}$, so that

$$\xi_p \cong \pi Y R^4 / 4 k_B T, \tag{2.69}$$

although we note that this expression is in error by a factor of two for microtubules ($R \sim 14$ nm and $R_i \sim 11.5$ nm; see Amos and Amos, 1991). Being raised to the fourth power, R must be known relatively well to make an accurate prediction with Eq. (2.69). Such is not always the case, and a somewhat gentler approach, which, in some sense averages over the bumpy atomic boundary of a molecule, replaces R^2 by the mass per unit length λ_p using the relationship $\lambda_p = \rho_m \pi R^2$ for a cylinder, where ρ_m is the mass per unit volume. Thus, we obtain

$$\xi_p \cong (Y/4\pi k_B T \rho_m^2) \lambda_p^2, \tag{2.70}$$

which implies that the persistence length should be proportional to the square of the mass per unit length, if Y and ρ_m are relatively constant from one filament to the next.

Data from Table 2.2 are plotted logarithmically in Fig. 2.20 as a test of the quadratic dependence of ξ_p on λ_p suggested by Eq. (2.70). With the exception of spectrin, which is a loosely intertwined pair of filaments, the data are consistent with the fitted functional form $\xi_p = 2.5 \times 10^{-5} \lambda_p^2$, where ξ_p is in nm and λ_p is in Da/nm. Because the data span so many orders of magnitude, the approximations (such as constant Y and ρ_m) behind the scaling law are supportable, and the graph can be used to find Young's modulus of a generic biofilament. Equating the fitted numerical

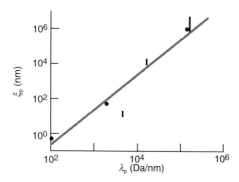

Fig. 2.20. Logarithmic plot of persistence length ξ_p against linear density λ_p for the data in Table 2.2. The straight line through the data is the function $\xi_p = 2.5 \times 10^{-5} \lambda_p^2$, where ξ_p is in nm and λ_p is in Da/nm.

factor 2.5×10^{-5} nm³/D² with $Y/4\pi k_B T \rho_m^2$ gives $Y = 0.5 \times 10^9$ J/m³ for $k_B T = 4 \times 10^{-21}$ J and $\rho_m = 10^3$ kg/m³ (which is the density of water, or roughly the density of many hydrocarbons). Although no more accurate than a factor of two, this value of Y is in the same range as $Y = 1–2 \times 10^9$ J/m³ found for collagen (linear density of 1000 Da/nm), but much smaller than that of dry cellulose (8×10^{10} J/m³) or steel (2×10^{11} J/m³). Of course, Eq. (2.70) can be applied to individual filaments if their radii and persistence lengths are sufficiently well-known, yielding $Y \sim (0.5–1.5) \times 10^9$ J/m³ for most filaments.

Filament geometry and elasticity

How should the cytoskeletal filaments, whose flexural rigidities we now know, behave in the cell? As one representative situation, we examine microtubules, which can easily be as long as a typical cell is wide (say 10 μm) and have a persistence length one hundred times the cell diameter. Microtubules should not display very strong thermal oscillations, and indeed Eq. (2.18) demonstrates that the root mean square angle of oscillation $\langle \theta^2 \rangle^{1/2}$ is about a tenth of a radian (or about 6°) for a sample microtubule with $L_c = 10$ μm if the persistence length is in the mid-range of the experimental values, say $\xi_p = 2 \times 10^3$ μm. This doesn't mean that microtubules in the cell behave quite like steel rods in a plastic bag, as can be seen by the image in Fig. 2.21(a), but a generous amount of energy is required to give our sample microtubule a substantial curvature: for example, $E_{arc} = (\xi_p/2L_c) k_B T = 100 \ k_B T$ if $R_c = L_c = 10$ μm, according to Eq. (2.14). Indeed, Elbaum et al. (1996) observe that a single microtubule with a length longer than the mean diameter of an artificial vesicle can cause the vesicle to deform into an ovoid.

Having a persistence length about one-tenth of its contour length, a spectrin tetramer should appear contorted on cellular length scales. In fact, its mean end-to-end displacement $\langle r_{ee}^2 \rangle^{1/2}$ is just 75 nm, which

Fig. 2.21. (a) Microtubules, with a persistence length in the millimeter range, are relatively stiff on cellular length scales (image is about 10 μm across, reprinted with permission from Osborn *et al.*, 1978; © 1978 by the Rockefeller University Press), in contrast to spectrin (b), which has a persistence length five orders of magnitude smaller (spectrin network of the human erythrocyte; image from Dr John Heuser, Washington University in St. Louis, as displayed in Steck, 1989; see also Heuser, 1983).

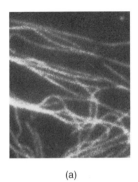

(a)

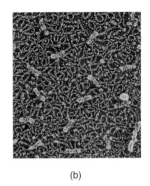

(b)

is only one-third of its contour length (200 nm), so the tetramers form a sinuous web when joined to form a network, as shown in Fig. 2.21(b). Thus, the network of spectrin tetramers in the human erythrocyte cytoskeleton can be stretched considerably to achieve a maximum area that is $(200/75)^2 \sim 7$ times its equilibrium area. As discussed in Section 2.4, highly convoluted chains such as spectrin resist extension, behaving like entropic springs with a spring constant $k_{sp} = 3k_B T/2\xi_p L_c$ in three dimensions, from Eq. (2.62). Our spectrin tetramer, then, has a spring constant of about 2×10^{-6} J/m^2, which, although not a huge number, provides the cytoskeleton with enough shear resistance to restore a red cell to its equilibrium shape after passage through a narrow capillary. Lastly, we recall from the definition $\xi_p = \kappa_f/k_B T$ that the persistence length should decrease with temperature, if the flexural rigidity is temperature-independent. The temperature-dependence of the elasticity of biofilaments has not been as extensively studied as that of conventional polymers, but Stokke *et al.* (1985a) do find that the persistence length decreases with temperature roughly as expected for an entropic spring.

2.6 Summary

Sometimes criss-crossing the interior of a cell, sometimes forming a mat or wall around it, the biological chains and filaments of the cell range in diameter up to 25 nm and have a mass per unit length covering more than three orders of magnitude, from ~ 100 Da/nm for alkanes to 160 000 Da/nm for microtubules. A simple mathematical representation of these filaments views them as structureless lines characterized by a position $\mathbf{r}(s)$ and tangent vector $\mathbf{t}(s) = \partial \mathbf{r}(s)/\partial s$, where s is the arc length along the line. The lowest-order expression for the energy per unit length of deforming a filament from its straight-line configuration is $(\kappa_f/2)(\partial \mathbf{t}(s)/\partial s)^2$, where κ_f is the flexural rigidity of the filament.

At non-zero temperature, filaments can exchange energy with their surroundings, permitting their shapes to fluctuate as they bend and twist. Their orientation changes direction with both position and time, such that the direction of the tangent vectors to the filament decays as $\langle \mathbf{t}(0) \cdot \mathbf{t}(s) \rangle = \exp(-s/\xi_p)$ due to thermal motion, where the persistence length $\xi_p = \kappa_f / k_B T$ depends upon the temperature T, where k_B is Boltzmann's constant. Also, because of fluctuations, the squared end-to-end displacement $\mathbf{r}_{ee}$ of a filament is less than its contour length L_c, having the form $\langle r_{ee}^2 \rangle = 2\xi_p L_c - 2\xi_p^2 [1 - \exp(-L_c/\xi_p)]$, which approaches the rigid rod limit $\langle r_{ee}^2 \rangle^{1/2} \sim L_c$ only when $\xi_p \gg L_c$. In comparison, long filaments with relatively short persistence lengths obey $\langle r_{ee}^2 \rangle = 2\xi_p L_c$, showing that the linear size of sinuous filaments grows only like the square root of the contour length, a property of all linear chains in which self-avoidance is neglected. When self-avoidance is enforced, the growth of $\langle r_{ee}^2 \rangle^{1/2}$ with L_c is dimension-dependent, achieving ideal behavior only in four dimensions. Branched polymers or linear chains with attractive interactions display still different scaling behavior.

At finite temperature, $\mathbf{r}_{ee}$ does not have a unique value but rather is distributed according to a probability per unit length of the form $\mathcal{P}(x) = (2\pi\sigma^2)^{-1/2} \exp(-x^2/2\sigma^2)$ for filaments in one dimension, where x is the displacement from one end of the chain to the other. For freely jointed chains with N identical segments of length b, the variance $\sigma^2 = Nb^2/d$, where d is the spatial dimension of the chain. This distribution, and its three-dimensional cousin, demonstrate that a chain is not likely to be found in its fully stretched configuration $r_{ee} = L_c$ because such a configuration is strongly disfavored by entropy. Thus, the free energy of a flexible chain rises as the chain is stretched from its equilibrium value of r_{ee}, and the chain behaves like an entropic spring with a force constant $k_{sp} = 3k_B T/2\xi_p L_c = 3k_B T/\langle r_{ee}^2 \rangle$ in three dimensions.

The persistence lengths of a variety of the cell's polymers and filaments have been measured, and they span the enormous range of 0.5 nm for alkanes up to a few millimeters for microtubules. This behavior can be understood by viewing the filament as a flexible rod of uniform density and cross section, permitting the flexural rigidity to be written as $\kappa_f = Y\mathcal{I}$, where Y is the material's Young's modulus. The moment of inertia of the cross section $\mathcal{I}$ has the form $\mathcal{I} = \pi(R^4 - R_i^4)/4$ for a hollow tube of inner and outer radii R_i and R, respectively, predicting that the persistence length has the form $\xi_p = \pi Y(R^4 - R_i^4)/4k_B T$. Treated as uniform rods, the filaments of the cell have Young's moduli in the range $(0.5\text{--}1.5) \times 10^9$ J/m^3, which is about two orders of magnitude lower than the moduli of conventionally "hard" materials such as wood or steel, but comparable to plastics.

2.7 Problems

Biological applications

2.1 Consider a large motor neuron running from the brain to the arm containing a core bundle of microtubules. Taking the persistence length of a microtubule to be 2 mm, what energy is required (in $k_B T$ at 300 K) to bend a microtubule of length 20 cm into an arc of radius 10 cm?

2.2 Let θ be the angle characterizing the change in direction of a filament along its length. For a tobacco mosaic virus of contour length 250 nm, determine the value of $\langle \theta^2 \rangle^{1/2}$ arising from thermal fluctuations. Quote your answer in degrees.

2.3 Consider a piece of spaghetti 2 mm in diameter. If Young's modulus Y of this material is 1×10^8 J/m³, what is its persistence length at $T = 300$ K? Is the result consistent with your everyday observations?

2.4 Flagella are whip-like structures typically about 10 μm long whose bending resistance arises from a microtubule core. Treating the flagellum as a hollow rod of inner radius 0.07 μm and outer radius 0.1 μm, find its persistence length at $T = 300$ K if its Young's modulus is 1×10^8 J/m³. Compare your result with the persistence length of a single microtubule. (*Note: this approximation is not especially trustworthy.*)

2.5 What are the structural advantages for a microtubule to be hollow? Calculate the mass ratio and the flexural rigidity ratio for a hollow microtubule with inner and outer radii 11.5 and 14 nm, respectively, compared to a solid microtubule with the same outer radius. What is the most efficient use of proteins to gain rigidity: one solid microtubule or several hollow ones?

2.6 The virus bacteriophage-λ contains a string of 97 000 base-pairs in its DNA. Find the length of this DNA strand at 0.34 nm/base-pair and compare it with the DNA persistence length. If the DNA is 2 nm in diameter, what is the radius of the smallest spherical volume into which it can be packed? If this DNA becomes a random chain once released into a host cell, what is its root mean square end-to-end distance? Compare your answers with the size of a typical bacterium.

2.7 Compare the root-mean-square end-to-end distance $\langle r_{ee}^2 \rangle^{1/2}$ of strands of spectrin, actin and microtubules 200 nm in contour length. What is the effective spring constant of each protein, for filaments with a contour length of 1 cm at 300 K? Use $\xi_p = 15$, 15×10^3, and 2×10^6 nm for spectrin, actin and microtubules, respectively.

2.8 Consider a 30 μm length of DNA, such as might be found in a virus. What force is required to stretch the DNA to an end-to-end displacement $x = 10$, 20 and 25 μm, according to the Gaussian approximation Eq. (2.62) and the worm-like chain model Eq. (2.66)? Assume the temperature is 300 K.

Formal development and extensions

2.9 Consider a polymer such as a linear alkane, where the bond angle between successive carbon atoms is a fixed value α, although the bonds are free to rotate around one another.

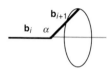

The length and orientation of the bond between atom i and atom $i+1$ defines a bond vector $\mathbf{b}_i$. Assume all bond lengths are the same, and that remote bonds can intersect.

(a) Show that the average projection of $\mathbf{b}_{i+k}$ on $\mathbf{b}_i$ is

$$\langle \mathbf{b}_i \cdot \mathbf{b}_{i+k} \rangle = b^2(-\cos\alpha)^k \qquad\qquad (k \geq 0).$$

[*Hint: start with* $\langle \mathbf{b}_i \cdot \mathbf{b}_{i+1} \rangle$ *and iterate.*]
(b) Write $\langle r_{ee}^2 \rangle$ in terms of $\langle \mathbf{b}_i \cdot \mathbf{b}_j \rangle$ to obtain

$$\langle r_{ee}^2 \rangle / b^2 = N[1 + (2 - 2/N)(-\cos\alpha) + (2 - 4/N)(-\cos\alpha)^2 + \ldots].$$

(c) Use your result from (b) to establish that, in the large N limit

$$\langle r_{ee}^2 \rangle = Nb^2(1 - \cos\alpha)/(1 + \cos\alpha).$$

(d) What is the effective bond length (in units of b) in this model at the tetrahedral value of 109.5°?

2.10 The backbones of polymers such as the polysiloxanes have alternating unequal bond angles:

even though the bond lengths are all equal.

(a) Show that, if self-intersections of this type of chain are permitted, its effective bond length is $B_{\text{eff}}^2 = b^2 (1 - \cos\alpha) \cdot (1 - \cos\beta)/(1 - \cos\alpha \cos\beta)$. [*Hint: follow the same steps as in Problem 2.9*]

(b) Confirm that this expression reduces to the fixed-angle rotating chain expression in Prob. 2.9(c) when $\alpha = \beta$.

(c) Evaluate B_{eff} when $\alpha = 109.5°$, $\beta = 130°$ and $b = 0.17$ nm.

2.11 Show that $\langle |\mathbf{r}_{ee}| \rangle = (8/3\pi)^{1/2} N^{1/2} b$ for ideal chains in three dimensions.

2.12 The radius of gyration, or root mean square radius, R_g of the $N + 1$ vertices in a linear chain, is defined by

$$R_g^2 = [\Sigma_{i=1,N+1} (\mathbf{r}_i - \mathbf{r}_{cm})^2]/(N+1)$$

where $\mathbf{r}_i$ is the position vector of each of the $N + 1$ vertices and $\mathbf{r}_{cm}$ is the center-of-mass position $\mathbf{r}_{cm} = \Sigma \mathbf{r}_i/(N+1)$. Show that $\langle R_g^2 \rangle = \langle r_{ee}^2 \rangle/6$ for ideal chains.

[*Hint: recast the problem to read $R_g^2 \propto \Sigma\Sigma r_{ij}^2$ and then use $\langle r_{ij}^2 \rangle = |j - i| b^2$ where $\mathbf{r}_{ij}$ is the displacement between vertices i and j. Justify!*]

2.13 Find $r_{ee, \text{most likely}}$ and $\langle |\mathbf{r}_{ee}| \rangle$ for ideal chains in two dimensions.

2.14 Suppose that a particle moves only in one direction at a constant speed v but changes direction randomly at the end of every time interval Δt.

(a) Find the diffusion constant D of the motion as a function of v and Δt, given the diffusion equation $\langle x^2 \rangle = 2Dt$.

(b) Find the temperature dependence of D if the kinetic energy of the particle is $k_B T/2$.

2.15 Consider a three-dimensional ideal chain of 50 segments, each with length 10 nm.

(a) What is $\langle |\mathbf{r}_{ee}| \rangle$ and $\langle r_{ee}^2 \rangle^{1/2}$?

(b) What is the effective spring constant at 300 K?

(c) If the chain has charges $+/-e$ on each end and is placed in a field of 10^6 V/m, what is the change in the end-to-end distance?

2.16 The results in the text for the distribution of $\mathbf{r}_{ee}$ for random chains in three dimensions can be generalized easily to random chains in d dimensions. For chains whose N segments have a uniform length b, show that:

(a) the distribution of end-to-end distances has the conventional
   Gaussian form, but with $\sigma^2 = Nb^2/d$,
(b) the effective spring constant is $k_{sp} = dk_B T/Nb^2$.

2.17 Compare the flexural rigidity of the solid cylindrical rod of Eq.
(2.11) with a solid rod with the cross section of a square. Take both
rods to have the same cross sectional area πR^2. Place the axis of
symmetry through the center of the square.

2.18 Determine the moment of inertia of the cross section for the
three regular-shaped rods with cross sections:

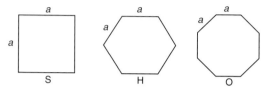

Take the length of the side in each case to be a. Find the ratio of these
moments to that of a cylinder of radius R ($\mathcal{I} = \pi R^4/4$), imposing the
condition that all shapes have the same cross sectional area πR^2 to
express a in terms of R.

2.19 Compare the two force–extension relations in Eqs. (2.64) and
(2.66) by plotting $(2\xi_p f/k_B T)$ against x/L_c. At what value of x/L_c is
the difference between these curves the largest?

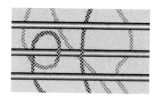

Chapter 3

Two-dimensional networks

With some notable exceptions such as DNA, most filaments in the cell are linked together as part of a network, which may extend throughout the interior of the cell, or be associated with a membrane as an effectively two-dimensional structure such as shown in Fig. 3.1. This chapter concentrates on the properties of planar networks, particularly the temperature- and stress-dependence of their geometry and elasticity. To allow sufficient time to develop the theoretical framework of elasticity, we focus in this chapter only on networks having uniform connectivity.

3.1 Soft networks in the cell

Two-dimensional networks arise in a variety of situations in the cell; they may be attached to its plasma or nuclear membrane, or be wrapped around a cell as its wall. Containing neither a nucleus nor other cytoskeletal components such as microtubules, for example, the human red blood cell possesses only a membrane-associated cytoskeleton. Composed of tetramers of the protein spectrin, the erythrocyte cytoskeleton is highly convoluted *in vivo* (see Fig. 2.21(b)), but can be stretched by about a factor of seven in area to reveal its relatively uniform four- to six-fold connectivity, as shown in Fig. 3.1(a) (Byers and

Fig. 3.1. (a) Membrane-associated cytoskeleton of the human erythrocyte. To show its connectivity, the cytoskeleton has been stretched, with a separation between junction complexes of about 200 nm (from Byers and Branton, 1985; courtesy of Dr Daniel Branton, Harvard University). (b) Section from the cortical lattice of an auditory outer hair cell; inset illustrates how the long, regularly spaced circumferential filaments are cross-linked. (Bar = 200 nm; reprinted with permission from Holley and Ashmore, 1990; © 1990 by the Company of Biologists; courtesy of Dr Matthew Holley, University of Bristol).

ankyrin

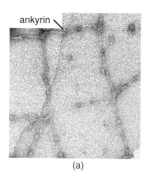

(a)

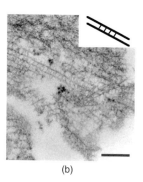

(b)

Branton, 1985; Liu *et al.*, 1987; Takeuchi *et al.*, 1998). Roughly midway along their 200 nm contour length, the spectrin tetramers are attached to the plasma membrane by the protein ankyrin (using another protein called band 3 as an intermediary). About 120 000 tetramers cover the 140 μm^2 membrane area of a typical erythrocyte, corresponding to a tetramer density of about 800 μm^{-2}. The tetramers are attached to one another at junction complexes containing actin segments 33–37 nm long; perhaps 35 000 junction complexes, with an average separation of about 75 nm, cover the erythrocyte.

An example of a two-dimensional biological network with lower connectivity than the red cell is the lateral cortex of the auditory outer hair cell, where the network junctions are frequently connected to just four of their neighbors. In guinea pigs, outer hair cells are roughly cylindrical in shape and about 10 μm in diameter, with the lateral cortex lying inside of, and attached to, the plasma membrane. As displayed in Fig. 3.1(b), the cortex consists of parallel filaments about 5–7 nm thick, spaced ~60 nm apart and wound around the axis of the cylinder. These filaments are cross-linked at intervals of ~30 nm (with a range of 10–50 nm) by thinner filaments just 2–3 nm thick (from Tolomeo *et al.*, 1996). Morphological and immunological studies suggest that the circumferential filaments are actin, and the crosslinks are spectrin, implying that the network is elastically anisotropic, being stiffer around the cylinder and more flexible along its axis.

The nuclear lamina is a further example of a network with extensive regions of four-fold connectivity (Aebi *et al.*, 1986; McKeon *et al.*, 1986). The nucleus is bounded by two membranes, and the lamina lies in the interior of the nucleus, adjacent to the inner nuclear membrane as shown in Fig. 3.2. Assembled into a 10–20 nm thick meshwork, the intermediate filaments of the lamina are 10.5 ± 1.5 nm in diameter and are typically separated by about 50 nm in the network. Composed of varieties of the protein lamin, the intermediate filaments form a dynamic network that can disassemble during cell division.

On a much shorter length scale than the inter-vertex separation of the networks described above, the few layers of peptidoglycan in the cell wall

Fig. 3.2. (a) Electron micrograph of a region about 2.5 μm in length from the two-dimensional nuclear lamina in a *Xenopus* oocyte, showing its square lattice of intermediate filaments (reprinted with permission from Aebi *et al.*, 1986; © 1986 by Macmillan Magazines Limited). Part (b) illustrates the approximate placement of the lamina adjacent to the inner nuclear membrane. For clarity, the endoplasmic reticulum has been omitted from the drawing. Although the pore is shown as a simple hole, in fact it is a 100 nm wide complex of proteins.

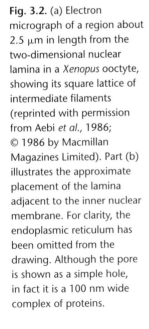

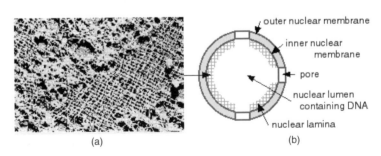

(a)

(b)

outer nuclear membrane
inner nuclear membrane
pore
nuclear lumen containing DNA
nuclear lamina

of Gram-negative bacteria form a tightly knit structure with three-fold coordinated junctions, having the approximate appearance of a planar honeycomb. As illustrated in Fig. 3.3, peptidoglycan consists of stiff chains of sugar molecules running in one direction, linked transversely by flexible chains of amino acids. However, the network connectivity may not be entirely two-dimensional: there are thought to be covalent links out of the plane formed by adjacent sugar chains, as indicated by the arrows in the figure. As with the erythrocyte cytoskeleton, the flexible amino-acid chains allow a peptidoglycan network to be stretched under stress, and the network area may increase by up to a factor of three compared to the *in vivo* area. The orientation of this anisotropic network is similar to that of the cortex of the auditory outer hair cell: the stiffer elements (sugars) run around the girth of the bacterium's cylindrical shape, while the softer elements lie parallel to its axis of symmetry.

The two-dimensional cellular structures described above have been presented as networks, rather than solid sheets, although these two materials display the same elastic behavior when measured on the appropriate length scale, as will be described in Chapter 6. Appropriately, the diameter of the filaments in our cellular networks is much less than the junction spacing, which ranges from a few nanometers in the bacterial cell wall to more than 50 nm in the erythrocyte cytoskeleton. The elastic properties of network filaments, both their resistance to bending and, for sinuous filaments, their resistance to stretching, are described in Chapter 2, where we emphasize the entropic contribution to their elasticity. What happens when we link filaments, both soft and stiff, into a network? How much of the network's behavior faithfully mimics that of its individual components, and how much represents collective effects arising from interactions among filaments?

To address these questions, we need to determine how a network changes in response to stress, in the same way as we characterized, in Chapter 2, the bending and stretching deformations of flexible filaments. In principle, two-dimensional networks, like their three dimensional cousins, have many deformation modes, although not all of the modes are independent. In two dimensions, isotropic materials (i.e., having an internal structure which is independent of direction), or networks with uniform six-fold connectivity, possess in-plane deformations involving only two fundamental modes – compression and shear – as illustrated in Fig. 3.4. The compression mode displayed in Fig. 3.4(a) preserves the internal angles of the network by uniformly scaling its linear dimensions, in contrast to the shear mode of Fig. 3.4(b), which preserves the area of the network, but not its internal angles. The resistance of two-dimensional, isotropic networks to these deformation modes is parametrized by just two elastic moduli, the area compression

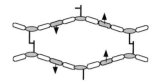

Fig. 3.3. Model for the connectivity of the peptidoglycan network of bacterial cell walls. The stiff chains of sugar molecules, represented by cylinders, are cross-linked both in- and out-of-plane by flexible chains of amino acids. Out-of-plane links are indicated by arrows (reprinted with permission from Koch and Woeste, 1992; © 1992 by the American Society for Microbiology).

Fig. 3.4. Fundamental deformation modes of a triangular network in two dimensions: (a) a compression mode which changes the network area without affecting its internal angles; (b) a shear mode which changes the angles, but leaves the area untouched.

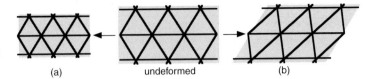

(a) undeformed (b)

modulus K_A and the shear modulus μ, although we emphasize again that networks without full rotational symmetry possess a larger number of independent deformation modes, and hence require a larger number of moduli.

As measures of the large-scale behavior of a system, elastic moduli can be related to the properties of its individual elements; for example, a uniform network of springs with force constant k_{sp} obeys

$$K_A = \sqrt{3}\, k_{sp}/2 \quad \mu = \sqrt{3}\, k_{sp}/4 \quad \text{(zero temperature and stress)}, \quad (3.1)$$

if the springs are connected at six-fold coordinated junctions as illustrated in Fig. 3.4. Note that, in two dimensions, the moduli have units of energy per unit area, the same as a spring constant. Similar relations to Eq. (3.1) can be obtained for the elastic moduli of networks with other symmetries, such as four-fold connectivity. Now, Eq. (3.1) applies only for small deformations of an unstressed network at zero temperature; as the temperature increases, or the network is subject to an external (two-dimensional) stress, the network may stray from its uniform appearance, as in Fig. 3.5, such that the elastic moduli change, and Eq. (3.1) becomes increasingly inaccurate. Approximate expressions for the temperature- and stress-dependence of the elastic moduli are available for several network structures of biological importance, and these can be used for interpreting the observed deformation of cellular networks.

We saw in Section 2.4 that a sinuous polymer can be approximately represented by a spring, so long as the polymer is not displaced too far from its equilibrium position. However, at large extensions, the simple spring description of a polymer fails, because a Hookean spring can be extended indefinitely, whereas real protein filaments cannot. Similar difficulties beset pure spring representations of networks: at zero temperature, for example, spring networks with uniform six-fold connectivity are predicted to expand without limit when subjected to a tensile stress exceeding $\sqrt{3}\, k_{sp}$, or collapse under moderate compression. Thus,

Fig. 3.5. Thermal fluctuations change a zero-temperature network, such as in part (a), into a more erratic configuration like (b), which may also be under a two-dimensional stress applied externally.

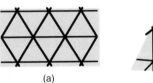
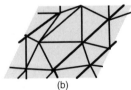

(a) (b)

in these large stress regimes, a successful description of a biological network requires a polymer representation that incorporates steric repulsion among the chain elements, and recognizes the maximal extension of the filament.

The mean shape adopted by a network under stress depends upon the elastic moduli as one might expect: a network with a large compression modulus undergoes only small deformations for a particular stress compared to the deformations experienced by a system with a small K_A. Equally interesting is the directional response of a network to a tensile stress applied along one direction only (referred to as a uniaxial stress). As illustrated in Fig. 3.6, when most materials are stretched in one direction, they shrink in the transverse direction. These changes in shape are characterized by the Poisson ratio, which is the fractional decrease in length along the transverse direction divided by the fractional increase in length along the longitudinal direction. Generally, the Poisson ratio is observed to be $+1/3$, meaning that the fractional shrinkage in the transverse direction is about 30% of the fractional extension along the direction of the stress. However, networks under certain ranges of stress, or two-dimensional networks fluctuating in three dimensions, may undergo little shrinkage in the transverse direction, or even expand transversely, when stretched longitudinally.

In the following three sections of this chapter, we review and apply the theory of elasticity to networks in two dimensions. Section 3.2 begins with an introduction to elastic moduli in two dimensions as applied to two networks in particular, namely those with uniform six-fold coordination (triangular networks) and four-fold coordination (square networks). The elastic moduli are obtained for spring networks both as an approximation useful for modest deformations, and as an illustration of the relationship between the macroscopic properties of a system and their microscopic components; networks with uniform six-fold coordination are treated in Section 3.3, while those with four-fold coordination are the subject of Section 3.4. Lastly, the formal results of Sections 3.2 to 3.4 are applied to a number of cellular networks in Section 3.5. Students unfamiliar with the tensor description of elastic deformations might consider reading part of Appendix D before tackling the proofs in the following three sections.

Fig. 3.6. Behavior of conventional materials when subject to a uniaxial stress: the object, whose original position is indicated by the shaded region, stretches in the direction of the stress, but contracts in the transverse direction.

3.2 Elastic moduli in two dimensions

The mechanical behavior of a membrane or a two-dimensional network depends upon whether it can undulate or is forced to lie flat in a plane. In this chapter, we consider only networks that are globally flat, and delay to Chapter 6 the more general question of out-of-plane motion. We begin our treatment of networks by introducing the strain tensor, which describes the deformation of two- or three-dimensional objects, similar to the way that the displacement from equilibrium, x, describes the one-dimensional deformation of a spring. For modest changes in shape, the free energy of deformation varies quadratically in the strain and involves a suite of elastic moduli analogous to the force constant k_{sp} of a spring. With this formalism in hand, our second task is finding the minimum number of elastic moduli required to describe the energetics of two specific systems, namely two-dimensional networks with four- and six-fold connectivity. Lastly, we provide a link between the macroscopic elastic properties of a network and its microscopic structure by determining the moduli of a six-fold network of springs at zero temperature.

Deformations and the strain tensor

How do we describe a deformation mathematically? Consider what happens to a square when it is deformed into a rectangle, as shown by the shaded region and wire outline in Fig. 3.7(a), respectively. Each point on the square, except the center in this particular deformation, moves from its original position $\mathbf{x}$ to a new position $\mathbf{x}+\mathbf{u}$, where $\mathbf{u}$ is a displacement vector. It is clear from Fig. 3.7(b) that $\mathbf{u}$ is not constant, but varies in magnitude and direction across the object. In fact, if $\mathbf{u}$ were constant over the object, it simply would be translated, unde-formed, to a new position. Hence, an important feature of the defor-mation is not just $\mathbf{u}$ by itself, but rather its rate of change with location on the object. Inspection of the two-dimensional deformation in Fig. 3.7 shows that $\mathbf{u}$ may have non-zero partial derivatives in any Cartesian direction.

The relevant description of a deformation lies in the strain tensor u_{ij}, related to the rate of change of $\mathbf{u}$ with position $\mathbf{x}$ by

$$u_{ij} = 1/2 \left[\partial u_i/\partial x_j + \partial u_j/\partial x_i + \sum_k (\partial u_k/\partial x_i)(\partial u_k/\partial x_j) \right], \qquad (3.2)$$

where the subscripts i, j, k refer to the Cartesian coordinate axes, such that there are four components of u_{ij} in two dimensions and nine com-ponents in three dimensions. Note that u_{ij} is unitless and is symmetric in

Fig. 3.7. (a) Movement of positions on a square (shaded region) as it changes to a rectangle (wire outline). (b) The deformation is described by the displacement vector $\mathbf{u}$, which varies locally on the object.

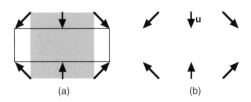

(a) (b)

indices i and j. For small deformations, the last term in Eq. (3.2) may be neglected, yielding

$$u_{ij} \cong 1/2 \, (\partial u_i / \partial x_j + \partial u_j / \partial x_i). \tag{3.3}$$

To illustrate the behavior of the strain tensor, consider two situations. First, if the object is simply translated by an applied force, then $\mathbf{u}$ is everywhere the same and all derivatives of the displacement vector vanish; hence, $u_{ij} = 0$. Second, suppose that the deformation is a modest version of Fig. 3.7, in which all positions are uniformly scaled, with their x-coordinates increasing from x to $1.1x$, and their y-coordinates decreasing from y to $0.9y$. Then $u_{xx} = (1.1 - 1.0)x/x = 0.1$ and $u_{yy} = (0.9 - 1.0)y/y = -0.1$. Since the magnitude of the scaling in a given Cartesian direction is independent of the transverse direction in this example, then $u_{xy} = u_{yx} = 0$. Note the important result here that u_{ij} is the same everywhere on the object in Fig. 3.7, even though $\mathbf{u}$ varies locally, so that u_{ij} describes the global behavior of this particular deformation. Further properties of the strain tensor can be found in Appendix D; alternate definitions of the strain tensor are described in Fung (1994).

Elastic moduli

In continuum mechanics, deformations such as that in Fig. 3.7 result from a force distributed across the surface of an object; for example, the deformation in the figure could arise from a pair of forces applied uniformly to the surfaces facing the vertical axis. What is important for these kind of deformations is not just the force vector itself, but also

- its orientation with respect to the surface: forces may result in shear or compression, depending on their orientation
- its magnitude per unit surface area (e.g., the pressure).

The stress tensor σ_{ij} is the force per unit area, taking into account its direction of application with respect to the surface. The component of the force in the i-direction, F_i, is given in terms of σ_{ij} by

$$F_i = \sum_j \sigma_{ij} a_j. \tag{3.4}$$

The sum is over the components of the surface area vector $\mathbf{a}$, which has a direction perpendicular to the surface and a magnitude equal to the surface area. The stress tensor has units of energy density and is symmetric in indices i and j.

How does an object respond to stress? For ideal springs in one dimension, we know that the restoring force f is proportional to the displacement from equilibrium x: $f = -k_{sp}x$, where k_{sp} is the spring constant. The corresponding relationship for continuous materials reads [*stress*] $\propto$ [*strain*], or

$$\sigma_{ij} = \sum_{k,l} C_{ijkl} u_{kl},$$ (3.5)

where the material-specific constants C_{ijkl} appearing in the summation are called the elastic stiffness constants or elastic moduli. Just as the potential energy of a Hooke's law spring is quadratic in the square of the displacement, the change in the free energy density $\Delta \mathcal{F}$ of a continuous object under deformation is quadratic in the strain tensor u_{ij}:

$$\Delta \mathcal{F} = 1/2 \sum_{i,j,k,l} C_{ijkl} u_{ij} u_{kl}.$$ (3.6)

The elastic constants have dimensions of energy per unit volume for three-dimensional materials, or energy per unit area for two-dimensional systems. Because a stress applied in one direction to an object may result in a deformation in several directions, as the experiment of squeezing jelly in one's hand easily demonstrates, the elastic moduli of two- or three-dimensional materials form a tensor, as opposed to the single k_{sp} of an isolated spring. However, the number of indices on C_{ijkl} should not intimidate the reader: symmetry considerations greatly reduce the number of independent components from the $3^4 = 81$ terms naively expected in three dimensions, or $2^4 = 16$ expected in two dimensions (see Landau and Lifshitz, 1986).

The actual number of independent moduli depends upon the symmetry of the system, with isotropic systems (equivalent in all directions) requiring the smallest number of moduli for their description. As a first step, we consider symmetry relations arising from Eq. (3.6) that can be applied to all materials. Because u_{ij} is symmetric under exchange of i and j, then according to Eq. (3.6), C_{ijkl} can be defined such that it is pairwise symmetric under exchange of i and j, or k and l; that is,

$$C_{ijkl} = C_{jikl} = C_{ijlk}.$$ (3.7)

Further, because the product $u_{ij} u_{kl}$ in Eq. (3.6) is symmetric under exchange of the pairs of indices ij and kl, the moduli must obey

$$C_{ijkl} = C_{klij}.$$ (3.8)

The set of conditions in Eqs. (3.7) and (3.8) decreases the number of independent values of C_{ijkl} to 21 in three dimensions, and 6 in two dimensions. Specifically, in two dimensions, the six independent moduli are

$$
\begin{aligned}
&C_{xxxx} \qquad C_{yyyy} \qquad C_{xxyy} = C_{yyxx} \\
&C_{xyxy} = C_{xyyx} = C_{yxyx} = C_{yxxy} \\
&C_{xxxy} = C_{xxyx} = C_{xyxx} = C_{yxxx} \\
&C_{yyyx} = C_{yyxy} = C_{yxyy} = C_{xyyy}.
\end{aligned}
$$ (3.9)

Symmetries of the material itself further reduce the number of elastic moduli, as we now demonstrate for two-dimensional networks with four-fold and six-fold connectivity, of which there are several biological examples (see Section 3.1). Although the word "network" is used to describe the materials in the following examples, the results apply to any system with the symmetries of the examples.

Six-fold networks in 2D

The two-dimensional network illustrated in Fig. 3.8(a) possesses six-fold axes of rotational symmetry through the vertices. To determine the effects of this symmetry, we follow Landau and Lifshitz (1986) and change from Cartesian coordinates x and y to complex coordinates ξ and η, where

$$\xi \equiv x + iy \qquad\qquad \eta \equiv x - iy, \qquad\qquad (3.10)$$

such that the free energy involves products of the form $C_{\xi\xi\eta\eta}u_{\xi\xi}u_{\eta\eta}$ and others. Now, a rotation about the xy coordinate origin by an angle θ changes the coordinates (x, y) of a given position to $(x\cos\theta - y\sin\theta, x\sin\theta + y\cos\theta)$ or, equivalently, changes the coordinates (ξ, η) to

$$\xi \rightarrow \xi\exp(i\theta) \qquad\qquad \eta \rightarrow \eta\exp(-i\theta). \qquad\qquad (3.11)$$

Specifically, six-fold symmetry demands that the moduli be invariant under rotations through $\theta = \pi/3$, or $\xi \rightarrow \xi\exp(i\pi/3)$ and $\eta \rightarrow \eta\exp(-i\pi/3)$. The only non-zero components of C_{ijkl} that remain unchanged by this transformation must contain ξ and η the same number of times, since $\exp(i\pi/3)\exp(-i\pi/3) = 1$. Hence only two components of C_{ijkl} are invariant under the symmetry of the network, namely the pair $C_{\xi\xi\eta\eta}$ and $C_{\xi\eta\xi\eta}$ and terms related by symmetry of the indices.

Expressed in the $\xi\eta$ representation, the change in the free energy density $\Delta\mathscr{F}$ from Eq. (3.6) becomes

$$\Delta\mathscr{F} = 2C_{\xi\eta\xi\eta}u_{\xi\eta}u_{\xi\eta} + C_{\xi\xi\eta\eta}u_{\xi\xi}u_{\eta\eta}, \qquad\qquad (3.12)$$

where the first term results from four combinations of $C_{\xi\eta\xi\eta}$ and the second term is from two combinations of $C_{\xi\xi\eta\eta}$; the expression includes the normalization factor of 1/2 from Eq. (3.6). To replace $u_{\xi\xi}$ etc. in Eq. (3.12) by the Cartesian components of the strain tensor, we use the fact that the components of a tensor transform as the products of the corresponding coordinates. That is, because $\xi^2 = (x + iy)^2 = x^2 - y^2 + 2ixy$, then

$$u_{\xi\xi} = u_{xx} - u_{yy} + 2iu_{xy}, \qquad\qquad (3.13)$$

and similarly with $u_{\xi\eta}$ and $u_{\eta\eta}$, permitting Eq. (3.12) to be written as

$$\Delta\mathscr{F} = 2C_{\xi\eta\xi\eta}(u_{xx} + u_{yy})^2 + C_{\xi\xi\eta\eta}\{(u_{xx} - u_{yy})^2 + 4u_{xy}^2\}. \qquad\qquad (3.14)$$

In Appendix D, we demonstrate how combinations of the strain tensor elements correspond to physical deformations; for example, the trace of u_{ij} (i.e., the sum of its diagonal elements) corresponds to a uniform compression or expansion, with no shear. Thus, we can replace C_{ijkl} by elastic moduli that are more directly related to the pure deformation modes of area compression (K_A) or shear (μ),

$$K_A = 4C_{\xi\eta\xi\eta} \qquad \mu = 2C_{\xi\xi\eta\eta}, \tag{3.15}$$

so that Eq. (3.14) becomes

$$\Delta\mathcal{F} = (K_A/2)(u_{xx} + u_{yy})^2 + \mu\,\{(u_{xx} - u_{yy})^2/2 + 2u_{xy}^2\}$$
$$\text{(six-fold symmetry).} \tag{3.16}$$

This is the same expression as that for isotropic materials in Appendix D; that is, in two dimensions, isotropic materials and those with six-fold symmetry are both described by two elastic moduli at small deformations.

Four-fold networks in 2D

As we established above from general symmetry considerations, two-dimensional materials with four-fold symmetry, as displayed in Fig. 3.8(b), have, at most, the six independent moduli given by Eq. (3.9). This number of independent moduli is reduced by additional symmetries, including invariance under each of the coordinate inversions $x \to -x$ and $y \to -y$ independently. Hence, any component of C_{ijkl} with an odd number of x or y indices changes sign under a single application of the inversions, because the components of a tensor transform as products of the corresponding coordinates. Now, the mechanical properties of a material should not change sign upon reflection, so that those components with an odd number of x or y indices must vanish, reducing the number of independent components to four. Lastly, clock-wise rotations by an angle of $\pi/2$ about the four-fold symmetry axis results in $x \to -y$ and $y \to x$, which implies that $C_{xxxx} = C_{yyyy}$. The moduli for pure deformations can be written in terms of these three C_{ijkl} by

$$K_A = (C_{xxxx} + C_{xxyy})/2$$
$$\mu_p = (C_{xxxx} - C_{xxyy})/2 \text{ (pure shear)} \tag{3.17}$$
$$\mu_s = C_{xyxy} \text{ (simple shear)},$$

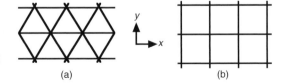

Fig. 3.8. Two-dimensional networks with six-fold (a) and four-fold symmetry (b).

(a) (b)

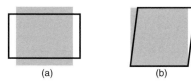

Fig. 3.9. Two potentially inequivalent shear modes in four-fold systems: pure shear (a) and simple shear (b). The shaded region shows the original configuration, while the wire outline is the deformed object.

so that the free energy density $\Delta\mathscr{F}$ in Eq. (3.6) becomes

$$\Delta\mathscr{F} = (K_A/2)(u_{xx} + u_{yy})^2 + (\mu_p/2)(u_{xx} - u_{yy})^2 + 2\mu_s u_{xy}^2$$
$$\text{(four-fold symmetry).} \tag{3.18}$$

There are two shear moduli present in Eq. (3.18), μ_p and μ_s, reflecting the resistance against the two shear modes (pure shear and simple shear, respectively) shown in Fig. 3.9. Both of these modes are area-preserving, and they can be related to each other through a coordinate transformation (see Fung, 1994, p. 132).

Networks of springs

In many cases, a system at low temperature exhibits approximate harmonic behavior and can be represented as a spring network having the appropriate symmetry. The elastic moduli of such a network can be expressed in terms of its microscopic characteristics, such as the spring constant, as we now illustrate for a network with six-fold connectivity at zero temperature (or constant spring length). Each spring has an unstretched length of s_o, and a potential energy V_{sp} of

$$V_{sp} = k_{sp}(s - s_o)^2/2, \tag{3.19}$$

where k_{sp} is the force constant. The relationship between microscopic and macroscopic descriptions is found by comparing two expressions for the change in free energy density under an infinitesimal deformation – one in terms of elastic constants (K_A and μ), and the other in terms of spring characteristics (k_{sp} and s_o).

We consider first the pure compression mode of Fig. 3.10(a). If each spring is stretched a small amount $\delta \equiv s - s_o$ away from its equilibrium length s_o, the change in potential energy per vertex ΔU is, according to Eq. (3.19),

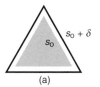

Fig. 3.10. Infinitesimal deformations (exaggerated for effect) used for calculating the compression modulus (a) and shear modulus (b) of a triangular network at zero temperature. In (a), all springs are stretched in length from s_o to $s_o + \delta$, while in (b), the top vertex is moved a distance δ to the right.

$$\Delta U = 3\, \Delta V_{sp} = 3 k_{sp} \delta^2/2, \tag{3.20}$$

where the factor of three arises because there are three times as many springs as there are vertices. Dividing Eq. (3.20) by the network area per vertex of $A_v = \sqrt{3}\, s_o^2/2$, for triangular networks, shows that the change in the free energy density $\Delta \mathcal{F}$ for small deformations around the equilibrium configuration is

$$\Delta \mathcal{F} = \Delta U/A_v = \sqrt{3}\, k_{sp} (\delta/s_o)^2. \tag{3.21}$$

Next, we write the strain tensor in terms of s_o and δ. Because the deformations are uniform in the x- and y-directions, it follows that $u_{xx} = u_{yy} = \delta/s_o$. Further, the off-diagonal elements vanish, $u_{xy} = 0$, because the displacement in the y-direction is independent of the position of the triangle in the x-direction. Substituting these elements of the strain tensor into Eq. (3.16) gives

$$\Delta \mathcal{F} = 2 K_A (\delta/s_o)^2, \tag{3.22}$$

for pure compression. Comparing Eqs. (3.21) and (3.22) yields

$$K_A = \sqrt{3}\, k_{sp}/2 \text{ (six-fold network).} \tag{3.23}$$

Note that the area compression modulus is independent of s_o at zero temperature.

An analogous method reveals an expression for the shear modulus, based upon the simple shear mode of Fig. 3.10(b). By moving the top vertex by an amount δ in the x-direction, the left hand spring is lengthened by $\delta/2$ and the right-hand spring is shortened by $\delta/2$, to lowest order in δ. Thus, the change in potential energy of either of the two stretched springs is $k_{sp} \delta^2/8$, again to lowest order in δ. Because the bottom spring exhibits no change in potential energy, the total change in potential energy per vertex is $\Delta U = k_{sp} \delta^2/4$, corresponding to a change in the energy density of

$$\Delta \mathcal{F} = \Delta U/A_v = k_{sp} (\delta/s_o)^2/(2\sqrt{3}). \tag{3.24}$$

The strain tensor describing this deformation is easy to evaluate. As both the y-coordinates of the vertices and the distance between vertices in the x-direction are unchanged in the simple shear mode, then $u_{xx} = u_{yy} = 0$. However, each successive row of vertices is displaced by δ in the positive x-direction for each increase $\sqrt{3} s_o/2$ in the y-direction, so that $u_{xy} = \delta/(\sqrt{3}\, s_o)$, where there is only one non-zero contribution to u_{xy} in Eq. (3.3). Thus, Eq. (3.16) yields

$$\Delta \mathcal{F} = (2\mu/3)(\delta/s_o)^2. \tag{3.25}$$

A comparison of Eq. (3.24) and (3.25) shows

$$\mu = \sqrt{3}\, k_{sp}/4 \qquad \text{(six-fold network).} \tag{3.26}$$

From Eqs. (3.23) and (3.26) we see that the ratio of K_A to μ is 2 for six-fold spring networks at zero temperature.

3.3 Spring networks with six-fold coordination

Not unexpectedly, the engineering specifications for a biological network vary from cell to cell. For example, a red blood cell needs a flexible network that permits it easily to deform in a capillary and then recover its equilibrium shape in a reasonable time frame. Thus, a cell may experience a variety of stresses during its everyday operation and adopts a network design to respond to these stresses in an appropriate way. How do the elastic properties of these networks depend upon quantities such as stress or temperature? Considering the stress-dependence for a moment, we know that the compression modulus of an ideal gas, for instance, is equal to its pressure (see end-of-chapter problems), indicating that a gas becomes more resistant to compression as its pressure rises. Do similar relationships exist for networks?

In this section, we determine the stress- and temperature-dependence of the elasticity and shape of a two-dimensional spring network with six-fold connectivity, similar to the cytoskeleton of the human erythrocyte. The properties of these networks are known from both analytical and computational treatments (Boal *et al.*, 1993; Discher *et al.*, 1997; Lammert and Discher, 1998; Wintz *et al.*, 1997). Four-fold networks are examined separately in Section 3.4, permitting us to observe the dependence of elastic properties on network connectivity. Although our primary interest is the behavior of networks under tension, there are situations in which a two-dimensional network can be compressed without buckling, encouraging us to consider both compression and tension.

Six-fold spring networks under stress

A network with six-fold connectivity, as illustrated in Fig. 3.11, can be viewed as a plane of triangular plaquettes, each plaquette bounded by three mechanical elements attached to neighboring vertices. In a network of identical springs at zero temperature and stress, all elements have the same length s_o and the plaquettes have the shape of equilateral triangles. The spring length changes from s_o to s_τ when the network is

(a)

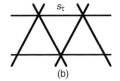
(b)

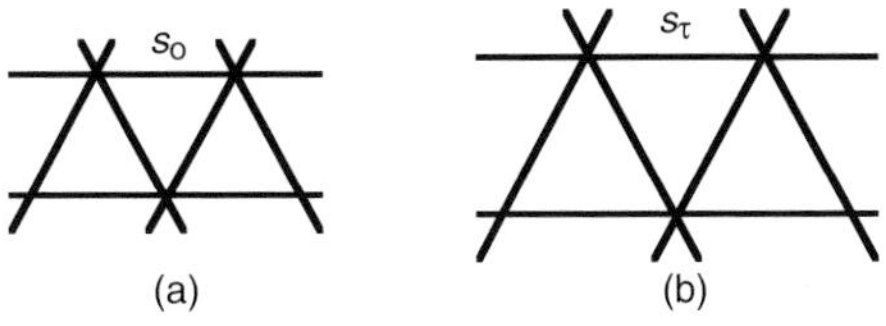

Fig. 3.11. The elementary plaquette in a planar network with six-fold connectivity has the shape of an equilateral triangle if all elements have the same length s. When the network is placed under a two-dimensional tension, s changes from its unstressed value s_o in (a) to a new value s_τ in (b).

placed under a two-dimensional isotropic tension τ, but the plaquettes remain equilateral triangles, albeit with different area. For such a network, s_τ can be calculated by minimizing the enthalpy H

$$H = E - \tau A, \tag{3.27}$$

where E is the energy of the springs and A is the area of the network. Here we adopt the sign convention that $\tau > 0$ corresponds to tension and $\tau < 0$ to compression. As there are three springs for each vertex in the network, the energy per vertex for springs of equal length s is $(3/2)k_{sp}$ $(s - s_o)^2$ and the area per vertex A_v is

$$A_v = \sqrt{3}\, s^2/2. \tag{3.28}$$

Hence, the enthalpy per vertex H_v is

$$H_v = (3/2)k_{sp}\, (s - s_o)^2 - \sqrt{3}\, \tau s^2/2, \tag{3.29}$$

for a six-fold spring network with equilateral plaquettes. The spring length s_τ that minimizes H_v for a particular stress can be found by equating the derivative $\partial H_v / \partial s$ to zero, yielding

$$s_\tau = s_o/(1 - \tau/[\sqrt{3}\, k_{sp}]) \qquad \text{(six-fold symmetry)}, \tag{3.30}$$

such that the minimum value of the enthalpy per vertex is

$$H_{v,min} = - (\sqrt{3}/2)\tau s_o^2/(1 - \tau/[\sqrt{3}\, k_{sp}])$$
$$\text{(equilateral plaquettes)}. \tag{3.31}$$

Eq. (3.30) shows that the network expands under tension ($\tau > 0$) and shrinks under compression ($\tau < 0$).

The behavior that we expect for triangular networks, including shapes other than equilateral triangles, is displayed graphically in Fig. 3.12, where the area per vertex is plotted against the reduced tension τ/k_{sp}. First, observe that the area increases without bound as the tension approaches a critical value τ_{exp},

$$\tau_{exp} = \sqrt{3}\, k_{sp} \qquad \text{(six-fold symmetry)}, \tag{3.32}$$

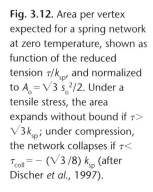

Fig. 3.12. Area per vertex expected for a spring network at zero temperature, shown as function of the reduced tension τ/k_{sp}, and normalized to $A_o = \sqrt{3}\, s_o^2/2$. Under a tensile stress, the area expands without bound if $\tau > \sqrt{3}k_{sp}$; under compression, the network collapses if $\tau < \tau_{coll} = - (\sqrt{3}/8)\, k_{sp}$ (after Discher et al., 1997).

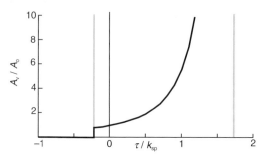

as can be obtained from Eq. (3.31) by inspection. The network area blows up at large tension because both the energy of the springs and the pressure term τA_v scale like s^2 at large extensions. However, the energy per vertex and τA_v enter the enthalpy with opposite signs if $\tau > 0$, so that if τ is sufficiently large, the tension term dominates over the spring energy, and the network expands without limit.

A second feature to note in Fig. 3.12 is the collapse of the network under compression, a behavior *not* present in networks of equilateral plaquettes. We have assumed, in deriving Eq. (3.31), that the enthalpy is minimized when its derivative with respect to s vanishes. However, Eq. (3.31) may not be the global minimum of H if shapes other than equilateral plaquettes are considered. Indeed, equilateral triangles enclose the largest area for a given perimeter, and therefore may not be the optimal plaquette shape under compression, where the tension term τA drives the system towards small areas. Clearly, the smallest τA contribution at $\tau < 0$ is given by plaquettes with zero area, of which those shaped like isosceles triangles have the lowest spring energy, namely $(k_{sp}/2) \cdot [(2s_1 - s_0)^2 + 2(s_1 - s_0)^2]$, where a zero-area triangle has two short sides of length s_1 and a long side of length $2s_1$. The minimum value of this energy as a function of s_1 is $k_{sp} s_0^2/6$ and occurs for $s_1 = 2s_0/3$. Thus, the enthalpy per vertex of the equilateral network rises with pressure ($\tau < 0$) according to Eq. (3.31) until it exceeds the enthalpy of a network of zero-area isosceles plaquettes ($k_{sp} s_0^2/6$) at a collapse tension τ_{coll} given by

$$\tau_{coll} = -(\sqrt{3}/8)\, k_{sp} \qquad \text{(six-fold symmetry)}, \qquad (3.33)$$

or, equivalently, at a spring length $s_\tau/s_0 = 8/9$. That is, the network collapses when subject to a two-dimensional pressure exceeding $|\tau_{coll}|$.

The elastic moduli of equilateral spring networks under stress can be obtained using the same technique as was used in Section 3.2 for unstressed networks. For a known deformation, two expressions are obtained for the change in the free energy density: one expression is in terms of the spring variables (k_{sp} and s_0) while the other is in terms of the elastic moduli and the strain tensor. As shown in Fig. 3.13, the deformation modes of choice are the same as those used in Section 3.2 except that, here, the small perturbations are performed on a state which is already under tension with sides of length s_τ.

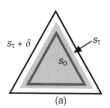 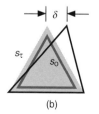

(a) (b)

Fig. 3.13. An unstressed equilateral triangle has sides of length s_0. An *equilateral* triangle under tension, with sides of length s_τ, is shown as the shaded region; this triangle is given an infinitesimal deformation as shown in outline. Mode (a) measures the area compression modulus, while mode (b) measures the shear modulus.

Table 3.1. *Change in free energy density $\Delta \mathcal{F}$ for deformation modes (a) and (b) in Fig. 3.13 of a triangulated spring network under stress. Column 2 (microscopic) refers to the enthalpy change expressed in spring variables, while column 4 (continuum) is the free energy change using Eq. (3.16) and the strain tensor in column 3*

Mode	$\Delta \mathcal{F}$ (microscopic)	Strain	$\Delta \mathcal{F}$ (continuum)
(a)	$\sqrt{3}\,k_{sp}\,(1-\tau/[\sqrt{3}\,k_{sp}]) \cdot (\delta/s_\tau)^2$	$u_{xx}=u_{yy}=\delta/s_\tau$ $u_{xy}=0$	$2K_A(\delta/s_\tau)^2$
(b)	$(k_{sp}/2\sqrt{3}) \cdot (1+\sqrt{3}\,\tau/k_{sp}) \cdot (\delta/s_\tau)^2$	$u_{xx}=u_{yy}=0$ $u_{xy}=(\delta/s_\tau)/\sqrt{3}$	$(2/3)\mu(\delta/s_\tau)^2$

Because the extraction of the elastic moduli from the deformation modes in Fig. 3.13 follows Section 3.2 very closely, the calculations are not repeated here. Rather, Table 3.1 summarizes the strain tensor and the changes in free energy density expected for the two deformation modes. Note that the deformations in Fig. 3.13 are with respect to the stressed state with spring length s_τ; i.e., the strain tensor measures a perturbation about a reference state which is itself deformed with respect to the unstressed state with spring length s_o. The elastic moduli can be obtained by equating columns 2 and 4 in Table 3.1, leading to the zero-temperature relations

$$K_A = (\sqrt{3}\,k_{sp}/2) \cdot (1-\tau/[\sqrt{3}\,k_{sp}]) \tag{3.34a}$$
$$\text{(six-fold symmetry, } T=0)$$
$$\mu = (\sqrt{3}\,k_{sp}/4) \cdot (1+\sqrt{3}\,\tau/k_{sp}), \tag{3.34b}$$

expressions that reduce to Eqs. (3.23) and (3.26) in the zero-stress limit. We remark that a more direct way of obtaining Eq. (3.34a) is to simply differentiate the area per vertex $\sqrt{3}\,s_\tau^2/2$ with respect to τ, and substitute the result into the definition of the areal compression modulus. Eqs. (3.34) show that the moduli do depend on the tension τ, becoming more resistant to shear as the network is stretched. The fact that K_A vanishes at $\tau = \sqrt{3}\,k_{sp}$ has already been anticipated, since the area per vertex grows without bound at this tension. Note that Eqs. (3.34) are not appropriate for collapsed networks.

The stress-dependence of the moduli has interesting implications for the Poisson ratio, which is a measure of how a material contracts in the transverse direction when it is stretched longitudinally. Specifically, the Poisson ratio σ_p is defined in two dimensions as the ratio of the strains

$$\sigma_p = -u_{yy}/u_{xx}, \tag{3.35}$$

where the stress is applied along the x-axis (not isotropically, as in Table 3.1). The negative sign is chosen so that conventional materials have a positive Poisson ratio: they shrink transversely when stretched longitudinally. In Appendix D, we show how to relate the Poisson ratio to the elastic moduli in three dimensions; the expression for the Poisson ratio in two dimensions can be obtained in a similar way, yielding

$$\sigma_p = (K_A - \mu)/(K_A + \mu). \tag{3.36}$$

Even though K_A and μ are both required to be positive (otherwise the material will spontaneously deform by compression or shear), there is no requirement that $K_A > \mu$ to ensure $\sigma_p > 0$. From Eqs. (3.23) and (3.26), a triangular network at zero temperature and stress has $K_A/\mu = 2$ and consequently $\sigma_p = 1/3$. However, the stress-dependence of the elastic moduli in Eq. (3.34) shows that the compression modulus decreases with tension, while the shear modulus increases with tension, leading to

$$\sigma_p = [1 - 5\tau/(\sqrt{3}\,k_{sp})]/[3 + \tau/(\sqrt{3}\,k_{sp})] \quad \text{(six-fold symmetry). (3.37)}$$

This expression for σ_p becomes negative for a six-fold spring network under tension for $\tau/k_{sp} > \sqrt{3}/5$, indicating that such a network should expand laterally when stretched longitudinally in this range of stress.

Just how well do our zero-temperature calculations reproduce the stress-dependence of a spring network at non-zero temperature? From the existence of the collapse transition alone, we know to be wary of severe approximations: the collapse transition only appears once we drop the assumption that all springs have the same length. The calculations presented here should be accurate for triangulated networks at low temperature and/or high tension, where the springs are expected to be of uniform length even if it is different from s_o. A Monte Carlo simulation has shown Eq. (3.30) to be accurate at the 90% level or better over the temperature range $k_B T = k_{sp} s_o^2/8$ to $k_{sp} s_o^2$ (Boal et al., 1993).

Six-fold spring networks at non-zero temperature

Our calculation of the stress-dependence of spring networks assumed that all springs have a common length, an assumption valid at low temperatures or high tensions relative to the scales set by $k_{sp} s_o^2$ and k_{sp}, respectively. However, as the temperature rises, the springs increasingly fluctuate in length. Consider an isolated spring, for which the probability of the displacement from equilibrium, $x = s - s_o$, is proportional to the

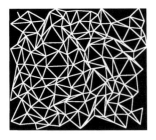

Fig. 3.14. Snapshot of a triangular network of springs as seen in a computer simulation. The network is held at zero tension and the relatively modest temperature of $k_B T = k_{sp} s_o^2/4$. The system is subject to periodic boundary conditions in the shape of a rectangle, whose location is indicated by the black background (after Boal *et al.*, 1993).

Boltzmann factor $\exp(-\Delta U/k_B T) = \exp(-k_{sp} x^2/2k_B T)$. Properly normalized, the probability $\mathcal{P}(x)dx$ that x lies between x and $x + dx$ is given by

$$\mathcal{P}(x)dx = dx \cdot \exp(-\beta k_{sp} x^2/2)/ \int_{-\infty}^{\infty} \exp(-\beta k_{sp} x^2/2) \, dx, \qquad (3.38)$$

where β is the inverse temperature $(k_B T)^{-1}$. Derived as an example in Appendix C, the mean square value of x obtained from this probability distribution is

$$\langle x^2 \rangle = \int_{-\infty}^{\infty} x^2 \, \mathcal{P}(x) \, dx = k_B T / k_{sp}, \qquad (3.39)$$

confirming that the fluctuation in the spring length increases with temperature. Thus, we expect spring networks to lose their regular appearance as the temperature grows, as can be seen in the simulated configuration displayed in Fig. 3.14.

It is not immediately clear from Fig. 3.14 how the network area should change with temperature, as there is considerable variation in the spring length and plaquette area. In fact, the thermal behavior of a network depends on its connectivity, as will be shown in Section 3.4 by comparing networks with four- and six-fold connectivity. As determined by computer simulation, the temperature-dependence of the area per vertex for the six-fold coordinated network is displayed in Fig. 3.15. Although one might expect the network to expand with temperature, as most conventional materials do, the area per vertex in the six-fold network surprisingly declines with temperature for $k_B T/k_{sp} s_o^2$ between 0 and 0.2. That is, the network has a negative coefficient of thermal expansion at low temperatures (Lammert and Discher, 1998). At higher temperatures, the area increases linearly with temperature with a slope close to 1/2. In contrast to the behavior of the area per vertex, the elastic moduli vary only weakly with temperature: both K_A

Fig. 3.15. Area per vertex $\langle A_v \rangle$ for a six-fold network of springs, shown as a function of the reduced temperature $k_B T/k_{sp} s_o^2$. The area is normalized to the area per vertex at zero stress and temperature, namely $A_o = \sqrt{3} \, s_o^2/2$, and is obtained by simulation.

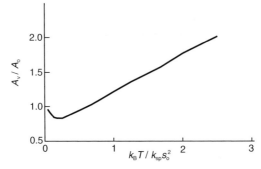

and μ are within 10–20% of their zero-temperature values ($\sqrt{3}\,k_{sp}/2$ and $\sqrt{3}\,k_{sp}/4$, respectively) up to a reduced temperature of $k_B T/k_{sp}s_o{}^2 = 0.3$. We return to the temperature-dependence of two-dimensional networks in Section 3.4.

3.4 Spring networks with four-fold coordination

Included among the simple two-dimensional networks of biological significance are those with uniform four-fold connectivity: for example, the nuclear lamina and the cortex of the auditory outer hair cell. In this section, we examine four-fold networks of identical springs, although we recognize that this system may represent only four-fold cellular networks of uniform composition. The free energy density $\Delta\mathcal{F}$ associated with the deformation of four-fold materials was shown in Section 3.2 to have a slightly more complicated form than that of six-fold or isotropic materials, namely, from Eq. (3.18)

$$\Delta\mathcal{F} = (K_A/2)(u_{xx}+u_{yy})^2 + (\mu_p/2)(u_{xx}-u_{yy})^2 + 2\mu_s u_{xy}{}^2$$
(four-fold symmetry),

an expression containing two shear moduli, μ_p and μ_s. If there is no restoring force between adjacent network elements, i.e., if the energy does not depend even implicitly upon the angle between neighboring bonds, the ground state is not unique and is said to be degenerate. For example, in the absence of angle-dependent forces, all of the plaquettes in Fig. 3.16 have the same energy, permitting a network to move among these configurations with no energy penalty. As a consequence, the simple shear modulus vanishes at zero temperature and stress, a behavior not seen in six-fold networks.

Rather than immediately address the implications of ground-state degeneracy, we first describe an approximation for the network that should be valid at simultaneously large tension and low temperature, namely that all plaquettes have the shape of squares. This is like the equilateral plaquette approximation for the six-fold network in Section 3.3, which describes the low temperature behavior of that network very well. For square plaquettes, all springs in the network have a length s_o when unstressed, and s_τ when subject to a two-dimensional tension τ. As there are two springs and one vertex per plaquette, the enthalpy per vertex in this approximation is $H_v = k_{sp}(s-s_o)^2 - \tau s^2$.

spring

Fig. 3.16. With no bending resistance, each of these spring configurations has the same energy, demonstrating that there is no energy penalty against simple shear deformations, to lowest order in the strain tensor.

Fig. 3.17. Sample deformation modes for a square: (a) is a pure compression or expansion, (b) is a pure shear, and (c) is a simple shear. The original shape is indicated by the shaded region, and the deformed shape is the wire outline. In the calculation leading to Eq. (3.43), the sides of the shaded square have length s_τ.

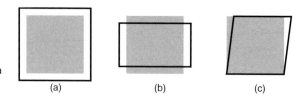

(a)   (b)   (c)

The spring length s_τ that minimizes H_v for square plaquettes is found to be

$$s_\tau = s_o/(1 - \tau/k_{sp}) \qquad \text{(four-fold symmetry, squares)}. \qquad (3.40)$$

From Eq. (3.40), we see that the spring length and, consequently, the area per vertex of square plaquettes both grow without bound at a tension given by

$$\tau_{exp} = k_{sp} \qquad \text{(four-fold symmetry, squares)}. \qquad (3.41)$$

This behavior occurs for the same reason that it does for six-fold spring networks: the energy $[k_{sp}(s-s_o)^2]$ and tension (τs^2) terms are both proportional to s^2 at large extension, and the tension dominates if τ/k_{sp} is sufficiently large. Under any positive pressure (negative tension) at zero temperature, four-fold networks collapse because all plaquettes have the same energy, including the plaquette with zero area:

$$\tau_{coll} < 0 \qquad \text{(four-fold symmetry)}. \qquad (3.42)$$

The elastic moduli of the square plaquette network can be found by performing small deformations on a plaquette of length s_τ to the side, as illustrated by the three deformation modes in Fig. 3.17. The calculation of the moduli follows the same procedure as that used for six-fold networks in Section 3.3, namely two expressions are generated for the free energy change of each deformation mode – one involving the elastic moduli and one involving spring variables – and the moduli are determined by equating the expressions for the free energy change. The method follows Section 3.3 so closely that we will not repeat it here; rather, we quote the results

$$K_A = (k_{sp} - \tau)/2 \qquad \mu_p = (k_{sp} + \tau)/2 \qquad \mu_s = \tau$$
$$\text{(four-fold symmetry, squares)}, \qquad (3.43)$$

and leave the proof to the problem set at the end of this chapter. These moduli are linear in stress and are found to describe computer simulations of four-fold spring networks at low temperature ($\beta k_{sp} s_o^2 \gg 1$) and high tension rather well (Tessier, Boal and Discher, personal communication). Of particular interest is the prediction that μ_s vanishes in the stress-free limit at low temperature. This behavior reflects the fact that the change in the spring energy of the simple shear deformation in Fig. 3.17(c) vanishes at zero tension, to leading order, and hence the plaquette is unstable against simple shear modes to lowest order in the energy (see problem set).

The square plaquette approximation also predicts that the area per vertex should be $A_v = s_o^2$ at $\tau = 0$, according to Eq. (3.40). This expression is certainly not accurate: the measured value from simulations at low temperature extrapolates to $0.62 s_o^2$ at $T = \tau = 0$. Further disagreements are found for the elastic moduli at low stress. The reason for the network area falling below s_o^2 is the degenerate ground state: the square configuration has the largest area for a given spring length, but the system can explore many other configurations with $0 < A_v < s_o^2$ with no energy penalty. That the network samples a wide array of plaquette shapes can be seen in Fig. 3.18, which displays a configuration from a computer simulation of a periodic network with four-fold connectivity. The temperature of this network is moderate, so the springs vary in length, as does the plaquette area. At low temperature, the springs should be close to s_o in length, and the plaquettes should be parallelograms. If the network samples these parallelogram shapes with equal probability, then the area per plaquette, or per vertex, is given by

Fig. 3.18. Configuration of a square network of springs taken from a Monte Carlo simulation at zero stress and a temperature of $k_B T = k_{sp} s_o^2/4$. The network is subject to non-orthogonal periodic boundaries whose location is indicated by the black background (from Tessier, Boal and Discher, personal communication).

$$\langle A_v \rangle = s_o^2 \int \sin\theta \, d\theta / \int d\theta = (2/\pi) s_o^2, \tag{3.44}$$

where θ is the angle between adjacent sides and covers the range 0 to $\pi/2$. This description predicts that the mean area per vertex is $(2/\pi) s_o^2 = 0.64 s_o^2$, in very good agreement with the simulation result of $0.62 \, s_o^2$.

Encouraged by our success in describing the network in terms of parallelogram plaquettes at vanishing stress and temperature, let us extend this approach to non-zero temperature at zero stress. Displayed in Fig. 3.19 is the area per vertex for four- and six-fold networks at zero stress. The areas are normalized to an area per vertex A_o, namely $(2/\pi) s_o^2$ for four-fold and $\sqrt{3} \, s_o^2/2$ for six-fold networks, respectively. The first feature to note in Fig. 3.19 is that the networks behave oppositely at low temperature: four-fold networks expand while six-fold ones contract, indicating that the thermal behavior of both bond lengths and angles must be important.

Fig. 3.19. Mean area per vertex for spring networks with four-fold (square symbols) and six-fold (circles) connectivity, normalized to $A_o = (2/\pi) s_o^2$ for four-fold networks and $A_o = \sqrt{3} \, s_o^2/2$ for six-fold networks, respectively. The grey line is the mean field calculation of Eq. (3.56), while the black line is a numerical integration of the corresponding expression for six-fold spring networks (from Tessier, Boal and Discher, personal communication).

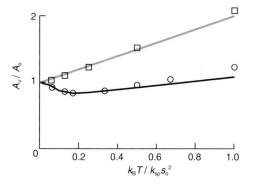

Fig. 3.20. (a) Parallelogram shapes generated by a fixed bond of length s_o and a freely moving vertex, indicated by the shading. The plaquettes can represent uniform networks with either four-fold (b) or six-fold (c) connectivity.

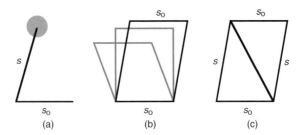

Is there a simple model that approximately captures the average behavior of the plaquettes in Fig. 3.18? In particular, is there a way of sampling the configurations of one plaquette that yields similar averages as found in the network as a whole? One can imagine many different approximations, but an approach that works fairly well is to sample the shapes of parallelograms with one side fixed at its zero-temperature length s_o. That is, one follows the shapes of parallelograms generated by a single vertex as it explores a two-dimensional coordinate space, as illustrated in Fig. 3.20. We refer to this as a mean field model, in the sense that all network plaquettes are assumed to be identical, such that the behavior of the network is the same as that of an individual plaquette. In reality, the shape of a plaquette is not independent of its neighbor, so that this simple model does not include local correlations in shape, nor does it include general quadrilateral shapes which proliferate as the temperature rises. Further, to be closer to the mean field approach, the length of the fixed side should be varied until the value that minimizes the free energy is found at each temperature. We introduce this model not so much because of its accuracy, as to illustrate how to treat a network at $T>0$.

Our mean field model can be evaluated analytically for four-fold networks, as we now show. The network as a whole has N plaquettes, N vertices, and $2N$ springs, so that each plaquette corresponds to a single vertex and two springs, one of length s and the other s_o. Hence, the potential energy of the plaquette is

$$E = (k_{sp}/2)(s - s_o)^2, \tag{3.45}$$

and the corresponding Boltzmann factor for a single vertex must be

$$\exp(-\beta E) = \exp(-\beta k_{sp}[s - s_o]^2/2) = \exp(-\alpha[\sigma - 1]^2), \tag{3.46}$$

where the dimensionless variables

$$\alpha = \beta k_{sp} s_o^2/2 \tag{3.47a}$$

$$\sigma = s/s_o, \tag{3.47b}$$

have been introduced.

The Boltzmann factor can be used to construct a probability distribution for plaquette shapes analogous to Eq. (3.38) for isolated springs. As the vertex moves in a two-dimensional plane, the probability distribution involves two Cartesian directions $dx \, dy$, which becomes $s \, ds \, d\theta$ in polar coordinates. For the four-fold plaquette in Fig. 3.20(b) (but not the six-fold version in part (c)), the energy does not depend upon θ, which can then be integrated separately from s. Thus, the two-dimensional generalization of Eq. (3.38) for the probability distribution in our mean field approximation is

$$\mathscr{P}(\sigma)\sigma d\sigma = \exp(-\alpha[\sigma-1]^2) \, \sigma d\sigma / \int \exp(-\alpha[\sigma-1]^2) \, \sigma d\sigma, \qquad (3.48)$$

where s has been rescaled according to Eq. (3.47b). The ensemble average of the area per vertex, $\langle A_v \rangle$, in the mean field approximation is proportional to an integral over the side length s through

$$\langle A_v \rangle = s_o \, \langle s \rangle \, \langle \sin\theta \rangle = (2/\pi) \, s_o^2 \, \langle \sigma \rangle, \qquad (3.49)$$

where

$$\langle \sigma \rangle = \int \sigma \mathscr{P}(\sigma)\sigma d\sigma = \int \sigma^2 \exp(-\alpha[\sigma-1]^2) \, d\sigma / \int \exp(-\alpha[\sigma-1]^2)\sigma d\sigma. \quad (3.50)$$

The range of σ in Eq. (3.50) runs from zero to infinity.

At low temperature, the integrands are concentrated around $\sigma = 1$, and it is convenient to change variables to

$$\varepsilon = \sigma - 1, \qquad (3.51)$$

so that Eq. (3.50) becomes

$$\langle \sigma \rangle = \int \exp(-\alpha\varepsilon^2) \, (\varepsilon+1)^2 d\varepsilon / \int \exp(-\alpha\varepsilon^2) \, (\varepsilon+1)d\varepsilon, \qquad (3.52)$$

where ε runs from -1 to $+\infty$. Expanding the polynomials in Eq. (3.52) leads to a number of integrals of the form

$$\int \exp(-\alpha\varepsilon^2)\varepsilon^n d\varepsilon = \alpha^{-(n+1)/2} \int \exp(-\tau^2)\tau^n d\tau \qquad (3.53)$$

where the integration limits on τ are from $-\sqrt{\alpha}$ to $+\infty$. Eq. (3.47a) shows that α tends to infinity at low temperature; as a consequence, the integrands are approximately symmetric about $\tau = 0$ at low temperature, and integrals with odd n must vanish. Thus, Eq. (3.52) becomes

$$\langle \sigma \rangle = 1 + \{\alpha^{-1} \int \exp(-\tau^2) \, \tau^2 d\tau / \int \exp(-\tau^2)d\tau \}. \qquad (3.54)$$

This expression appears in the mean square displacement of a spring in one dimension (see Appendix C), and can be evaluated using $\int \exp(-\tau^2) \tau^2 d\tau = \sqrt{\pi}/2$ and $\int \exp(-\tau^2)d\tau = \sqrt{\pi}$, where the integration limits are $-\infty$ and $+\infty$. Hence, we find

$$\langle \sigma \rangle = 1 + 1/(2\alpha), \qquad (3.55)$$

and, finally, the area per vertex is

$$\langle A_v \rangle = (2/\pi)s_o^2(1 + k_B T / k_{sp}s_o^2). \tag{3.56}$$

This mean field result for the area per vertex is plotted in Fig. 3.19, and the agreement with the simulations of a complete network is very good, even for temperatures as high as $k_B T = k_{sp}s_o^2$. Somewhat more impressive are the predictions of this approach for the six-fold spring network, also shown in Fig. 3.19. The methodology for six-fold networks is the same as that for four-fold, except that Eq. (3.45) for the energy per vertex now contains the energy of the second spring running along a diagonal of the plaquette. Unfortunately, this extra energy depends on the angle of the "moving" vertex with respect to the fixed base. The resulting probability distribution involves both s and θ, but is straightforward to integrate numerically, predicting that A_v initially decreases with temperature, as observed for the network simulations. The predictions of this approximation lie within a few percent of the simulated networks up to $k_B T = k_{sp}s_o^2/2$.

Our mean field approach assumes that one spring in the plaquette has a fixed length, and that the shape fluctuations of the parallelogram plaquettes mimic those of the network as a whole. Although this model is clearly useful in determining averages of some simple observables, it is less accurate for observables that probe the distribution of spring lengths. For example, the parallelogram approximation to four-fold networks at low temperature predicts that the area compression modulus, which is a measure of the fluctuations in plaquette area, should be (see problem set)

$$\beta K_A s_o^2 = 4\pi/(\pi^2 - 8). \tag{3.57}$$

Eq. (3.57) is about 50% higher than what is found from simulations of four-fold networks, demonstrating that the area fluctuations are not well-described by the mean field approach (Tessier, Boal and Discher, personal communication).

3.5 Membrane-associated networks

Before we describe and interpret measurements of network elasticity in two dimensions, let us quickly review some relevant results from Sections 3.2–3.4. When an object deforms in response to an applied stress, an element of the object moves from its original position x to a new position $x + u$, where u is a displacement vector whose magnitude and direction vary locally across the object. The nature and severity of the deformation can be characterized by the strain tensor u_{ij}, a unitless quantity proportional to the rate of change of u with respect to x. To lowest order, the change in the free energy density associated with the deformation is proportional to the squares of various combinations of components of the strain tensor, and each combination can be made to represent a physical deformation mode such as compression or shear. The forces on an object can be represented by a stress tensor σ_{ij}, which has units of energy per unit area in two dimensions or energy per unit volume in three. For Hooke's law materials, the stress is proportional to the strain, $\sigma_{ij} = \Sigma_{k,l} C_{ijkl} u_{kl}$, much like $f = -k_{sp} x$ of a one-dimensional spring. Although the number of elastic moduli C_{ijkl} is large in principle, symmetry considerations show that isotropic or six-fold materials in two dimensions can be represented by just two moduli (the area compression and shear moduli, K_A and μ, respectively), while two-dimensional materials with four-fold symmetry can be represented by three moduli (K_A, μ_p and μ_s, where the two shear moduli correspond to pure and simple shear modes, respectively).

The moduli depend upon both temperature and stress. For example, two-dimensional spring networks under an in-plane tension τ approximately obey Eqs. (3.34) and (3.43)

$$K_A = (\sqrt{3}\, k_{sp}/2) \cdot (1 - \tau / [\sqrt{3}\, k_{sp}]) \qquad \mu = (\sqrt{3}\, k_{sp}/4) \cdot (1 + \sqrt{3}\, \tau / k_{sp})$$
$$\text{(six-fold symmetry),} \tag{3.58a}$$

and

$$K_A = (k_{sp} - \tau)/2 \qquad \mu_p = (k_{sp}/2) \cdot (1 + \tau/k_{sp}) \qquad \mu_s = \tau$$
$$\text{(four-fold symmetry),} \tag{3.58b}$$

if the springs have uniform length in the deformed state, where $\tau > 0$ ($\tau < 0$) corresponds to tension (compression). The temperature-dependence of the moduli is weaker than the stress-dependence, and varies with the connectivity of the network. For instance, the behavior of the area per vertex at low temperature and zero stress is opposite for the two spring models investigated here: six-fold networks initially shrink with rising temperature while four-fold networks expand.

Measured network elasticity

The elastic characteristics of many different cell types have been determined, although not all cells are amenable to theoretical analysis because of their complex cytoskeleton or inhomogeneous structure. Here, we concentrate on just a few cells with a simple architecture: red blood cells, auditory outer hair cells and bacteria. One set of techniques for measuring elasticity can be broadly categorized as micromechanical manipulation: a cell or extracted network is subject to a known stress – by means of a micropipette, for instance – and the response of the cell

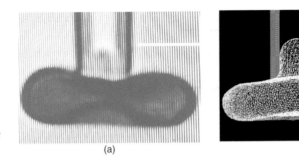

(a)   (b)

Fig. 3.21. (a) Micropipette aspiration of a red blood cell. The diameter of the cell is about 8 μm, and the inner diameter of the pipette is roughly 1 μm. In line with the horizontal bar, a small segment of the membrane can be seen inside the pipette (from Evans and Waugh, 1980; courtesy of Dr Evan Evans, University of British Columbia). (b) Computer simulation of an aspiration experiment, using a computational blood cell about 4 μm in diameter (from Discher et al., 1998).

is observed by optical microscopy or another imaging method. The stress–strain curves found in these experiments are interpreted using conventional mechanics to yield one or more elastic moduli. A second experimental approach records the thermal fluctuations in a system's shape; as demonstrated for filaments in Chapter 2, the amplitude of these fluctuations is inversely proportional to the deformation resistance of the system.

One of the most thoroughly studied two-dimensional networks in the cell is the membrane-associated cytoskeleton of the human erythrocyte. A classic procedure for investigating the mechanical properties of this cytoskeleton is to apply a suction pressure to a red cell through a micropipette, pulling a segment of its plasma membrane into the pipette, as illustrated in Fig. 3.21 (Evans, 1973a, 1973b; Waugh and Evans, 1979). The figure shows a flaccid human red blood cell, with a small segment of its plasma membrane and cytoskeleton being drawn up into the pipette. The length of the aspirated segment L is related to the applied pressure P and the two-dimensional shear modulus μ through

$$P = (\mu / R_p) [((2L / R_p) - 1) + \ln(2L / R_p)], \tag{3.59}$$

where R_p is the pipette radius (see Section 7.4). Several independent experiments based upon this technique yield a shear modulus of $6{-}9 \times 10^{-6}$ J/m^2 for human erythrocytes (Waugh and Evans, 1979, and references therein). In fact, the shear moduli of red cells from several vertebrates have been measured by aspiration, and their reported values generally lie not far from 10^{-5} J/m^2 for red cells without a nucleus, and about an order of magnitude higher for nucleated cells (Waugh and Evans, 1976). Deforming a cell with optical tweezers also allows the extraction of μ for the red cell cytoskeleton: one experiment obtains $2.5 \pm 0.4 \times 10^{-6}$ J/m^2 (Hénon et al., 1999) while another finds about 2×10^{-4} J/m^2 (Sleep et al., 1999). Further, an analysis of the thermal fluctuations of red cell shape yields a value for μ that may be more than an order of magnitude smaller than the micromanipulation results

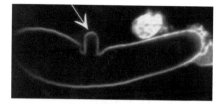

Fig. 3.22. Fluorescence image of an auditory outer hair cell deformed by micropipette aspiration. About 80 μm long and 10 μm wide, the cell is from a guinea pig; the position of the aspirated segment is indicated by the arrow (reprinted with permission from Oghalai *et al.*, 1998; © 1998 by the Society for Neuroscience; courtesy of Dr William Brownell, Baylor College of Medicine).

(Peterson *et al.*, 1992). It is not clear why these techniques give such different results.

The variation of the red cell cytoskeleton elasticity with temperature has been measured in two experiments. Stokke *et al.* (1985a) observe that the spectrin persistence length decreases with increasing temperature, as expected for flexible filaments. Evans and Waugh (1979) find that the shear modulus decreases from 8×10^{-6} J/m^2 to 5.5×10^{-6} J/m^2 over the temperature range of 5–45°C. We return to the significance of these results after describing related measurements on other simple cells. We should also point out that the measured value of μ decreases with spectrin density, roughly as expected from Eq. (3.60), although there are subtleties associated with network behavior at low density that will be described further in the treatment of percolation in Chapter 4 (Waugh and Agre, 1988). Lastly, we mention that the area compression modulus as observed in aspiration experiments is close to twice the shear modulus (Discher *et al.*, 1994).

Auditory outer hair cells are relatively large, and can be aspirated in the same way as erythrocytes to determine the elastic properties of the lateral cortex, as displayed in Fig. 3.22. Under the assumption that it is isotropic, the effective shear modulus of the cortex is found to be $1.5 \pm 0.3 \times 10^{-2}$ J/m^2, which is about 1000 times larger than that of the erythrocyte (Sit *et al.*, 1997). However, the lateral cortex is known to be anisotropic, as seen in Fig. 3.1(b): it is stiffer in the circumferential direction than in the axial direction. On the basis of the response when an outer hair cell is gently bent by an external probe, the circumferential filaments are estimated to have a Young's modulus of 1×10^7 J/m^3 while the axial cross links have a smaller value of $Y = 3 \times 10^6$ J/m^3 (Tolomeo *et al.*, 1996). These estimates are significantly lower than Y estimated for actin in Section 2.5, perhaps reflecting different experimental conditions. Lastly, we mention that a shear modulus has been obtained for fibroblasts on a substrate, where the deformation is generated by applying a magnetic field to microscopic magnetic beads attached to the plasma membrane. The cytoskeleton of a fibroblast is more complex than the cells we have discussed so far, and the measured shear resistance is correspondingly large, $\mu \sim 2\text{–}4 \times 10^{-3}$ J/m^2 (Bausch *et al.*, 1998).

Fig. 3.23. Scanning tunneling microscope image of a bacterial sheath suspended across a 0.3 μm gap in a Ga–As substrate. The sheath can be deformed by an atomic force microscope to determine an elastic modulus (reprinted with permission from Xu *et al.*, 1996, © 1996 by the American Society for Microbiology).

Elastic properties of several thin structural components of bacteria have been measured, although the analysis has been in terms of three-dimensional moduli, rather than the two-dimensional quantities of this chapter. For example, the cell walls of Gram-positive bacteria, containing a thick layer of peptidoglycan, can be extracted and stretched, allowing the determination of a Young's modulus from the resulting stress/strain curve. As applied to *Bacillus subtilis* (Thwaites and Surana, 1991), the technique shows that the Young's modulus of a dry cell wall is $1.3 \pm 0.3 \times 10^{10}$ J/m^3, decreasing by several orders of magnitude to about 3×10^7 J/m^3 when the wall is wet. These values for Y, corresponding to a deformation along the axis of the bacterium, can be compared to the generic $Y \sim 10^9$ J/m^3 of biofilaments (see Section 2.5).

In some cases, individual bacteria are aligned end-to-end like sausages, surrounded by a hollow cylindrical sheath which can be isolated and manipulated as shown in Fig. 3.23 (Xu *et al.*, 1996). A Young's modulus can be extracted from the stress/strain curve obtained when a known force is delivered to the sheath by the tip of an atomic force microscope. For the bacterium *Methanospirillum hungatei*, Y lies between 2×10^{10} J/m^3 and 4×10^{10} J/m^3, a value which is larger than that of typical biopolymers but similar to collagen ($1–2 \times 10^{10}$ J/m^3). The AFM approach has also been applied to the Gram-negative bacterium *Escherichia coli* by Xu *et al.* (1999), yielding an average $Y = 2.5 \times 10^7$ J/m^3 for a hydrated cell wall, where Y changes by a factor of two depending upon direction.

Interpretation of measurements

The values of the elastic moduli for many of the two-dimensional networks summarized above are much lower than conventional materials, including plastics. The softness of these networks reflects both the low density and the high flexibility of their filamentous components. The erythrocyte cytoskeleton demonstrates both of these effects, as can be seen in Fig. 3.24. Part (a) of the figure is a drawing of a stretched cytoskeleton, as imaged by an electron microscope, showing once again the approximate six-fold connectivity of the network (the authors of this study find 3% and 8% of the junctions have five- or seven-fold connectivity, respectively; Liu *et al.*, 1987). The distance between junctions in the figure is close to the tetramer contour length of 200 nm, in contrast to the average separation of 75 nm *in vivo*. Unless there are local attractive interactions that tighten up the tetramers when they are in a cell (Ursitti and Wade, 1993; see also McGough and Josephs, 1990), the cytoskeleton should appear something like Fig. 3.24(b), with the tetramers being moderately convoluted and making a large entropic contribution to the elasticity of the network.

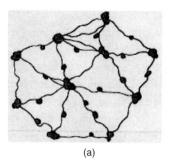 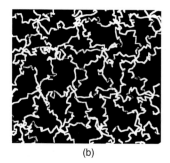

(a)   (b)

Fig. 3.24. (a) Drawing of a spread erythrocyte cytoskeleton, copied from an electron microscope image; the distance between junctions is close to 200 nm (reprinted with permission from Liu *et al.*, 1987, © 1987 by the Rockefeller University Press). (b) Computer simulation of the cytoskeleton, in which each tetramer is represented by a single polymer chain, linked at six-fold vertices and attached to the bilayer at its mid-point. In this unstretched configuration, elements closer to the viewer are shaded lighter (from Boal, 1994).

The elastic moduli of a network of floppy chains can be estimated from the results we have obtained so far in Chapters 2 and 3. Let's consider six-fold networks as an example. Although the chains are contorted, their junctions, which appear lighter in Fig. 3.24(b), undergo only limited excursions from their equilibrium positions, and behave like the junctions in a spring network at low temperature. Regarded as entropic springs, the chains would have an effective spring constant $k_{sp} = 3k_B T/\langle r_{ee}^2\rangle$ according to Eq. (2.62), where Eq. (2.34) has been used for the value of the mean square end-to-end distance $\langle r_{ee}^2\rangle$. Now, the area per vertex A_v of an equilateral plaquette in a six-fold network is $A_v \sim \sqrt{3}\langle r_{ee}^2\rangle/2$, so that the spring constant could alternatively be written $k_{sp} \sim 3\sqrt{3} k_B T/2A_v$. Recognizing that the two-dimensional density of chains ρ is given by $\rho = 3/A_v$ (three chains per vertex), the effective spring constant is $k_{sp} = (\sqrt{3}/2)\rho k_B T$. Lastly, the shear modulus for a six-fold spring network at low temperature is $\mu = \sqrt{3}\, k_{sp}/4$, so that we expect $\mu = (3/8)\rho k_B T$. Now, the implicit assumptions in this derivation, such as the freedom of the chain to move without restrictions imposed by other chains or the bilayer, challenge the accuracy of the factor of 3/8 in this expression. Nevertheless, as obtained on more general grounds in Chapter 4, it is true that μ approximately obeys

$$\mu \sim \rho k_B T. \tag{3.60}$$

Eq. (3.60) provides a guide to the entropic contribution to the elasticity of the erythrocyte cytoskeleton. With a spectrin tetramer density of $\rho \sim 800\ \mu m^{-2}$ (see Section 3.1), the cytoskeleton would be expected to have a shear modulus of 3×10^{-6} J/m^2, which is at the low end of the experimental determinations. Computer simulations that include steric interactions of the polymer chains (Boal, 1994) confirm the approximate results, although the predicted shear modulus rises to 1×10^{-5} J/m^2. What about the area compression modulus K_A? If the picture of the erythrocyte cytoskeleton as a low temperature network of entropic or conventional springs is correct, then we would expect $K_A/\mu = 2$ from Eq. (3.34) at zero stress. Experiments using micropipette aspiration find K_A/μ

close to 2 as well (Discher *et al.*, 1994). These experiments have been further discussed in terms of effective models of the cytoskeleton by Discher *et al.*, (1998).

The temperature-dependence of the network geometry and elasticity provides a test of the importance of entropy in the cytoskeleton. Here, experiment appears to be divided. On the one hand, $\langle r_{ee}^2 \rangle$ of spectrin tetramers falls with rising temperature, and is found to be quantitatively consistent with $\langle r_{ee}^2 \rangle \propto \xi_p$ in Eq. (2.31) and the variation of the persistence length $\xi_p \propto 1/T$ in Eq. (2.19) (Stokke *et al.*, 1985a). On the other hand, if spectrin tetramers do behave like entropic springs, we expect that the elastic moduli should rise roughly linearly with temperature, according to Eq. (3.26). Over the temperature range of 5–45°C, this change should correspond to an increase of 13%. However, the shear modulus is observed to decrease by about 30% instead (Waugh and Evans, 1979). The reasons for this discrepancy are not yet apparent.

With its composite structure of soft spectrin and somewhat stiffer actin, the cortical lattice of outer hair cells is strongly anisotropic (the flexural rigidity of F-actin is a thousand times that of spectrin). The resistance of this network against simple shear modes should vanish in the absence of angle-dependent interactions at its four-fold junctions. One can obtain a crude estimate of the entropic contribution to the pure shear modulus through the density of polymer chains, namely two chains per 25×65 nm rectangular plaquette. For a network of floppy chains, this density corresponds to a shear modulus of 5×10^{-6} J/m², using $\mu = \rho k_B T$. In fact, the measured shear modulus is three orders of magnitude larger (Sit *et al.*, 1997; Tolomeo *et al.*, 1996), indicating that the shear resistance is of energetic, rather than entropic origin, reflecting the stiffness of the network filaments. Lastly, we note for comparison that a rough estimate of the two-dimensional moduli of the bacterial cell wall may be obtained by multiplying the three-dimensional moduli by the thickness of the wall, as explained further in Chapter 5. With $Y \sim 10^9$ J/m³ and a thickness of 5 nm, the two-dimensional moduli should be around 5 J/m², illustrating the strength of the cell wall compared to the loose networks of the red cell or hair cell. Cell walls are analyzed in more detail in Section 7.5.

3.6 Summary

The soft filaments of a cell can be knitted into a variety of two- or three-dimensional networks, some of which have surprisingly regular connectivity. To within a factor of two larger or smaller, the typical diameter of a network filament is 10 nm, whereas the common spacing between filaments is 50–100 nm. However, some networks, such as the bacterial cell wall, are much denser than this, consisting of thin polymers less than 1 nm wide, connected at intervals of less than 10 nm. Most networks in the cell, with the exception of the cell wall, are moderately flexible and deform in response to ambient forces.

The mathematical description of a deformation involves the dimensionless strain tensor u_{ij}, related to the displacement of a given element of an object at position $\mathbf{x}$ by an amount $\mathbf{u}$ according to $u_{ij} = 1/2\ [\partial u_i/\partial x_j + \partial u_j/\partial x_i + \Sigma_k\ (\partial u_k/\partial x_i)(\partial u_k/\partial x_j)]$, where the subscripts i, j, k refer to the Cartesian coordinate axes. For small deformations, the last term may be safely neglected. The forces experienced by an object are described by the stress tensor σ_{ij}, such that the component F_i of the net force is given by $F_i = \Sigma_j\ \sigma_{ij} a_j$, where a_j is a component of the vector $\mathbf{a}$ having magnitude equal to the surface area element and a direction perpendicular to that element. For Hooke's law materials, the stress is proportional to the strain, or $\sigma_{ij} = \Sigma_{k,l}\ C_{ijkl} u_{kl}$, where the proportionality constants C_{ijkl} form a tensor and are the analogues of the scalar spring constant k_{sp} of Hooke's law. To lowest order, the free energy density $\Delta\mathscr{F}$ associated with the deformation is proportional to the squares of different combinations of the strain tensor. The symmetry of the strain tensor, and the internal symmetries of the material, relate different components of its elastic moduli, reducing the number of independent moduli to just a few, in many cases. For instance, the free energy density of two-dimensional materials with six-fold (or full rotational) symmetry is given by $\Delta\mathscr{F} = (K_A/2)\cdot(u_{xx} + u_{yy})^2 + \mu\cdot\{(u_{xx} - u_{yy})^2/2 + 2u_{xy}^2\}$, where K_A and μ are the compression and shear moduli in two dimensions; the corresponding Young's modulus is $Y_{2D} = 4K_A\mu/(K_A + \mu)$. For four-fold materials in two dimensions, $\Delta\mathscr{F} = (K_A/2)\cdot(u_{xx} + u_{yy})^2 + (\mu_p/2)\cdot(u_{xx} - u_{yy})^2/2 + 2\mu_s u_{xy}^2$, where the two shear moduli, μ_p and μ_s, reflect the resistance against pure shear and simple shear, respectively.

The elastic moduli depend upon both tension τ and temperature T. Considering regular networks of ideal springs, we find that at low temperature, six-fold networks obey $K_A = (\sqrt{3}\ k_{sp}/2)\cdot(1 - \tau/[\sqrt{3}\ k_{sp}])$ and $\mu = (\sqrt{3}\ k_{sp}/4)\cdot(1 + \sqrt{3}\ \tau/k_{sp})$, where $\tau > 0$ corresponds to tension. Four-fold networks have a similar form, namely $K_A = (k_{sp} - \tau)/2$, $\mu_p = (k_{sp}/2)\cdot(1 + \tau/k_{sp})$ and $\mu_s = \tau$, but the relations are valid only at larger values of τ. The low temperature behavior of network geometry at zero stress is intriguing: six-fold networks initially shrink, while four-fold networks expand, as the temperature rises. At modest temperatures, the mean area per vertex A_v of four-fold networks is described by $\langle A_v \rangle = (2/\pi)s_o^2 (1 + k_B T/2k_{sp}s_o^2)$.

The contribution of energy and entropy to the elastic properties of a network depends, of course, on the flexibility of its filaments. If the filaments are linked together at intervals much less than their persistence length, energy considerations may dominate the elasticity, and the shear resistance is not too far below that of conventional materials. However, if the distance between network junctions is many times the persistence length, the shear modulus should be that expected for a network of floppy chains, namely $\mu \sim \rho k_B T$ where ρ is the chain density in two dimensions.

3.7 Problems

Biological applications

3.1 A triangular network of identical springs is expected to display a negative Poisson ratio σ_p over a range of tensions.

(a) Verify the range of tension where $\sigma_p < 0$, based on Eq. (3.37).
(b) Calculate the range of the area per junction vertex where $\sigma_p < 0$. Quote your answer as a ratio to the area per vertex of an unstressed network.
(c) Find the minimum tension needed for $\sigma_p < 0$ in the cytoskeleton of the human red blood cell. Determine the equivalent spring constant from the expression for the shear modulus using $\mu = 5 \times 10^{-6}$ J/m^2 at zero tension.

3.2 Triangular networks, such as the spectrin cytoskeleton of the human erythrocyte, are predicted to shrink initially as their temperature rises from zero. From Fig. 3.19, determine the (reduced) temperature above which a triangular spring network is predicted to expand. Is this temperature above the (dimensionless) operating temperature of the red cell cytoskeleton, viewed as a triangular network of ideal springs with $\mu = 5 \times 10^{-6}$ J/m^2 and unstressed spring length 75 nm?

3.3 Taking 5×10^{-6} J/m^2 for the shear modulus of the erythrocyte cytoskeleton, predict the areal compression modulus K_A, assuming that the cytoskeleton behaves like a triangular network of ideal springs (at zero tension). Calculate the relative change in area $\langle \Delta A^2 \rangle^{1/2}/A_o$ arising from thermal fluctuations at 300 K, where $\Delta A = A - A_o$ and A_o is the unstressed area. Choose two values of A_o:

(i) the area of a triangular plaquette 75 nm to the side
(ii) 50 μm^2 (roughly one-third the surface area of the human erythrocyte).

(*Hint: use the definition* $(\beta K_A)^{-1} = \langle \Delta A^2 \rangle / \langle A \rangle$; *see Appendix D for the three-dimensional analogue*).

3.4 Evaluate the thermal expansion of a square network like the lateral cortex of the auditory outer hair cell. As a simplification, assume that the network is uniform with an elementary square plaquette of length $s_o = 40$ nm to the side and an areal compression modulus $K_A = 2 \times 10^{-2}$ J/m^2 at $T = 0$ and $\tau = 0$. Treat the network elements as springs, and plot the relative area A_v / s_o^2 from $T = 0$ to 300 K.

3.5 Actin can be crosslinked by the proteins α-actinin or fimbrin to form bundles of parallel filaments, whose structures are schematically represented by

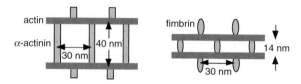

Treating the cross-links as inextensible, calculate the area compression modulus of these two networks using the results of Problem 3.15(c). Establish first that the flexural rigidity is given by $\kappa_f = b \cdot k_{bend}$ using Eq. (2.14) at small θ (b, k_{bend} and θ are defined in Problem 3.15). Assume the persistence length of actin is 15 μm at 300 K to set κ_f.

3.6 The basal lamina is a complex mat which, early in its development, is composed largely of the protein laminin. Laminin has the structure of an asymmetric cross about 140 nm long and 70 nm across, with arms of average cross-sectional radius R of roughly 5 nm. Consider a network knit with the rectangular structure

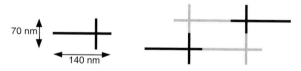

(a) Find the effective spring constant k_{sp} of the network elements using the result from Problem 3.17 that k_{sp} of a uniform rod equals $\pi R^2 Y/L$, where Y is the three-dimensional Young's modulus and L is the length of the filament (assume $Y = 0.5 \times 10^9$ J/m^3).
(b) Calculate the areal compression modulus for this network using the model system in Problem 3.16.

3.7 The cell wall of a Gram-negative bacterium is composed of a thin peptidoglycan network having the approximate two-dimensional structure (see Fig. 3.3):

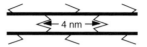

The filaments running horizontally in the diagram are sugar rings with a mean separation between successive in-plane crosslinks of ~4 nm; the crosslinks are floppy chains of amino acids having a contour length of about 4 nm, but a much smaller mean end-to-end displacement of 1.3 nm.

(a) What is the persistence length of the amino-acid chain?
(b) Assuming that the glycan filaments are inextensible rigid rods, calculate K_A using results from Problem 3.16 with $k_x \gg k_y$.

3.8 The lateral cortex of the auditory outer hair cell in guinea pigs is composed of two different filaments linked to form rectangular plaquettes. The mean dimensions of the network are, with 20% variability:

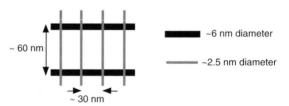

(a) Obtain the effective spring constant of both network elements from $k_{sp} = \pi R^2 Y/L$ (see Problem 3.17 for definitions), assuming $Y = 0.5 \times 10^9$ J/m^3.
(b) Calculate the areal compression modulus from the expression in Problem 3.16(b). For comparison, the observed *shear* modulus of this network is $(1.5 \pm 0.3) \times 10^{-2}$ J/m^2.

Formal development and extensions

3.9 The area compression modulus K_A is related to the change of area with respect to (two-dimensional) pressure P by $K_A^{-1} = -(\partial A/\partial P)/A$. For an ideal gas obeying the two-dimensional version of $PV = Nk_B T$, show that $K_A = P$.

3.10 For an isotropic material in two dimensions, the stress is related to the strain by

$$\sigma_{ij} = (K_A - \mu)\,\delta_{ij}\,\mathrm{tr}\,u + 2\mu u_{ij}.$$

(a) Invert this expression to obtain the strain in terms of the stress (an analogous expression for three dimensional materials is derived in Appendix D).

(b) Evaluate u_{xx} and u_{yy} for a uniaxial stress in the x-direction: $\sigma_{xx} \neq 0$, $\sigma_{yy} = 0$.

(c) Using your results from part (b), show that the Poisson ratio $\sigma_p = -u_{yy}/u_{xx}$ is given by $\sigma_p = (K_A - \mu)/(K_A + \mu)$.

(d) The two-dimensional Young's modulus is defined by $Y_{2D} = [stress]/[strain] = \sigma_{xx}/u_{xx}$. Use your results from part (b) to prove that $Y_{2D} = 4K_A\mu/(K_A + \mu)$.

3.11 Suppose that a two-dimensional isotropic material is stretched uniformly by a fixed scale factor in the x-direction, such that $u_{xx} = \delta$.

(a) Find the corresponding u_{yy} in terms of K_A, μ and δ by minimizing the free energy of the material.

(b) Show that the Poisson ratio $\sigma_p = -u_{yy}/u_{xx}$ is given by $\sigma_p = (K_A - \mu)/(K_A + \mu)$ using your results from part (a).

3.12 In Section 3.3, the area compression modulus for a triangular network of springs under stress was obtained by comparing two expressions for the free energy density, one in terms of spring variables and one in terms of the elastic moduli. As an alternative approach, use the stress-dependent expression for the spring length s_τ to obtain the area compression modulus by differentiation: $K_A^{-1} = (\partial A/\partial \tau)/A$.

3.13 Find the tension-dependence of the elastic moduli of a spring network with four-fold connectivity. Under tension τ, the springs are stretched from their unstressed length s_0 to a stretched length of s_τ. Consider the following three deformations about a square plaquette (shaded) under tension with spring length s_τ:

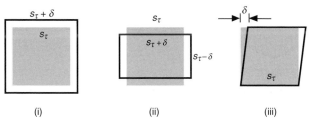

(i)   (ii)   (iii)

(a) Establish that $s_\tau = s_o/(1 - \tau/k_{sp})$ by minimizing the enthalpy per vertex, which is

$$H_v = k_{sp}(s - s_o)^2 - \tau s^2 \text{ for a square plaquette.}$$

(b) Show that the change in H_v/A_v for each mode in terms of spring variables k_{sp}, s_τ and δ is, to order δ^2:

$$(\Delta H_v/A_v)_i = (k_{sp} - \tau)\cdot(\delta/s_\tau)^2, \quad (\Delta H_v/A_v)_{ii} = (k_{sp} + \tau)\cdot(\delta/s_\tau)^2,$$
$$(\Delta H_p/A_p)_{iii} = (\tau/2)\cdot(\delta/s_\tau)^2,$$

where A_v is the area per vertex.

(c) By calculating the strain tensor for each mode, show that the free energy density is

$$(\Delta \mathcal{F})_i = 2K_A(\delta/s_\tau)^2, \quad (\Delta \mathcal{F})_{ii} = 2\mu_p(\delta/s_\tau)^2, \quad (\Delta \mathcal{F})_{iii} = (\mu_s/2)\cdot(\delta/s_\tau)^2,$$

(d) Compare your results from parts (b) and (c), to obtain the elastic constants

$$K_A = (k_{sp} - \tau)/2, \quad \mu_p = (k_{sp} + \tau)/2, \quad \mu_s = \tau.$$

3.14 Consider a network of rigid rods of length s_o joined together at four-fold junctions, around which the rods can rotate freely, so long as they don't intersect. In the mean field limit where all plaquettes of this network have the same shape (as in configuration (iii) of Problem 3.13), show that the area compression modulus is given by

$$K_A = 4\pi (k_B T/s_o^2)/(\pi^2 - 8).$$

Compare this expression for K_A to $\rho k_B T$, where ρ is the density of rods (Hint: use the definition $(\beta K_A)^{-1} = \langle\Delta A^2\rangle/\langle A\rangle$; see Appendix D for the three-dimensional analogue.)

3.15 The two-dimensional honeycomb network is an example of a network with three-fold connectivity with unequal bond lengths a and b. Let's assume that the bonds are inextensible, so that the only contribution to the network energy is $k_{bend}\theta^2/2$ at each vertex.

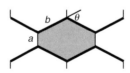

(a) Find the area of the plaquette in grey as a function of θ. How many vertices are there per plaquette (not 6!)? What is the area per vertex A_v of this network?

(b) Defining E_{v} as the bending energy per vertex, show that the enthalpy per vertex, $H_{\mathrm{v}} = E_{\mathrm{v}} - \tau A_{\mathrm{v}}$, for this network under tension τ is minimized at an angle

$$\theta_{\min} = \tau b^2 / (2k_{\mathrm{bend}} + \tau a b / 2)$$

if the deformation of the network is small.

(c) Find A_{v} at small tensions from your value of $\theta_{\min}$, then prove that the areal compression modulus at zero tension is

$$K_{\mathrm{A}} = 4ak_{\mathrm{bend}} / b^3,$$

by differentiating A_{v} [as in $K_{\mathrm{A}}^{-1} = (\partial A / \partial \tau) / A$].

3.16 The spring network with four-fold symmetry in Section 3.4 can be generalized to a rectangular network of inequivalent springs, as in

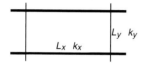

L_y k_y

L_x k_x

where L_x, L_y are the unstretched spring lengths in the x and y directions and k_x, k_y are the corresponding spring constants.

(a) Apply an isotropic tension τ to this network and find the displacements from equilibrium x and y by minimizing the enthalpy per vertex $H_{\mathrm{v}} = k_x x^2 / 2 + k_y y^2 / 2 - \tau (L_x + x)(L_y + y)$.

(b) Show that the areal compression modulus $[K_{\mathrm{A}}^{-1} \equiv (\partial A / \partial \tau) / A]$ is given by

$$K_{\mathrm{A}}^{-1} = L_y / k_x L_x + L_x / k_y L_y,$$

at small tensions.

(c) Confirm that your result from (b) reduces to Eq. (3.43) when $k_x = k_y = k$ and $L_x = L_y$.

3.17 A force f is applied uniformly across each end of a cylindrical rod, changing its length by an amount x.

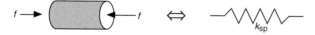

k_{sp}

Viewing the rod as a spring, show that its effective spring constant is $k_{\mathrm{sp}} = \pi R^2 Y / L$, where R and L are the radius and length of the rod, respectively, and Young's modulus Y is defined *via* $[stress] = Y[strain]$.

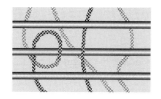

Chapter 4

Three-dimensional networks

The filaments of the cell vary tremendously in their bending resistance, having a visual appearance ranging from ropes to threads if viewed in isolation on the length scale of a micron. Collections of these biological filaments have strikingly different structures: a bundle of stiff microtubules may display strong internal alignment, whereas a network of very flexible proteins may resemble the proverbial can of worms. Thus, the elastic behavior of multicomponent networks containing both stiff and floppy filaments may include contributions from the energy and the entropy of their constituents. In this chapter, we first review a selection of three-dimensional networks from the cell, and then establish the elastic properties of four different model systems, ranging from entropic springs to rattling rods. In the concluding section, these models are used to interpret, where possible, the measured characteristics of cellular networks.

4.1 Networks of biological rods and ropes

The filaments of the cytoskeleton and extracellular matrix collectively form a variety of chemically homogeneous and heterogeneous structures. Let's begin our discussion of these structures by describing two networks of microtubules found in the cell. The persistence length of microtubules is of the order of millimeters, such that microtubules bend only gently on the scale of microns and will not form contorted networks (see Section 2.5). For example, the microtubules of the schematic cell in Fig. 4.1(a) are not cross-linked, but rather extend like spikes towards the cell boundary, growing and shrinking with time. However,

Fig. 4.1. Cartoon of microtubule configurations conventionally found in cells. In part (a), microtubules radiate from the centrosome near the nucleus, while in part (b), they lie parallel to one another along the axon of a nerve cell (not drawn to scale).

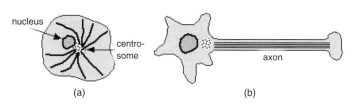

Fig. 4.2. In these newborn mouse keratinocytes, keratin intermediate filaments are seen to form an extended network attached to both the plasma membrane and the surface of the nucleus (courtesy of Dr Pierre A. Coulombe, Johns Hopkins University School of Medicine).

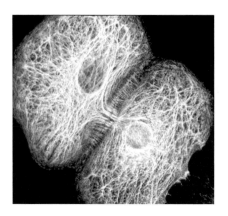

there exist varieties of microtubule-associated proteins (or *MAPs*) that can form links between microtubules, resulting in a bundled structure like Fig. 4.1(b): linking proteins are spaced along the microtubule like the rungs of a ladder, tying two microtubules parallel to one another. Such microtubule bundles are present in the long axon of a nerve cell (see Fig. 1.1) or in the whip-like flagella extending from some cells. Neither of these configurations resembles the regular meshwork of the erythrocyte cytoskeleton or the nuclear lamina described in Section 3.1. The function of the microtubules in both the radial and parallel structures includes transport: different motor proteins can walk along the tubule, as described further in Chapter 9.

The cell's intermediate filaments form a variety of networks that may run throughout the cell, or be sheet-like. In some cases, the filaments terminate at a desmosome at the cell boundary, which is attached to a desmosome on an adjoining cell (see Fig. A.2); in these and other ways, extended networks of filaments can bind the cells into collective structures. For example, the immunostained keratin filaments in the cells displayed in Fig. 4.2 extend throughout the cytoplasm and are attached to both the cell nucleus and the plasma membrane. Of the proteins that can form intermediate filaments, several lamins are found in the nuclear laminae of eucaryotic cells, while more than 20 different keratins are present in human epithelial cells alone. The intermediate filament vimentin, discussed in Section 4.5, is observed in a wide range of cells, including fibroblasts and many others.

With a persistence length of $10-20$ μm, actin filaments appear moderately flexible on cellular length scales, as can be seen in the micrograph of a reconstituted actin network in Fig. 4.3 (see Section 4.5 for procedure). These thin filaments may join together as part of a network, although actin filaments, like microtubules, require linking proteins in order to form a network or composite structure. There are a variety of such actin-binding proteins, α-actinin, filamin, and fimbrin being common

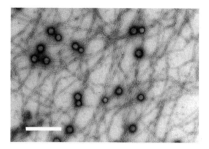

Fig. 4.3. Electron micrograph of a network of actin extracted from rabbit muscle cells and purified before being reconstituted; bar is 500 nm. The dark disks are 100 nm diameter latex spheres introduced in the experimental study of the network (reprinted with permission from Schmidt *et al.*, 1989; © 1989 by the American Chemical Society; courtesy of Dr Erich Sackmann, Technische Universität München).

examples, and they combine with actin to form networks of varying appearance and mechanical properties. Cartoons of composite actin structures are displayed in Fig. 4.4. Fimbrin and α-actinin link actin filaments into parallel bundles, with an average spacing between strands of 14 and 40 nm, respectively. The relatively large spacing between filaments linked by α-actinin permits motor proteins such as myosin to slide between adjacent strands. Filamin is a hinged dimer having roughly the shape of a V, whose individual arms lie length-wise along distinct actin filaments, resulting in an irregular, gel-like network. Actin also can be linked via spectrin: the auditory outer hair cell is thought to be an actin/spectrin network (see Section 3.1), and the spectrin tetramers of the erythrocyte cytoskeleton are joined, not to each other, but to short stubs of actin, each stub having the same average coordination as the network itself.

While actin, intermediate filaments and microtubules are the principal components of the internal networks of the cell, other proteins are present in the two- and three-dimensional networks surrounding the cell, such as the plant cell wall or the extracellular matrix. These networks tend to be composite structures, with a heterogeneous molecular composition and organization, and only a few of their components have yet been isolated and reconstituted into chemically homogeneous networks whose properties can be studied in isolation.

The mechanical description of three-dimensional materials in general requires a larger number of elastic moduli than are needed for

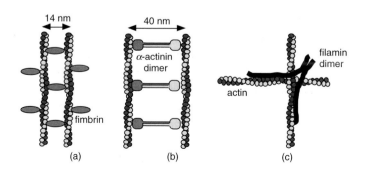

Fig. 4.4. Schematic representation of three structures formed by different actin-binding proteins (after Alberts *et al.*, 1994; reproduced with permission; © 1994 by Routledge, Inc., part of The Taylor and Francis Group). (a) Fimbrin has two actin-binding sites and links actin into parallel strands with a separation of 14 nm, while (b) α-actinin, a dimer with one binding site per monomer, results in a looser actin bundle with a separation of 40 nm. (c) In contrast, filamin produces a gel-like network.

Fig. 4.5. (a) Above the connectivity percolation threshold, a continuous path exists across the network along its elements, as indicated by the thick dark line. (b) Below the percolation threshold, a network is not completely connected and its shear resistance vanishes.

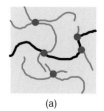
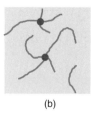

(a)   (b)

Fig. 4.5. (a) Above the connectivity percolation threshold, a continuous path exists across the network along its elements, as indicated by the thick dark line. (b) Below the percolation threshold, a network is not completely connected and its shear resistance vanishes.

two-dimensional materials, although materials with high symmetry are simple to describe in any dimension. Both two- and three-dimensional isotropic systems involve only two independent moduli, and a simple cubic material needs three independent moduli in three dimensions. However, the macroscopic elastic parameters of a material may have a slightly more complicated dependence on the characteristics of its microscopic elements in three dimensions than in two. A network of identical springs with spring constant k_{sp} and equilibrium length s_o, for example, has moduli proportional to just k_{sp} in two dimensions but k_{sp}/s_o in three dimensions, with a numerical prefactor that depends upon the network connectivity.

Our treatment of biological networks must recognize that there may or may not be permanent cross-links between its filamentous elements. If the density of filaments and cross-links is sufficiently high, the network possesses both compression and shear resistance, as in Fig. 4.5(a). For instance, in networks composed of floppy chains, the elasticity of the network near its equilibrium configuration is entropic and the moduli are proportional to $\rho k_B T$, where ρ is the density of chains. Of course, if the density of chains is so low that there are few junctions per chain, the network may not be completely connected, and the shear modulus vanishes as in Fig. 4.5(b). There is a well-defined threshold for the average connectivity, referred to as the connectivity percolation threshold, above which network elements are sufficiently linked to their neighbors that a continuous path can be traced along the elements from one side of the network to the other, as in Fig. 4.5(a). At temperatures above zero, the elastic moduli also are believed to be non-vanishing immediately above the connectivity percolation threshold, although this is not true at zero temperature for certain classes of forces.

Suppose that there are no cross-links at all. Clearly, such collections of rods and chains must behave like fluids if they are observed over long time scales: even if the filaments are wrapped around each other forming knots and loops, they can wiggle past each other owing to thermal motion such that the system as a whole ultimately deforms in response to an applied shear. However, on short time scales, the system may respond somewhat elastically, a behavior referred to as viscoelasticity. Thus, the elastic moduli for uncross-linked chains may be time-dependent, with the

dilute   semi-dilute   concentrated

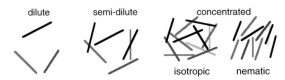

isotropic   nematic

Fig. 4.6. Polymer solutions can be classified according to concentration as dilute, semi-dilute and concentrated. In addition, concentrated solutions of stiff filaments may display a disordered isotropic phase or, at higher density, an ordered nematic phase, in which the filaments possess long-range orientational order.

effective shear modulus decreasing as time increases. The viscoelastic properties of a polymer solution depend upon its density, as illustrated in Fig. 4.6 for stiff filaments. In the dilute regime, the filaments are sufficiently separated in space that they can rotate largely unencumbered and the solution's viscosity is close to that of its solvent. At the opposite extreme, in the concentrated regime the filaments are so dense that they are in frequent contact with each other. In between lies the semi-dilute regime, characteristic of many biopolymer solutions.

The behavior of a polymer solution at moderate to high concentrations depends in part on the stiffness of the polymer. Floppy chains, with an average contour length L_c much longer than the polymer persistence length ξ_p, form entangled configurations with random orientations of the polymers. Individual stiff rods, with $L_c \ll \xi_p$, each have a well-defined orientation, which may or may not be alligned with those of its neighbors. Thus, a solution of polymeric rods may be disordered and isotropic at modest concentrations, but ordered at higher density into a liquid crystalline state called the nematic phase, in which individual rods or molecules display long-range orientational order (see Vertogen and de Jeu, 1988).

One way of characterizing the time-dependence of the effective elastic moduli of viscoelastic materials is to apply a strain to the material at a fixed driving frequency, as shown in Fig. 4.7. The relationship between the applied strain u_{xy} and the resulting stress σ_{xy} developed within the system involves two functions of the angular frequency ω of the driving strain, namely $\sigma_{xy} = G'(\omega)u_{xy}(t) + G''(\omega)\cdot(du_{xy}/dt)/\omega$, where du_{xy}/dt is the rate of strain. The functions $G'(\omega)$ and $G''(\omega)$ are called the shear storage and shear loss moduli, respectively, and are measures of the energy stored (and recovered) or lost during a cycle of the system. At very long times, or, equivalently when ω vanishes, G' becomes the shear modulus and G''/ω becomes the viscosity. Both of these moduli have been measured for a variety of cytoskeletal filaments.

In Section 4.2, we review the static elastic properties of several three-dimensional systems, and relate the elastic constants of a regular spring

Fig. 4.7. Time evolution of a system being driven by an oscillating shear strain, permitting the frequency-dependence of the strain response of a viscoelastic material to be determined.

network to the microscopic characteristics of the springs themselves. In Section 4.3, we develop an approximate expression for the free energy density of a cross-linked network of floppy chains under deformation, and we also describe the behavior of networks near the connectivity percolation threshold. The formalism for representing the mechanical properties of viscoelastic materials is presented in Section 4.4. Lastly, our analytical results are summarized and applied to cytoskeletal networks in Section 4.5.

4.2 Elastic moduli in three dimensions

The three-dimensional materials of interest in our treatment of the cell are not so much conventional bulk solids (bone is an obvious exception to this statement) as semiflexible networks having more complex elastic characteristics than networks of floppy chains. In this section, we review the formalism for describing the generic elastic properties of three-dimensional materials, and present the free energy of deformation for both isotropic materials and those with four-fold symmetry. We then discuss the compression modulus of one particular system, namely a spring network with cubic symmetry, as a demonstration of how the microscopic characteristics of a three-dimensional network are reflected in its elastic moduli.

The general formalism for describing the deformation of a system under stress, as laid out in Section 3.2 and Appendix D, involves elastic moduli C_{ijkl} whose number rises rapidly with dimension d as d^4 (16 for $d = 2$ and 81 for $d = 3$) if the symmetries of the tensor are ignored. Yet even the simplest symmetry operations arising from the definition of the elastic moduli and of the strain tensor, namely $C_{ijkl} = C_{jikl} = C_{ijlk}$ and $C_{ijkl} = C_{klij}$, sharply reduce the number of independent moduli to 21 in three dimensions and 6 in two dimensions. Other symmetry considerations may further decrease the number of independent moduli.

We begin our discussion of elasticity in three dimensions by considering isotropic materials. As explained in Appendix D, only two quadratic combinations of the strain tensor u_{ij} are invariant under arbitrary rotations, namely $\Sigma_{i,j} u_{ij}^2$ and tru, where tr is the trace operator. Hence, only two elastic moduli are needed to represent an isotropic Hooke's law material in three dimensions, the same number needed for two-dimensional isotropic materials, as we saw in Chapter 3. Using combinations of the strain tensor corresponding to physically simple deformations, the free energy density of deformation $\Delta\mathcal{F}$ can be written as

$$\Delta\mathcal{F} = \mu \sum_{i,j} (u_{ij} - \delta_{ij} \, \text{tr}u \, /3)^2 + 1/2 \, K_V \, (\text{tr}u)^2, \tag{4.1}$$

where K_V is the volume compression modulus and μ is the (three-dimensional) shear modulus. As the strain tensor is unitless in any system, K_V and μ must carry the same units as the free energy density, or energy per unit volume in three dimensions.

Most deformation modes involve a combination of K_V and μ. For example, a stress may be applied along one direction only, say σ_{zz}, and the resulting strain is measured in the same direction, u_{zz}. Such deformations measure Young's modulus Y, defined by

$$Y = \sigma_{zz}/u_{zz}. \tag{4.2}$$

As shown in Appendix D, Young's modulus for isotropic materials can be expressed in terms of the elastic moduli K_V and μ by

$$Y = 9K_V\mu/(3K_V + \mu). \tag{4.3}$$

The corresponding relation for two-dimensional materials is obtained in the problem set in Chapter 3. Another quantity measured by uniaxial stress is the Poisson ratio σ_p, which is the negative ratio of the strain in a transverse direction, e.g. u_{xx}, to the strain in the longitudinal direction u_{zz}; that is, $\sigma_p = -u_{xx}/u_{zz}$. For isotropic materials in three dimensions, the Poisson ratio can be rewritten as

$$\sigma_p = (3K_V - 2\mu)/(6K_V + 2\mu). \tag{4.4}$$

Eqs. (4.3) and (4.4) permit the combination Y and σ_p to be used as an alternative to the pair K_V and μ for describing the elastic characteristics of isotropic materials. Many three-dimensional materials display values for σ_p in the range 1/4 to 1/3.

In two dimensions, the deformation free energy has the same functional form for both isotropic materials and those with six-fold symmetry. However, this is not the case in three dimensions, wherein hexagonal systems have five independent elastic moduli, compared with the two moduli of isotropic systems. Rhombohedral and other symmetries require even more moduli. However, cubic systems, as illustrated in Fig. 4.8, require only three moduli, just as four-fold systems do in two dimensions. In fact, the expression for the free-energy density is a straightforward generalization of the two-dimensional result, becoming (Landau and Lifshitz, 1986, p. 35)

$$\Delta \mathcal{F} = (C_{xxxx}/2) \cdot (u_{xx}^2 + u_{yy}^2 + u_{zz}^2) + C_{xxyy}(u_{xx}u_{yy} + u_{xx}u_{zz} + u_{yy}u_{zz})$$
$$+ 2C_{xyxy}(u_{xy}^2 + u_{xz}^2 + u_{yz}^2). \tag{4.5}$$

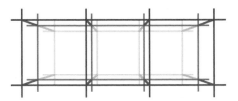

Fig. **4.8.** Example of a network with cubic symmetry in three dimensions.

Although this equation clearly expresses its historical roots in Eq. (3.6), it does not directly reflect simple deformations. By taking combinations of elastic moduli and strain tensor components, Eq. (4.5) can be rewritten

$$\Delta \mathcal{F} = (K_V/2) \cdot (u_{xx} + u_{yy} + u_{zz})^2 + (\mu'/3)[(u_{xx} - u_{yy})^2 + (u_{xx} - u_{zz})^2 + (u_{yy} - u_{zz})^2] + 2\mu_s(u_{xy}^2 + u_{xz}^2 + u_{yz}^2), \tag{4.6}$$

where

$$K_V = (C_{xxxx} + 2C_{xxyy})/3 \quad \mu' = (C_{xxxx} - C_{xxyy})/2 \quad \mu_s = C_{xyxy}. \tag{4.7}$$

With this definition of the elastic moduli, Eq. (4.6) reduces to the isotropic expression (4.1) when $\mu' = \mu_s = \mu$.

To see how these moduli are related to the microscopic structure of a material, we turn to our favorite example, a network of identical springs, each having a potential energy $V(s) = (k_{sp}/2) \cdot (s - s_o)^2$, where k_{sp} is the spring constant and s_o is the unstretched spring length. Suppose the springs are linked at six-fold junctions to form a network with cubic symmetry, as in Fig. 4.8. This is a somewhat dangerous example because, as we saw for two-dimensional spring networks with four-fold connectivity in Section 3.4, the ground state of this network is degenerate unless interactions between non-adjacent network nodes are introduced to provide rigidity. However, the simple cubic network is easy to evaluate and yields a qualitatively correct relationship between k_{sp}, s_o, and the elastic moduli. To find the relation between an elastic modulus of the network and its microscopic description, we compare the change in the free energy according to Eq. (4.6) with an equivalent expression involving the spring variables k_{sp} and s_o. Specifically, let's consider the deformation mode illustrated in Fig. 4.9, wherein all springs change in length from s_o to $s_o + \delta$, which is a pure compression. For this mode, the off-diagonal elements of the strain tensor vanish, leaving only the diagonal elements $u_{xx} = u_{yy} = u_{zz} = \delta/s_o$. The change in the free energy density is, according to Eq. (4.6),

Fig. 4.9. Sample deformation mode of a spring network with cubic symmetry. The springs change in length from s_o to $s_o + \delta$. The original shape of a unit cell of the network is shaded, while the deformed shape is indicated by the wire outline.

$$\Delta \mathcal{F} = (9K_V/2) \cdot (\delta/s_o)^2. \tag{4.8}$$

Now, a cubical unit cell of the network contains a single vertex and three springs, such that the change in the potential energy of this cell upon deformation is $3k_{sp}\delta^2/2$. Dividing the change in potential by the volume of the unit cell s_o^3, yields

$$\Delta \mathcal{F} = (3k_{sp}/2) \cdot (\delta^2/s_o^3). \tag{4.9}$$

The compression modulus can be found by comparing Eqs. (4.8) and (4.9)

$$K_V = k_{sp}/3s_o \qquad \text{(rigid cubic symmetry)}, \tag{4.10}$$

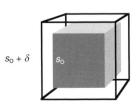

$s_o + \delta$ s_o

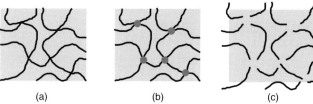

Fig. 4.10. Construction of a random network of random chains. Six long chains in (a) are welded at five points to produce the shear-resistant network in (b) with $n=16$ segments, shown as isolated units in (c).

(a)   (b)   (c)

showing that the elastic moduli depend upon both the spring constant and the equilibrium spring length in three dimensions. Two other regular spring networks are treated in the end-of-chapter problems, also yielding $K_V \propto k_{sp}/s_o$, with a proportionality constant of about 1/2.

4.3 Entropic networks

We showed in Section 3.5 for a specific two-dimensional network of floppy chains that its elastic moduli are roughly equal to $\rho k_B T$, where T is the temperature and ρ is the chain density. A similar relation holds in three dimensions, with ρ becoming the three-dimensional density of chains. We now derive this result more generally for a network of chains joined at random locations.

Construction of random chain networks

The model investigated here is one developed some time ago for interpreting the properties of vulcanized rubber (Flory, 1953; Treloar, 1975). To construct the model network, flexible chains are packed together with sufficient density that a given chain is close to its neighbors at many positions along its contour length. The network is then given rigidity by welding different chains to one another at points where they are in close proximity, as illustrated in Fig. 4.10. After welding has occured, there are n chain segments, each a random chain in its own right (having been cut from a larger random chain) with an end-to-end displacement $\mathbf{r}_{ee}$ obeying the conventional Gaussian probability distribution, Eq. (2.52). Despite our freezing the positions of the chains during network construction, the network nodes are not fixed in space at non-zero temperature and the instantaneous end-to-end displacement of a segment may change, even though the contour length is fixed. When the network deforms in response to an external stress, the average value of $\mathbf{r}_{ee}$ changes as well.

Deformation of the network

To describe quantitatively the deformation of the network as a whole, we introduce the scale factors Λ_x, Λ_y, Λ_z, such that a system in the shape

Fig. 4.11. Under deformation, the rectangle of sides L_x and L_y in (a) changes to a different rectangle of sides $\Lambda_x L_x$ and $\Lambda_y L_y$ in (b), where the scale factors satisfy $\Lambda_x > 1$ and $\Lambda_y < 1$ for the deformation shown.

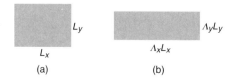

L_y $\quad\quad$ $\Lambda_y L_y$

L_x $\quad\quad\quad\quad$ $\Lambda_x L_x$

(a) $\quad\quad\quad\quad\quad$ (b)

of a rectangular prism of sides L_x, L_y, L_z before the deformation becomes $\Lambda_x L_x$, $\Lambda_y L_y$, $\Lambda_z L_z$ after the deformation, as illustrated in Fig. 4.11. Thus, an extension (compression) of the network in a given Cartesian direction corresponds to $\Lambda > 1$ ($\Lambda < 1$). We follow Chapter XI of Flory (1953) to determine how the entropy of the network depends upon the scale factors.

As is apparent in Fig. 4.10(c), the chains possess a distribution of end-to-end displacement vectors that changes with deformation as the chains are stretched, compressed and re-oriented. The probability that a given chain i has a particular displacement vector $\mathbf{r}_i = (x_i, y_i, z_i)$ *after* deformation can be found from the probability that it had a displacement vector $(x_i/\Lambda_x, y_i/\Lambda_y, z_i/\Lambda_z)$ in the unstressed state *before* the deformation. To be a little more precise, the probability that the displacement of a chain lies in the range $\mathbf{r}_i$ to $\mathbf{r}_i + \Delta\mathbf{r}$ *after* deformation is given by

$$\mathcal{P}(x_i/\Lambda_x, y_i/\Lambda_y, z_i/\Lambda_z)\,(\Delta x/\Lambda_x)\cdot(\Delta y/\Lambda_y)\cdot(\Delta z/\Lambda_z), \tag{4.11}$$

where the three-dimensional probability density $\mathcal{P}$ is given by Eq. (2.55). The actual number of chains n_i lying in this range of $\mathbf{r}_i$ is equal to the product of this probability and the total number of chains n in the system, resulting in

$$n_i = n\,(2\pi\sigma^2)^{-3/2}\,\exp\{-[(x_i/\Lambda_x)^2+(y_i/\Lambda_y)^2+(z_i/\Lambda_z)^2]/2\sigma^2\}$$
$$\Delta x \Delta y \Delta z / \Lambda_x \Lambda_y \Lambda_z, \tag{4.12}$$

after substitution of Eq. (2.55), where $\sigma^2 = Nb^2/3$ for a three dimensional chain of N equal segments of length b. We will use n_i in determining the change in the free energy during deformation.

Before evaluating the entropy of the network, let us review the method of constructing one of its configurations. To generate a specific configuration such as Fig. 4.10(b), two conditions must be satisfied:

(a) the distribution of n individual segments must have the correct number of chains n_i for each range of $\mathbf{r}_i$, and
(b) each weld site must be within the appropriate distance of a site on a neighboring chain.

The probability of the chains satisfying conditions (a) and (b) are defined as $\mathcal{P}_a$ and $\mathcal{P}_b$, respectively; the combined probability for constructing the configuration is $\mathcal{P}_a \mathcal{P}_b$.

To evaluate $\mathcal{P}_a$, we need to know two quantities: a number and a probability. First, the number of chains available in the original configuration to change to the new configuration is n_i of Eq. (4.12). Second, the probability ρ_i that any one of these n_i chains has the correct $\mathbf{r}_i$ is

$$\rho_i = (x_i, y_i, z_i)\, \Delta x\, \Delta y\, \Delta z, \tag{4.13}$$

so that the probability of all n_i chains in this range have the correct $\mathbf{r}_i$ is $\rho_i^{n(i)}$, ignoring permutations. For all n chains in the network, the total probability of the correct position involves the product of the individual terms, or $\Pi_i\, \rho_i^{n(i)}$. However, the individual chains can be permuted a total of $n!/\Pi_i n_i!$ ways, so that the probability of satisfying criterion (a) is

$$\mathcal{P}_a = (\Pi_i\rho_i^{n(i)})\cdot(n!/\Pi_i n_i!) = n!\; \Pi_i(\rho_i^{n(i)}/n_i!). \tag{4.14}$$

Let's review the welding process before determining $\mathcal{P}_b$. How many welds are there for n segments? Each weld links four chain segments; however, the number of welds is not $n/4$, but rather $n/2$, since each segment is defined by two welds (we neglect dangling chains; see Chapter XI of Flory, 1953, for a discussion of this approximation). The number of welding sites on the chains must be twice the number of welds, or n, because each weld involves two separate weld sites, usually on different chains. Thus, there are n segments, n welding sites and $n/2$ welds. Now, what is the probability of finding a welding site in the correct position with respect to neighboring chains? Starting with site number 1, the probability that one of the $n-1$ remaining welding sites lies within a volume δV of site number 1 is just $(n-1)\delta V/V$, where V is the volume of the entire network. After the first weld, two sites have been removed from the list of potential weld locations. For the second weld, the probability that one of the remaining $(n-3)$ welding sites lies within a volume δV of site number 3 is then $(n-3)\delta V/V$. We repeat this argument for all $n/2$ welds, to obtain

$$\mathcal{P}_b = (n-1)(n-3)(n-5)\ldots 1\, (\delta V/V)^{n/2}$$
$$\cong (n/2)!\, (2\,\delta V/V)^{n/2}, \tag{4.15}$$

where we have used the approximation $(n/2-1/2)(n/2-3/2)\ldots 1/2 \cong (n/2)!$. In this expression, V is the deformed volume $\Lambda_x\Lambda_y\Lambda_z V_o$, where V_o is the original volume of the network.

Entropy change under deformation

The probability for the network to have a particular shape is $\mathcal{P}_a\mathcal{P}_b$, permitting its entropy to be written as $S = k_B\ln\mathcal{P}_a + k_b\ln\mathcal{P}_b$ (see Appendix C). Our next task is explicitly to evaluate $\ln\mathcal{P}_a$ and $\ln\mathcal{P}_b$. The logarithm of Eq. (4.14) can be simplified by using Stirling's approximation for factorials, $\ln n! = n \ln n - n$, such that

$$\ln \mathcal{P}_a = n \ln n + \sum_i n_i \ln(\rho_i / n_i)$$
$$= \sum_i n_i \ln(n \rho_i / n_i). \tag{4.16}$$

The product $n \rho_i$ can be obtained from Eq. (4.13), and the n_i themselves are given by Eq. (4.12). The calculation is performed in the problem set, and leads to

$$\ln \mathcal{P}_a = -(n/2)[\Lambda_x^2 + \Lambda_y^2 + \Lambda_z^2 - 3 - 2\ln(\Lambda_x \Lambda_y \Lambda_z)]. \tag{4.17}$$

Mercifully, it takes considerably less effort to obtain $\ln \mathcal{P}_b$. From Eq. (4.15), we find

$$\ln \mathcal{P}_b = \ln (n/2)! + (n/2)\ln(2 \, \delta V / V_o) - (n/2)\ln(\Lambda_x \Lambda_y \Lambda_z) \tag{4.18}$$

after substituting $V = \Lambda_x \Lambda_y \Lambda_z V_o$. Combining $\ln \mathcal{P}_a$ and $\ln \mathcal{P}_b$, we obtain the entropy S of the deformation

$$S = -(k_B n/2)[\Lambda_x^2 + \Lambda_y^2 + \Lambda_z^2 - 3 - \ln(\Lambda_x \Lambda_y \Lambda_z) - \ln (n/2)!$$
$$- (n/2)\ln(2 \, \delta V / V_o)]. \tag{4.19}$$

The last two terms in this expression are independent of the deformation and will not appear in the entropy difference ΔS with respect to a particular reference state. Taking the reference state to be $\Lambda_x = \Lambda_y = \Lambda_z = 1$, Eq. (4.19) leads to

$$S = -(k_B n/2)[\Lambda_x^2 + \Lambda_y^2 + \Lambda_z^2 - 3 - \ln(\Lambda_x \Lambda_y \Lambda_z)]. \tag{4.20}$$

Because there is no internal energy scale associated with the chains, the free energy ΔF is just $-T\Delta S$, and

$$\Delta F = (k_B T n/2)[\Lambda_x^2 + \Lambda_y^2 + \Lambda_z^2 - 3 - \ln(\Lambda_x \Lambda_y \Lambda_z)]. \tag{4.21}$$

The appeal of having an analytical expression for ΔF should not cause us to forget the physical and numerical approximations in its derivation. In particular, opinion is divided regarding the presence of the logarithmic term (see Flory, 1976; Deam and Edwards, 1976).

Elastic moduli

Elastic moduli can be obtained from Eq. (4.21) by imposing specific deformations. Imposing volume conservation on the system requires $\Lambda_x \Lambda_y \Lambda_z = 1$, which removes the logarithmic term and leaves

$$\Delta F = (k_B T n/2) \cdot (\Lambda_x + \Lambda_y + \Lambda_z - 3). \tag{4.22}$$

We can extract the shear modulus from ΔF by performing a pure shear, with $\Lambda_x = \Lambda = 1/\Lambda_y$ and $\Lambda_z = 1$, yielding

$$\Delta F = (k_B T n/2) \cdot (\Lambda^2 + 1/\Lambda^2 - 2) \qquad \text{(pure shear)}. \tag{4.23}$$

The expression in the brackets can be rewritten as $(\Lambda - 1/\Lambda)^2$, which becomes $4\delta^2$ when $\Lambda = 1 + \delta$ and δ is small. Dividing Eq. (4.23) by the network volume, which has not changed from its undeformed value V_o by shear, the free energy density $\Delta\mathcal{F}$ for the pure shear mode is

$$\Delta\mathcal{F} = 2\delta^2 \rho k_B T, \tag{4.24}$$

where $\rho = n/V$ is the density of chains. To find the shear modulus, we evaluate $\Delta\mathcal{F}$ for isotropic materials in Eq. (4.1) under pure shear conditions corresponding to $\Lambda = 1 + \delta$, namely $u_{xx} = \delta$, $u_{yy} = -\delta$, $u_{zz} = 0$, to obtain

$$\Delta\mathcal{F} = 2\delta^2 \mu. \tag{4.25}$$

A comparison of Eqs. (4.24) and (4.25) gives the long-awaited result

$$\mu = \rho k_B T. \tag{4.26}$$

Although our expression for the free energy of random networks qualitatively describes their behavior under compression and tension, we should not lose sight of the fact that several approximations were used in its derivation so that the expression may not be completely accurate. For example, our approach omits self-avoidance within a chain, repulsive interactions between chains, and the effects of chain entanglement (see references in Everaers and Kremer, 1995). Further, the free energy of Eq. (4.22) is not a minimum at the reference state $\Lambda_x = \Lambda_y = \Lambda_z = 1$. Although computer simulations of random networks (for example, see Plischke and Joos, 1998) confirm the density-dependence of the shear modulus predicted by Eq. (4.26), they have not yet verified the numerical prefactor. Nevertheless, the approach to random networks presented above does capture their most important features at modest chain densities.

Percolation

In constructing our network, we assume that each initial chain in Fig. 4.10(a) is sufficiently long that it is welded at many points to neighboring chains. What happens when the density of welds decreases to the point where there is less than a single weld per chain, on average? Clearly, if the density is so low that many chains are physically separated and sustain no welds, then we may not even have a connected network. This phenomenon has been studied in a variety of contexts, including the gelation of polymers (see Flory, 1953, Chapter IX) and goes under the generic name of percolation (a good introductory text is Stauffer and Aharony, 1992).

Consider the two-dimensional square lattice in Fig. 4.12, in which bonds have been placed between lattice sites in a random fashion. The

Fig. 4.12. Percolation phenomena on a square lattice in two dimensions. For the purpose of this example only, we ignore bonds along the edge of the lattice, so that the maximum allowable number of bonds is 24. Configuration (a), with bond occupation probability $p = 8/24 = 1/3$, is below the connectivity percolation threshold p_C, while configuration (b), with $p = 16/24 = 2/3$, is above p_C, resulting in at least one path completely traversing the network.

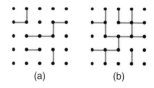

(a)   (b)

Table 4.1. *Percolation threshold for a selection of regular lattices in two and three dimensions*

Lattice	z	p_C (bond)	p_C (site)	p^* (bond)
Two dimensions				
honeycomb	3	0.653	0.696	1
square	4	0.500	0.593	1
triangular	6	0.347	0.500	2/3
Three dimensions				
simple cubic	6	0.249	0.312	1
body-centered cubic	8	0.180	0.246	3/4
face-centered cubic	12	0.119	0.198	1/2

Notes:
"Bond" and "site" refer to different measures of defects in the lattice, as described in the text; z is the maximum coordination of a site and p^* is the rigidity percolation threshold (for bonds) as calculated from Eq. (4.29). Data from Chapter 2 of Stauffer and Aharony (1992).

population of bonds on the lattice as a whole can be described by a parameter p, which is the probability that two nearest-neighbor sites are connected by a bond. If the lattice has a full complement of bonds, $p = 1$ and each site on the lattice has four-fold connectivity. If no bonds are present, obviously p vanishes. The sample configurations in the figure have values of p intermediate between 0 and 1.

One attribute of a configuration is whether a continuously connected path traverses the lattice. Such a path is absent in configuration (a) of Fig. 4.12 (with $p = 1/3$), but present in configuration (b) (with $p = 2/3$). For infinite systems, the existence of a connecting path across the lattice is a discontinuous function of p and there is a well-defined value of p, called the connectivity percolation threshold p_C, below which there is no connecting path. The connectivity threshold is lattice-dependent, as can be seen from the sample of values in Table 4.1. The two columns in the table labelled "bond" and "site" refer to which element of the lattice is used to parametrize its defects. We have so far discussed bond depletion, where p is the fraction of occupied bonds. However, one can also use site occupation as a parameter, bonds then being added between all occupied nearest neighbor sites. The occupation probabilities at percolation for these two measures are similar, but different, as they should be. Note that the percolation threshold decreases with increasing z, the maximum number of bonds that can be connected to each site.

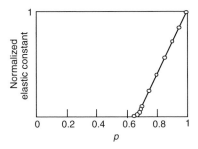

Fig. 4.13. Behavior of an elastic constant for a bond-diluted triangular network. For this network, $p_C = 0.325$, well below the apparent value of p_R (data reprinted with permission from Feng *et al.*, 1985, © 1985 by the American Physical Society).

The elastic moduli also depend upon p: for $p < p_C$, the network is so dilute as to be disconnected, so that both μ and K_A vanish. However, the presence of a connecting path at $p > p_C$ may not guarantee that the network can resist deformation. Even though p is safely above p_C in configuration (b), there is only one connecting path from left to right – which is not a lot of paths to provide shear resistance. In fact, even at $p = 1$, square networks do not resist simple shear, as proven in Chapter 3. It is found (Feng and Sen, 1984) that the elastic moduli vanish below a well-defined threshold, referred to as the rigidity percolation threshold p_R. At zero temperature, p_R may be considerably larger than p_C if the forces between network nodes are *central* forces, i.e. they are directed only along the bonds and do not depend upon bond angles (Feng *et al.*, 1985). Once the percolation threshold has been crossed from below, the elastic moduli rise with a power-law dependence on $p - p_R$ towards their values for a fully connected network, as shown in Fig. 4.13. As $p \to 1$, an elastic modulus C is approximately described by

$$C(p)/C(p=1) \cong (p - p_R)/(1 - p_R) \quad (p \to 1), \tag{4.27}$$

which vanishes at p_R and is equal to unity at $p = 1$.

As we are reminded by Thorpe (1986), the study of rigidity goes back more than a century to Maxwell (1864), who developed a simple model for the rigidity percolation threshold based on the bond-induced constraints on the motion of lattice points. Let's consider a regular lattice with N sites embedded in a d-dimensional space. Each site can be linked to its z nearest neighbors with a probability p. If they are unconstrained, the N sites have Nd variables describing their possible motion. The presence of a bond between sites provides one constraint on the number of ways the sites can move; we will say that the constraint reduces the number of floppy modes of the system. For example, two particles in a plane may move in four ways if unconstrained, but only three ways if they are joined by a rigid bar (two translations in the plane, and one rotation). How many constraints are present at an occupation p? The number of bonds linked to a given vertex is zp on average; however, the total number of bonds is not Nzp, because each bond is shared by two sites. Thus, the number of bonds or constraints is $Nzp/2$, reducing the number of floppy modes to

$$[\text{floppy modes}] \cong Nd - Nzp/2. \tag{4.28}$$

This equation is an approximation because some bonds are redundant, i.e. their presence does not reduce the number of floppy modes. The structure becomes rigid when the number of floppy modes vanishes, which occurs at a probability p^* of

$$p^* = 2d/z, \tag{4.29}$$

according to Eq. (4.28). As indicated by the right-hand column in Table 4.1, p^* is always larger than the observed p_C, consistent with the idea that rigidity requires more bonds than does connectivity. Further, Eq. (4.29) is in good agreement with p_R obtained by simulations at zero temperature (Feng and Sen, 1984; Feng et al., 1985). However, at finite temperature, nodes can depart from their zero-temperature lattice sites and interact entropically, generating a resistance to deformation. Computer studies of defective networks at finite temperature argue that entropic repulsion is enough to restore elasticity for $p > p_C$, such that $p_R = p_C$ even for central-force networks at non-zero temperature (Plischke and Joos, 1998).

4.4 Semiflexible polymer solutions

The networks that we have investigated thus far have permanent connection points linking together different structural elements such as polymer chains. If the number of connection points per chain is high enough, at least some of the elastic moduli are non-vanishing, depending on the nature of the connectivity. Consider now a situation like Fig. 4.14, in which there is a reasonable density of chains, but no permanent cross-links. Subjected to a shear, this network behaves like a fluid if allowed to relax completely, in the sense that the filaments ultimately slide around each other, and the network does not recover its original shape when the stress is released. Raising the filament density causes the network to become ever more entangled, such that its fluid behavior emerges only at ever longer time scales. One can imagine that, at sufficiently high density, the fluid response of the network takes so long that it behaves like an elastic solid on time-scales of interest.

Fig. 4.14. Application of a shear stress (from left to right) to a network of short filaments. If the filaments are not cross-linked the network responds like a fluid, when observed on long enough time-scales.

Solutions of polymer chains are examples of viscoelastic materials, behaving like elastic solids on short time-scales and viscous fluids on long time-scales. Their mechanical properties depend upon such characteristics as the polymer contour length and density, as we discuss momentarily. The importance of viscoelasticity to biofilaments in the cell is the subject of current research: chemically pure actin, intermediate filaments and microtubules can be studied in solution at concentrations comparable to those in the cell; the properties of such solutions are presented in Section 4.5. In this section, we describe universal features of viscoelastic materials, such as frequency-dependent elastic moduli, and then report without proof several predictions of viscoelastic behavior from model systems.

Classification of polymer solutions

The viscoelastic characteristics of a polymer solution depends upon its concentration c and the ratio of the polymer contour length L_c to its persistence length ξ_p. For simplicity, we present the generic features of two polymeric systems: one with very flexible chains ($L_c \gg \xi_p$) and a second with moderately stiff rods ($L_c \ll \xi_p$). Their solutions can be classified as dilute, semi-dilute and concentrated, although the boundaries between these categories are not necessarily sharp. The behavior of a solution is uniform within a given concentration regime, and may change smoothly or abruptly in crossing a transition region. Cartoons of polymer configurations expected in the three concentration regions are drawn in Fig. 4.15; many biological systems of interest fall into the semi-dilute category.

Individual polymers can rotate without obstruction in the dilute regime, which spans a concentration range determined by the effective volume per polymer. For a random coil, the effective volume is about $4\pi R_g^3/3$, where R_g is the radius of gyration, defined as $R_g^2 = N^{-1} \Sigma_i (\mathbf{r}_i - \mathbf{r}_{cm})^2$ for a discrete chain of N segments located at positions $\mathbf{r}_i$ (the center-of-mass position is at $\mathbf{r}_{cm} = N^{-1} \Sigma_i \mathbf{r}_i$). For instance, if the chain is uniformly distributed within a sphere of radius R, the radius of gyration

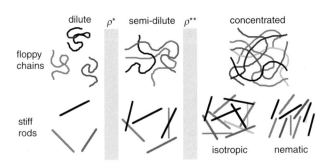

dilute ρ^* semi-dilute ρ^{**} concentrated

floppy chains

stiff rods

isotropic   nematic

Fig. 4.15. Generic behavior of a solution of polymer chains as a function of chain density, which increases from left to right. Densities at the transitions between regions are indicated by ρ^* and ρ^{**}.

is $(3/5)^{1/2}R$. In contrast, to rotate freely of its neighbors, a stiff rod of length L_c must have an effective volume much larger than that of a random chain with the same L_c: the centers of mass of two hard rods are excluded from a volume of the order L_c^3. In the dilute regime, then, the average volume occupied by a single filament (stiff or floppy) must be greater than the effective volume that we have just estimated. Equivalently, the concentration must be less than a transition number density ρ^* of about

$$\rho^* \sim (4\pi R_g^3/3)^{-1} \quad \text{(floppy)} \quad \text{or} \quad \rho^* \sim L_c^{-3} \quad \text{(stiff)}. \tag{4.30}$$

At the other extreme are concentrated solutions, wherein the individual polymeric units strongly overlap and intertwine. For flexible polymers, this regime may have the appearance of a polymer melt and occurs at densities above a transition density ρ^{**} of

$$\rho^{**} \sim v_{ex}/b^6 \quad \text{(floppy)}, \tag{4.31}$$

where v_{ex} is the approximate volume occupied by a chain segment, and b is the average segment length. The proof of this result is somewhat involved; we refer the interested reader to Section 5.1 of Doi and Edwards (1986). The persistence lengths of the filaments discussed in Section 4.5 are all sufficiently long ($\xi_p > 1$ µm) that our primary interest here is in the properties of semi-flexible rods, rather than floppy chains.

The transition concentration to the high-density region for rigid rods depends upon their diameter D_f. Suppose that one rod, A, lies along the y-axis while another rod, B, is free to move, so long as its axis is parallel to the x-axis, as illustrated in Fig. 4.16. These rods intersect each other if the center of rod B lies within the box shown in the figure. That is, the centers are excluded from a volume equal to $2D_fL_c^2$ if the rods are forbidden from intersecting. Now, angular averages of rod orientations must be performed to obtain an exact expression for the excluded volume, but the result must be of order $D_fL_c^2$, corresponding to a transition number density of

$$\rho^{**} \sim (D_fL_c^2)^{-1} \quad \text{(stiff)}. \tag{4.32}$$

For densities not too far above ρ^{**}, the rods are randomly oriented, and the system is isotropic. However, rigid rods may order into a so-called

Fig. 4.16. Configurations of two identical rods of length L_c and diameter D_f. Rod A is fixed along the y-axis, while rod B may move parallel to the x-axis. If the rods cannot overlap, the center of rod B is excluded from the region indicated by the box of volume $2D_fL_c^2$ (plus terms higher order in D_fL_c; after Doi and Edwards, 1986; © 1986 by Oxford University Press).

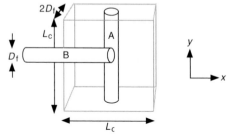

nematic phase at high concentrations, in which they display long-range orientational order even though their positions are not regularly spaced as they might be on a crystal lattice. The differences between isotropic and nematic phases are illustrated in Fig. 4.15. Whether the nematic phase is achievable depends upon the anisotropy of the interaction between rods. Considering hard rods of diameter D_f and length L_c, Onsager (1949) showed that the nematic phase exists only for L_c/D_f greater than approximately 3 (see also Vertogen and de Jeu, 1988, Chapter 13). Note that the highest concentration permitted for hard rods is roughly $(L_c D_f^2)^{-1}$.

Storage and loss moduli

How do we formally describe a material whose response to an imposed stress or strain is time-dependent? Let's consider a fixed-strain measurement in which a shear strain u_{xy} is imposed upon an object, creating a stress σ_{xy}. If the system were an elastic solid, the stress would be proportional to the strain, $\sigma_{xy} = \mu u_{xy}$, with the shear modulus μ fixed for the duration of the measurement. However, for a viscoelastic material like an uncross-linked polymer solution, the stress slowly decays to zero from its initial value as the polymers untangle and the network relaxes. Thus, the effective elastic modulus is time-dependent, and the stress-strain relationship can be written as

$$\sigma_{xy} = \int_{-\infty}^{t} G(t - t')\,(du_{xy}/dt')\,dt', \qquad (4.33)$$

where du_{xy}/dt is the shear rate and $G(t-t')$ is the relaxation modulus, having the same units as the shear modulus (see Ferry, 1980, for alternative formulations of this equation). Eq. (4.33) implies that the response of the system is additive in time: small changes in stress result from small changes in strain, which are integrated to yield the total stress.

A variety of experimental configurations can be employed to measure the relaxation modulus. Commonly, the frequency-dependence of the elastic response is measured, rather than its time-dependence. In experiments illustrated in Fig. 4.17, the system is subjected to an oscillating strain of the form

$$u_{xy}(t) = u_{xy}^{\circ} \sin\omega t, \qquad (4.34)$$

where u_{xy}° is the amplitude of the strain, ω is its frequency, and the initial value of the strain has been set to zero for convenience. The corresponding strain rate is simply

$$du_{xy}/dt = \omega\, u_{xy}^{\circ} \cos\omega t. \qquad (4.35)$$

Fig. 4.17. Application of an oscillating shear strain $u_{xy}(t)$ to a three-dimensional system. The undisturbed system is indicated by the shaded region, and the deformation is shown by the wire outline.

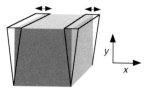

Eqs. (4.34) and (4.35) can be substituted into the definition of the relaxation modulus to obtain

$$\sigma_{xy}(t) = \int_0^\infty G(\tau)\omega\, u_{xy}^{\;\circ}\cos[\omega(t-\tau)]\, d\tau$$

$$= u_{xy}^{\;\circ}\, [\omega \int_0^\infty G(\tau)\sin\omega\tau\, d\tau]\sin\omega t$$

$$+ u_{xy}^{\;\circ}\, [\omega \int_0^\infty G(\tau)\cos\omega\tau\, d\tau]\cos\omega t, \qquad (4.36)$$

where the integration variable has been changed to $\tau = t - t'$ and we have used the identity $\cos(\alpha - \beta) = \sin\alpha\,\sin\beta + \cos\alpha\,\cos\beta$. The terms in the square brackets are functions of frequency, but their explicit time-dependence has been integrated away. We replace these terms with two new moduli by writing

$$\sigma_{xy} = u_{xy}^{\;\circ} G'(\omega)\sin\omega t + u_{xy}^{\;\circ} G''(\omega)\cos\omega t, \qquad (4.37)$$

where $G'(\omega)$ is the shear storage modulus and $G''(\omega)$ is the shear loss modulus.

The physical meaning of the moduli can be seen by considering their phase with respect to the applied strain, $u_{xy}(t) = u_{xy}^{\;\circ}\sin\omega t$. First, we recast Eq. (4.37) in terms of the applied strain and its rate of change, Eq. (4.35),

$$\sigma_{xy} = G'(\omega)u_{xy}(t) + G''(\omega)\cdot(du_{xy}/dt)/\omega. \qquad (4.38)$$

If only the storage modulus is non-vanishing (i.e., $G''=0$), the stress is directly proportional to the strain in Eq. (4.42) and G' appears like a frequency-dependent shear modulus μ. That is, G' is a measure of the elastic energy stored in and retrieved from the system. Conversely, if only the loss modulus is non-vanishing ($G'=0$), the stress is proportional to the rate of strain and has the same generic form as Newton's law for a viscous fluid, $\sigma_{xy} = \eta(du_{xy}/dt)$ where η is the viscosity. Thus, G''/ω appears as a frequency-dependent dynamic viscosity and G'' should vanish linearly with ω as $\omega \to 0$.

If the material is truly a fluid, then the storage modulus should decrease as the frequency decreases, ultimately vanishing at $\omega=0$. While the observed behavior of G' is monotonic with frequency, it is not without structure, as the schematic plot in Fig. 4.18 illustrates. Dilute solutions display the smallest storage moduli, as one might expect from their near-fluid behavior. At the other extreme, concentrated polymeric solutions have storage moduli typical of hard plastics, of the order 10^9 J/m^3. Semi-dilute solutions range between these extremes, displaying little resistance to shear at small ω (long times) but a high resistance at

Fig. 4.18. Schematic representation of the storage modulus $G'(\omega)$ as a function of frequency ω for three sample concentrations: dilute, semi-dilute and concentrated. Physical systems may range between the examples shown here.

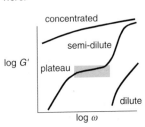

large ω (short times). At intermediate frequencies, they may display slowly varying behavior associated with polymer entanglements, referred to as the plateau region.

At each frequency range, different shear relaxation modes are allowed during the oscillation of the applied strain. At short times, only local molecular rearrangements may take place, while over long times, large-scale motion of the polymer as a whole is permitted. Between these time frames, the polymers move on intermediate length scales, which may be comparable to the distances between entanglement points for semi-dilute solutions. By entanglements, we mean loops and even knots that might occur as polymers thread their way around one another. Thus, over a range of frequencies, entanglements behave almost as fixed cross-links, which become unstuck only when viewed on longer time scales. This results in a plateau at intermediate frequency, where the solution behaves like a network of (temporarily) fixed cross-links, resulting in a frequency-independent shear resistance determined by the density of cross-links.

Model results for $G'(\omega)$

The formalism that we have developed above employs two frequency-dependent moduli to represent the viscoelasticity of a polymeric solution at fixed concentration. A number of simple mechanical models have been advanced for understanding the measured behavior of $G'(\omega)$ and $G''(\omega)$, as reviewed in Ferry (1980) (see Elson, 1988 for biological applications). For example, the spring-and-dashpot representation shown in Fig. 4.19 permits the ω-dependence of the moduli to be interpreted in terms of effective spring constants and the viscosity of the system as a whole. However, our interest lies in relating the experimental measurements of G' and G'' to the microscopic structure of the material, and this has been the subject of theoretical investigation for several years now.

The plateau region of G' at intermediate frequencies tends to cover time scales of importance in the cell. Now, we have shown for networks of floppy chains with permanent cross-links that the shear modulus μ is proportional to the density ρ (or concentration c) of chains: $\mu \sim \rho k_B T$. The plateau region is a frequency range in which entanglements behave like quasi-permanent cross-links. Thus, if a cellular network consists of floppy

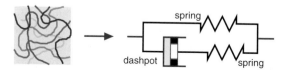

Fig. 4.19. Effective representation of a polymer solution in terms of a simple mechanical model involving springs and viscous dashpots (see Elson, 1988).

Fig. 4.20. Cartoon of a polymer confined to an effective tube created by its neighbors (indicated by disks). Referred to as the tube model (Doi and Edwards, 1986), this representation has been used to predict the frequency and concentration-dependence of G' (Isambert and Maggs, 1996; Morse, 1998).

chains with a persistence length much shorter than the distance between entanglements, we would expect G' to be proportional to c^1 or ρ^1 (see also Ferry, 1980, Chapter 13). However, for most cellular filaments except spectrin, the persistence length is comparable to, if not much larger than, the diameter of a typical cell, and the filaments appear as gently curving ropes or rods on cellular length scales. Is the concentration-dependence of G' in the plateau region affected by the rigidity of the filament?

Several studies have addressed this question. Some years ago, Kirkwood and Auer (1951) showed that in a particular representation of a rod, the storage modulus obeyed

$$G'(\omega) = (3\rho k_B T/5)\, \omega^2\tau^2/(1+\omega^2\tau^2), \tag{4.39}$$

where τ is a relaxation time depending on the rod geometry and solution viscosity (see also Ferry, 1980, Chapter 9). Although this expression does not contain an intermediate plateau, nevertheless G' rises with ω to reach an asymptotic value of $(3/5)\rho k_B T$ at high frequency; i.e., in the elastic regime, G' scales like ρ^1. More recently, Isambert and Maggs (1996) and Morse (1998) find that G' is proportional to $\rho^{7/5}$ for moderately flexible filaments whose appearance in solution has the snake-like form of Fig. 4.20 (see also Gittes and MacKintosh, 1998). MacKintosh et al. (1995) find that G' in the plateau region grows faster than the square of the density, namely $\rho^{11/5}$ for polymer solutions and $\rho^{5/2}$ for densely cross-linked gels, by considering the effects of tension on rods that are close to straight.

These three calculations emphasize the entropic contribution to network elasticity. In contrast, Satcher and Dewey (1996) examine the elasticity of networks where flexible network elements are joined at right angles to form T-junctions. Deformations of the network bend the elements perpendicular to their length, and the elasticity of the network reflects the energy required to bend the elements. In this approach, the shear modulus is predicted to grow as ρ^2. The concentration-dependence of G' predicted by the above models can be tested experimentally, as we describe in the following section.

4.5 Rheology of cytoskeletal components

In this section, we investigate the rheology of polymer networks and solutions in three dimensions (rheology is the study of the flow and deformation of matter). The systems on which we report are the principal filaments of the cytoskeleton – actin, intermediate filaments, and microtubules. In the case of actin, both pure and cross-linked networks have been investigated. We also provide references to measurements made on the cytoplasm of several cell types, although the findings do not have a complete interpretation owing to the chemical and structural heterogeneity of the cytoplasm. We first review some results from Sections 4.2–4.4 on viscoelasticity, and provide a brief synopsis of several techniques used in rheological studies, before turning our attention to the experimental findings themselves.

Storage and loss moduli

Isotropic materials in three dimensions are characterized by the compression modulus K_V and shear modulus μ. When constructed with permanent connections, networks of floppy polymers obey $\mu \sim \rho k_B T$, where ρ is the chain number density (i.e., the number of chains per unit volume), a single chain being defined as a chain segment running between two junctions. In the absence of permanent cross-links, polymer solutions display a time-dependent elasticity characteristic of viscoelastic materials. When subject to an oscillating strain u_{xy} of angular frequency ω, the stress σ_{xy} developed in the solution obeys $\sigma_{xy}(t) = G'(\omega)u_{xy}(t) + G''(\omega)\cdot(du_{xy}/dt)/\omega$, where $G'(\omega)$ and $G''(\omega)$ are the shear storage and shear loss moduli, respectively. As $\omega \to 0$, G' becomes the shear modulus and G''/ω becomes the viscosity.

The behavior of G' and G'' at intermediate ω depends upon the number density ρ. Solutions of moderately inflexible filaments of contour length L_c and diameter D_f can be categorized according to concentration as dilute, semi-dilute or concentrated, the regions being separated by their transition densities $\rho^* \sim L_c^{-3}$ and $\rho^{**} \sim (D_f L_c^2)^{-1}$. In the dilute regime, G' is close to that of the solvent, while in the concentrated regime, it is similar to the shear modulus of a hard plastic. In semi-dilute solutions, $G'(\omega)$ first increases with ω until reaching a plateau, where polymer entanglements act as transient connection points. Beyond the plateau region, $G'(\omega)$ rises again to reach its asymptotic value in the 10^9 J/m^3 range. The density-dependence of G' at the plateau depends on the nature of the network: floppy chains with permanent connections increase as ρ^1, while semiflexible filaments with no cross-links rise like $\rho^{7/5}$. Solutions of rigid rods do not exhibit a plateau at intermediate frequency, but rather rise smoothly towards their high-frequency limit, where G' is proportional to ρ^1.

Measurements of viscoelasticity

Several methods for measuring the viscoelastic properties of polymer solutions can be applied to biofilaments, although care must be taken to ensure that the technique does not disrupt the fragile biological network being measured. The procedures can be assigned to one of two broad

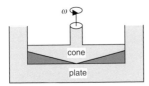

Fig. 4.21. Schematic cross section of a cone and plate device for measuring viscoelastic properties. The sample occupies the shaded region between the externally driven cone, and the plate (see Tran-Son-Tay, 1993).

categories involving either the observation of a response to an externally applied stress, or the measurement of thermal fluctuations in a sample. Methods of applying a stress or strain to a system include conventional viscometers; the cone-and-plate device in Fig. 4.21 is but one of many geometries that are in use (see Ferry, 1980, Chapter 5). The cone is driven at a frequency ω to impose a strain on the sample, which fills the gap between the cone and the plate. A variation of this method has been developed for observing the shear response of cells in a solution: the cone and plate are counter-rotating and made of optically transparent materials, permitting the deformation of the cells to be observed through a microscope (see review by Tran-Son-Tay, 1993).

Viscometers such as the cone-and-plate device measure the bulk properties of systems and require sample sizes much larger than a typical cell volume. In contrast, deformations of an individual cell and its internal components can be made visible by the injection of magnetic or fluorescent beads into its cytoplasm. Such beads might be simply passive markers aiding the analysis of strain, or may contain magnetic material permitting the bead to be manipulated by a constant or oscillating external magnetic field (Crick and Hughes, 1950; Ziemann *et al.*, 1994, and references therein). Analysis of video recording of the beads through a microscope fixes their position to a precision of better than about 0.05 μm, compared to a bead diameter of about 2–3 μm. Although such beads are not tiny, they are certainly small enough to be sensitive to local inhomogeneities of the cytoskeleton.

Earlier, we introduced the relation between thermal fluctuations and elasticity in the context of filaments and two-dimensional networks. The elastic characteristics of the cytoskeleton also can be determined by observing thermal fluctuations in the position of marker beads in the network, thus reducing disturbances to the material caused by the measuring device. Several techniques based on fluctuations have been developed, such as the recording of single bead motion (Schnurr *et al.*, 1997) or measuring interference of scattered light from an ensemble of beads (Palmer *et al.*, 1999). These procedures permit the study of oscillations at higher frequency, into the kilohertz range, than may be possible with mechanical rheometers. Bead diameters of 0.5–5 μm have been used, permitting the study of micron-size domains of a sample.

Chain densities in actin networks

Filament-forming proteins such as actin can be extracted from cells, purified, and then repolymerized into filaments and networks in a controlled fashion. Let us determine the filament density of a reference solution of 1 mg/ml F-actin, lying midway in the range often studied in the laboratory; from this reference solution, the chain density of a solution at arbitrary concentration can be obtained by multiplying by the ratio of their concentrations. At 370 monomers per micron of contour length, each monomer having a molecular mass of 42000 Da, a filament of F-actin possesses a mass per unit length λ_p of 16×10^6 Da/μm (see Table 2.2). A solution of 1 mg/ml (equal to 24 μM) corresponds to $c = 6.1 \times 10^{26}$ Da/m^3, leading to a density of filament length $\rho_L = c/\lambda_p = 3.8 \times 10^{19}$ μm/m^3; i.e., a cubic meter of this solution contains 3.8×10^{19} μm of F-actin filaments. The chain density ρ can be determined from ρ_L knowing the average filament length.

Cross-linked actin networks

Our discussion of actin rheology begins with cross-linked networks, whose properties have been studied for more than two decades with conventional viscometers (see, for example, Maruyama *et al.*, 1974, and Brotschi *et al.*, 1978). At our reference F-actin concentration of 1 mg/ml, the minimum ratio of cross-linking proteins to actin monomers for gel formation is observed to be roughly 1:100 or lower, depending on the cross-linker (Brotschi *et al.*, 1978). To estimate the shear modulus expected for such gels, let us assume a 1:100 ratio of cross-linker to actin monomer. If the cross-linking protein creates a four-fold junction, we then assign a quarter of a cross-linker per segmented end. As a segment has two ends, there is half a cross-linker per segment, or 50 monomers per segment at a ratio of 100 monomers per cross-linker. At 370 F-actin monomers per micron, the mean distance between network nodes must then be $50/370 = 0.14$ μm, corresponding to a filament density of 2.7×10^{20} filaments/m^3 if $\rho_L = 3.8 \times 10^{19}$ μm/m^3 at 1 mg/ml. Our estimate of the shear modulus from $\mu \sim \rho k_B T$ for this model network is thus 1.1 J/m^3 with $k_B T = 4.0 \times 10^{-21}$ J at 300 K.

The observed mechanical properties of cross-linked networks depend upon the stability of the cross-link, among other things. Janmey *et al.* (1990) used an actin-binding protein (ABP-280) to provide robust cross-links at concentration ratios to actin monomers of 1:100 or less. The shear storage modulus $G'(\omega)$ was observed to be almost independent of frequency in the range examined, having a value of 16–18 J/m^3 for an actin concentration of 1.6 mg/ml and ABP:actin monomer ratio of 1:133. This value is about an order of magnitude larger than expected from $\mu \sim \rho k_B T$: the actin concentration is 60% higher than our reference concentration, but the number of cross-links per monomer is 30% lower, yielding $\rho k_B T \sim 1.3$ J/m^3. A different study of strongly cross-linked actin (Xu *et al.*, 1998a) used 2% biotinylated actin (i.e., the protein biotin was attached to 2% of the actin monomers), yielding a gel in the presence of avidin, a protein capable of linking with 4 biotins per avidin. Even the small concentration of avidin used in this study (1:500 actin monomers) raises G' by almost two orders of magnitude compared to that of pure 2% biotinylated actin, illustrating the rigidity imparted by the cross-links. Assuming that all available biotins are linked to an avidin, actin concentrations of 0.65 mg/ml correspond to $\rho k_B T \sim 0.7$ J/m^3, which is more than an order of magnitude smaller than the measured G' in this experiment, namely ~20 J/m^3, independent of ω. The main result of both of these studies is that G' is observed to be largely independent of frequency, like a permanently cross-linked network; further, G' is about an order of magnitude larger than the nominal $\rho k_B T$.

The elastic behavior of actin networks is markedly different when α-actinin (from *Acanthamoeba*) is the cross-linker. At the relatively high ratio of α-actinin to actin monomer of 1:15 in a 1 mg/ml solution, G' is observed to be about 1.4 J/m^3 at low frequency, rising to more than 100 J/m^3 at frequencies exceeding 1 Hz (Sato *et al.*, 1987). At these concentrations, the average distance between crosslinks is a scant 20 nm (if all available α-actinin proteins form elastically active four-fold junctions) and the nominal value of $\rho k_B T$ is 7 J/m^3, this time much higher than the observed value. The origin of this discrepancy is believed to lie in the weaker affinity of α-actinin to actin, a conclusion deduced in part from the temperature dependence of $G'(\omega)$ for these networks. Over a broad frequency range, G' drops by about an order of magnitude as the temperature rises from 8 °C to 25 °C, implying that the linkage is susceptible to rupture by thermal energies at 25 °C. In contrast, the elasticity of actin networks with biotin/avidin junctions displays no change from 8 to 25 °C. At physiological temperatures, then, actin networks crosslinked by α-actinin tend to be stiff at short time-scales and soft at long time-scales, typical of a viscoelastic material (see also Tempel *et al.*, 1996, for a percolation approach to this network).

Actin solutions

The behavior of actin networks without permanent crosslinks has been the subject of intense study for some years (see Xu *et al.*, 1998b; Hinner *et al.*, 1998; Palmer *et al.*, 1999; Tang *et al.*, 1999; and references therein). Many studies have been made of the semi-dilute regime with concentrations around 0.5–2 mg/ml (approaching the concentration of the actin cortex of some cells) finding that the shear storage modulus is independent of frequency at $10^{-3} < \omega < 10$ rad/s, as shown in Fig. 4.22. The measured values of G' in the low-frequency range depend on preparation conditions: at concentrations of 1 mg/ml, fresh actin has G' of the order 0.1–1 J/m^3, with several measurements close to 1 J/m^3 (Hinner *et al.*, 1998;

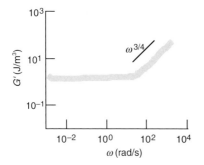

Fig. 4.22. Approximate shear storage modulus $G'(\omega)$ of a 1 mg/ml actin solution, as measured by light scattering (redrawn with permission, after Palmer *et al.*, 1999; © 1999 by the Biophysical Society). At small ω, G' is frequency-independent, but scales like $\omega^{3/4}$ at higher frequency.

Palmer *et al.*, 1999; Xu *et al.*, 1998b), although G' rises by an order of magnitude as the actin ages. If the actin filaments have a mean length of 2 μm, for example, the corresponding $\rho k_B T$ is just under 0.1 J/m³. These observations suggest that the filaments are sufficiently entangled to behave like a cross-linked network on time-scales long compared with many cellular time-scales even though the linkages are weak. However, the behavior of the plateau modulus is more subtle than $\rho k_B T$, as G' increases with chain density like $\rho^{1.2-1.4}$ (Xu *et al.*, 1998b; Hinner *et al.*, 1998; Palmer *et al.*, 1999). This concentration dependence agrees with theoretical expectations of semi-dilute solutions of semi-flexible filaments which predict a scaling of $\rho^{7/5}$ (Isambert and Maggs, 1996; Morse, 1998). At high frequency, $G'(\omega)$ rises like $\omega^{3/4}$ (Hinner *et al.*, 1998; Palmer *et al.*, 1999; Schnurr *et al.*, 1997), again as expected on theoretical grounds (Isambert and Maggs, 1996; Morse, 1998; Gittes and MacKintosh, 1998).

Actin filaments at high density

Depending on location within the cell, actin concentrations may be 5 mg/ml or larger, corresponding to filament densities of 1.9×10^{20} chains/m³ for filaments of length 1 μm (see Podolski and Steck, 1990, for actin filament lengths). Such densities lie in the concentrated solution regime, whose lower limit from Eq. (4.32) is $\rho^{**} \sim (D_f L_c^2)^{-1} \sim 1.2 \times 10^{20}$ chains/m³ if $L_c = 1$ μm and $D_f = 8$ nm. Is the actin concentration in cells sufficiently high to form a nematic phase? Onsager (1949) predicted that the threshold for nematic ordering of hard, rigid rods of uniform length occurs at a density (see also de Gennes and Prost, 1993, Sec. 2.2)

$$\rho_N = 4.25/D_f L_c^2. \tag{4.40}$$

This is a factor of four larger than ρ^{**} and has a value of 5.3×10^{20} chains/m³ (at $L_c = 1$ μm, $D_f = 8$ nm), clearly larger than ρ at actin concentrations of 5 mg/ml (if $L_c = 1$ μm). However, studies of actin with a distribution of filament lengths indicate that the nematic phase may

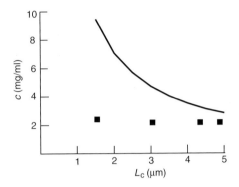

be reached at densities lower than predicted by ρ_N at the mean filament length (Suzuki *et al.*, 1991; Coppin and Leavis, 1992; Furukawa *et al.*, 1993). Fig. 4.23 illustrates how Eq. (4.40) exceeds the observed transition density to the nematic phase found by Furukawa *et al.* (1993), although the filament length was not measured directly in the experiment. Käs *et al.* (1996) observe an onset of nematic ordering at concentrations exceeding 2.5 mg/ml in experiments with a measured distribution of filament lengths; for this experiment, Eq. (4.40) predicts the onset should occur at $\sim 3 \times 10^{18}$ filaments/m^3 or 1.0 mg/ml (using $\langle L_c \rangle = 14$ μm), which is lower than the experimental threshold.

Intermediate filament and microtubule solutions

A comparative rheological study has been made of microtubules and vimentin (a common intermediate filament), both of which are more rigid than actin (Janmey *et al.*, 1991). In the experiment, both protein solutions were held at concentrations of 2 mg/ml, which corresponds to a lower density of contour length than actin solutions owing to the higher mass per unit length λ_p of the filament. Repeating the same calculation as performed for actin, the density of contour length ρ_L is 7.5×10^{18} μm/m^3 for microtubules ($\lambda_p = 16 \times 10^7$ Da/μm from Table 2.2) and 3.4×10^{19} μm/m^3 for vimentin (assuming $\lambda_p = 3.5 \times 10^7$ Da/μm, typical of intermediate filaments) at a concentration $c = 2$ mg/ml. The measured shear and storage moduli of both protein solutions show little frequency-dependence, and are in the range 2–3 J/m^3 over an angular frequency range $10^{-2} < \omega < 10^2$ rad/s. Not unexpectedly, these values are about a factor of five lower than that of F-actin networks measured in the same experiment; $\rho_L = 5.7 \times 10^{19}$ μm/m^3 at $c = 1.5$ mg/ml for actin, which is 1.7 times that of vimentin and 7.6 times that of microtubules at 2 mg/ml. The concentration-dependence of the shear modulus of the microtubule solutions was found to obey $c^{1.3}$, not inconsistent with the behavior expected for rigid or semiflexible filaments, given the uncertainty of the measurement. However, vimentin solutions obey $c^{0.5}$, an unusually low exponent that may arise if the average length of the vimentin filaments is concentration-dependent (Janmey et al., 1991). Just for comparison, the study quotes an average vimentin filament length of $L_c = 3.5$ μm, giving $\rho = \rho_L / L_c \sim 1 \times 10^{19}$ chains/m³ and $\rho k_B T = 0.04$ J/m³.

Other measurements of cell elasticity

The networks discussed in this text are chosen on the basis of their structural simplicity, permitting the microscopic origin of their elastic properties to be investigated. The characteristics of other homogeneous networks, such as collagen (Harley et al., 1997) and reconstituted spectrin (Stokke et al., 1985b), have also been examined. Achieving a moderately complete picture of single-component networks aids the interpretation of multicomponent networks present in the cell; for example, Griffith and Pollard (1982) have determined the viscosity of actin/microtubule networks. Highly heterogenous systems have been investigated as well, although their interpretation may be necessarily qualitative (for a review of pioneering work in this area, see Elson, 1988). The techniques employed include micropipette aspiration (Hochmuth and Needham, 1990), deformation by atomic force microscope (Radmacher et al., 1996) or the use of internal markers to drive or record deformation of the cytoplasm. For example, magnetic beads (Bausch et al., 1999) or fluorescent markers can be injected into the cell cytoplasm to follow the local strain it develops in response to a stress (Ragsdale et al., 1997, and references therein). Not unexpectedly, such measurements made on the cytoplasm of a particular macrophage show that the shear storage modulus is in the range quoted above for networks of pure filaments such as actin, ranging from $10-10^2$ J/m³. Further, the elastic response may vary by a factor of two within a given cell, and much more between different cells of a given population (Bausch et al., 1999).

The mechanical properties of the bacterial cell wall are treated in some detail in Sections 3.5 and 7.5. The measured values of a typical elastic modulus of hydrated wall material is a few times 10^7 J/m³ (see, for example, Yao et al., 1999), which is well below the value of 10^9 J/m³ often found for dense polymers. However, this is not unexpected given that the molecular structure of peptidoglycan, the material of the wall, has stiff longitudinal elements linked by floppy transverse elements. We present a simple model of this structure in the problem set, and show that its elastic moduli should be in the 10^6 J/m³ range for thick walls.

Systematic studies also have been made of defective networks in cells, particularly the two-dimensional network of the human erythrocyte (Waugh and Agre, 1988). Spectrin-depleted cytoskeletons exhibit a reduced shear resistance, consistent with their rigidity emerging at the connectivity percolation threshold p_C, although their elasticity is not completely described by percolation (see problem set).

4.6 Summary

We have focused in this chapter on the three principal components of the cytoskeleton – actin, intermediate filaments and microtubules. With persistence lengths greater than or equal to the diameter of a typical cell, these filaments have gentle curvature when viewed in their cellular environment. Solutions of chemically pure filaments of each of these proteins can be studied in the laboratory. In the cell, binding proteins may be present that link the filaments into three-dimensional networks with fixed connection points.

The static elastic properties of three-dimensional materials with high symmetry can be summarized in just a few elastic moduli. For materials with cubic symmetry, three moduli appear in the free energy of deformation, Eq. (4.6); for isotropic materials, only two moduli (K_V and μ) are present: $\Delta \mathcal{F} = \mu \sum_{i,j} (u_{ij} - \delta_{ij} \, \mathrm{tr} u /3)^2 + K_V \, (\mathrm{tr} u)^2/2$, where u_{ij} is the strain tensor. The moduli are rooted in the microscopic characteristics of the material; for example, at zero temperature the moduli of a network of springs with force constant k_{sp} and equilibrium spring length s_o are proportional to k_{sp}/s_o. The elastic moduli of floppy chain networks with fixed connection points are close to $\rho k_B T$, where ρ is the number of chains per unit volume. However, when the average number of connection points per chain is less than the connectivity percolation threshold, the shear modulus vanishes. Under certain circumstances (e.g., central forces at zero temperature) the shear modulus vanishes below a separate rigidity percolation threshold, at a density higher than the connectivity threshold. For a network on a lattice in d dimensions, the bond occupation probability at the rigidity percolation threshold is not far from $2d/z$, where z is the number of nearest neighbors to a lattice site.

Without permanent cross-links, polymers in solution may behave like fluids over very long time-scales. However, on short time-scales, their mechanical elements may be sufficiently dense or entangled that they move past each other only very slowly, and exhibit shear resistance. Such viscoelastic materials are characterized by time- (or frequency-) dependent shear moduli: subject to an oscillating strain of angular frequency ω, they develop a stress $\sigma_{xy} = G'(\omega)u_{xy}(t) + G''(\omega) \cdot (du_{xy}/dt)/\omega$, where $G'(\omega)$ and $G''(\omega)$ are the shear storage modulus and shear loss modulus, respectively. In the zero-frequency limit, G' becomes the shear modulus and G''/ω becomes the viscosity. In general, G' increases with frequency, although its specific behavior depends upon concentration. Of interest in the cell is the behavior of a solution of stiff filaments, say with contour length L_c held at a number density ρ. In the dilute regime, $\rho < \rho^* \sim L_c^{-3}$, the polymer solution behaves like a simple fluid with small values of G'. In contrast, at high filament concentrations, $\rho > \rho^{**} \sim (D_f L_c^2)^{-1}$, the solution lies in the concentrated regime and G' approaches 10^9 J/m^3. Nematic ordering may occur at such high concentrations if L_c/D_f exceeds approximately 3. Less uniform behavior is seen at concentrations between ρ^* and ρ^{**}, where G' first rises with frequency, then becomes roughly constant through a plateau region, and finally rises again towards an asymptotic value. The plateau reflects the presence of chain entanglements, and is predicted to scale with concentration like ρ^1 for floppy chains or $\rho^{7/5}$ for semi-flexible filaments. Solutions of rigid rods also display a high-frequency regime that increases like ρ^1.

Networks of biofilaments in the cell display most of the theoretical expectations described above; in fact, biofilaments represent an excellent laboratory for testing some concepts from polymer physics because, unlike alkanes, they can be fluorescently labelled and studied individually by microscopy. Filamentous networks with strong permanent cross-links are found to have frequency-independent storage moduli G' that are an order of magnitude larger than $\rho k_B T$, depending on the stiffness of the filament. Semidilute solutions of actin and microtubules individually show that G' in the plateau region scales with concentration like $\rho^{7/5}$, and with frequency like $\omega^{3/4}$ at high frequency, in agreement with theoretical predictions. At high concentrations, actin solutions are observed to form a nematic phase, although the concentration at the onset of ordering may be different than expected theoretically for systems of rigid rods of uniform length.

4.7 Problems

Biological applications

4.1 Suppose that a spherical cell of radius 5 μm contains a solution of actin filaments at a concentration of 10 mg/ml.

(a) Calculate the nominal value of $\rho k_B T$ of the filaments if they have a length of 5 μm (where $T = 300$ K).

(b) Taking the volume compression modulus K_V to be $\rho k_B T$, find the fluctuations in the volume occupied by the network $\langle \Delta V^2 \rangle^{1/2} / V$ (assuming that the cell membrane permits these fluctuations).

4.2 A simplification of the peptidoglycan network of the bacterial cell wall treats the glycan strands as rigid rods and the string of amino acids as entropic springs, as in the figure in Problem 4.13. There, we show that the compression modulus of this model at zero tension is $K_V = k_{sp}/(8b)$, with symbols defined in the figure.

(a) If the contour length of the peptide chains is 4 nm, but their mean end-to-end displacement is $a = 1.3$ nm, what is their persistence length and effective spring constant k_{sp} (at $T = 300$ K)?
(b) If $b = 1.0$ nm, what is K_V using your results from part (a)?
(c) Compare the result from (b) with $\rho k_B T$ at $T = 300$ K using the bond density (transverse plus longitudinal) calculated in Problem 4.13(a).

4.3 Hereditary spherocytosis is a red blood cell disorder in which a reduced spectrin content of the cytoskeleton is associated with a reduced shear modulus. A sample of data taken from a study by Waugh and Agre (1988) is shown in the table below (all quantities have been normalized to a control sample). Plot these data and compare them with two predictions from percolation theory: $\mu/\mu_0 = (p - p_{perc})/(1 - p_{perc})$ where p_{perc} is either p_C or p^* from Eq. (4.29) for a triangular lattice. Interpreting the measurements in terms of a randomly depleted network, with which value of p_{perc} are the data more consistent?

Spectrin content	Shear modulus	Spectrin content	Shear modulus
0.54	0.62 ± 0.04	0.52	0.62 ± 0.03
0.97	0.82 ± 0.09	0.68	0.73 ± 0.05
0.73	0.77 ± 0.04	0.55	0.61 ± 0.04
0.65	0.61 ± 0.03	0.71	0.73 ± 0.05
0.43	0.62 ± 0.09	0.43	0.63 ± 0.07

4.4 What is the lowest density of bonds required for rigidity percolation for the three-dimensional lattices in Table 4.1 – simple cubic, body-centered cubic and face-centered cubic? Take the bonds in each lattice to have the same length b and assume that the rigidity percolation threshold p_R is given by Eq. (4.29).

4.5 In one experiment (Abraham *et al.*, 1999), actin filaments at the leading edge of a migrating cell were found to have concentrations as high as 1600 ± 600 μm of filament per cubic micron of cell volume.

(a) If the filaments were 1 μm in length, in what concentration regime would this system fall? Would you expect the filaments to form a nematic phase?

(b) From the density you obtain in (a), what is the pressure of the filaments, regarding them as non-interacting objects (at $T = 300$ K)? Quote your answer in J/m^3 and atmospheres.

4.6 The bacterium *E. coli* has the approximate shape of a cylinder capped at each end by hemispheres. Use the image in Fig. 1.1 to obtain a typical length and diameter of this bacterium.

(a) Evaluate the densities ρ^*, ρ^{**} separating the dilute, semi-dilute and concentrated regions of bacterial solutions. Quote your answer in bacteria per cubic meter.

(b) Does the shape of the bacterium satisfy Onsager's condition for the existence of a nematic phase?

4.7 The tobacco mosaic virus (TMV) has the shape of a cylindrical rod 260 nm in length and 18 nm in diameter; its mass per unit length is given in Table 2.2.

(a) Does the shape of the virus satisfy Onsager's condition for the existence of a nematic phase?

(b) Find the transition concentrations separating the four possible phases of this virus in solution. Calculate ρ^*, ρ^{**} and ρ_N in filaments/m^3, and c^*, c^{**} and c_N in mg/ml.

4.8 Under certain conditions, disk-like objects can form an ordered phase in which the disks display long-range orientational order.

Characterizing a disk by its diameter D and thickness T, it is found that:

    condition I: the disks may form an ordered phase if $T/D < 0.2$
    (with some uncertainty, see Veerman and Frenkel, 1992);

condition II: an ensemble of very thin disks ($T \ll D$) may form an ordered phase if its density satisfies $\rho > 4/D^3$ (see Eppenga and Frenkel, 1984).

(a) Find the approximate diameter and thickness of a human red cell, and compare the dimensions with condition I.
(b) Find the approximate concentration of red cells in the blood (see Chapter 22 of Alberts *et al.*, 1994) and compare with condition II.

Formal development and extensions

4.9 Suppose that we have a three-dimensional isotropic material whose compression resistance is much greater than its shear resistance: $K_V \gg \mu$.

(a) Find the strain tensor for the deformation $x \to \Lambda_- x$, $y \to \Lambda_- y$, $z \to \Lambda_+ z$ in terms of δ, where $\Lambda_+ = 1 + \delta$ and there is no volume change.
(b) Predict the Poisson ratio from your results in (a) using the definition $\sigma_p = -u_{xx}/u_{zz}$.
(c) Compare your prediction from (b) with Eq. (4.4) in the appropriate limit.

4.10 Consider a network of springs connected in the shape of a cube, as in Fig. 4.8, such that the springs are forced to lie at right angles to one another. Each spring has the same unstretched length s_o and force constant k_{sp}.

(a) By minimizing the enthalpy H of the network under pressure ($H = E + PV$), find the spring length s_p at pressure P, and show that $s_p = s_o(1 - Ps_o/k_{sp})$ at small P (where $P > 0$ is compression).
(b) Using $V = s_p^3$ of a single cube, determine the compression modulus K_V at $P = 0$ from your results in part (a).

4.11 A spring network with the symmetry of a body-centered cubic lattice has the structure:

Each vertex is connected to its eight nearest neighbors by springs of unstretched length s_o and spring constant k_{sp}.

(a) Find the volume per vertex in terms of s_o.
(b) Find the strain tensor and the change in energy density if all springs are stretched from s_o to $s_o + \delta$.
(c) Show that the compression modulus is given by $K_V = (1/\sqrt{3}) \cdot (k_{sp}/s_o)$.

4.12 A spring network with the symmetry of a face-centered cubic lattice has the structure:

Each vertex is connected to its twelve nearest neighbors by springs of unstretched length s_o and spring constant k_{sp}.

(a) Find the volume per vertex in terms of s_o.
(b) Find the strain tensor and the change in energy density if all springs are stretched from s_o to $s_o + \delta$.
(c) Show that the compression modulus is given by $K_V = (2\sqrt{2}/3) \cdot (k_{sp}/s_o)$.

4.13 The box in the diagram below represents a section of an extended network that we will use as a simplified model of peptidogly-can, the material of the bacterial cell wall. The multisegmented rods with monomer length b represent the glycan strands, while the five short bonds of length a perpendicular to the glycan are strings of amino acids. All bonds (indicated by the thick lines) and vertices (indi-cated by the disks) are shared with neighboring sections of the wall (we use the word "section" to represent a specific volume of the cell wall with a specific structure; here, the volume of a section is $4a^2b$).

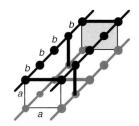

(a) How many vertices and bonds (both transverse and longitudinal) are there in this section? Show that the volume per vertex is a^2b. (*If you haven't done this type of problem before, determine what "frac-tion" of a bond or vertex belongs to each section.*)

(b) Show that the enthalpy per vertex H_V for this network under tension τ is

$$H_V = (k_{sp}/4) \cdot (a - a_o)^2 - \tau a^2 b,$$

where k_{sp} is the effective spring constant of the peptide chain and a_o is its unstressed (but not contour) length. Assume that the glycan strand (b) is a rigid rod, but a can vary about a_o.

(c) Show that H_V is minimized by the bond length

$$a_\tau = a_o / (1 - 4\tau b/k_{sp}).$$

(d) Find the volume per vertex, and show that the volume compression modulus is

$$K_V = (k_{sp}/8b) \cdot (1 - 4\tau b/k_{sp}).$$

4.14 Let us use Eq. (4.21) as a model free energy density $\Delta\mathcal{F}$ for a network of random chains under compression. Consider only configurations with $\Lambda_x = \Lambda_y = \Lambda_z = \Lambda$.

(a) Show that the minimum value of $\Delta\mathcal{F}$ occurs at $\Lambda = 1/\sqrt{2}$, not $\Lambda = 1$.

(b) What is $\Delta\mathcal{F}$ at $\Lambda = 1/\sqrt{2}$ (call this $\Delta\mathcal{F}_{min}$)?

(c) What is $\Delta\mathcal{F}$ at the uniform dilation $\Lambda = (1/\sqrt{2}) + \delta$ (call this $\Delta\mathcal{F}_{def}$)?

(d) From $\Delta\mathcal{F}_{def}$ and $\Delta\mathcal{F}_{min}$ above, find the compression modulus K_V at $\Lambda = 1/\sqrt{2}$?

4.15 Consider a three-dimensional cage formed from 12 inextensible rigid rods of length s_o. If the rods join at right angles, the cage is a cube of volume s_o^3, as in Fig. 4.8.

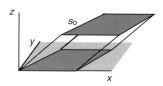

Find the average volume of the cage if the rods freely pivot at the junctions (You may find it calculationally helpful to fix one rod to lie along the x-axis and a second rod to lie in the xy-plane).

4.16 Determine the mean value of the squared volume $\langle V^2 \rangle$ for the freely jointed cage in problem 4.15, and extract the compression modulus of the cage from

$$(\beta K_V)^{-1} = (\langle V^2 \rangle - \langle V \rangle^2)/\langle V \rangle.$$

4.17 Complete the mathematical proof leading from Eq. (4.16) to (4.17). The steps are as follows:

- substitute n_i and n_{f_i} into Eq. (4.16); the resulting expression contains summations of the form $\Sigma_i f(x_i)\Delta x$, which can be converted to integrals, yielding $\ln\mathscr{P}_a = n(I_1 + I_2)$, where

$$I_1 = (\Lambda_x\Lambda_y\Lambda_z)^{-1} (2\pi\sigma^2)^{-3/2} \int\int\int \exp\{-[(x/\Lambda_x)^2 + (y/\Lambda_y)^2$$
$$+ (z/\Lambda_z)^2]/2\sigma^2\} \cdot [x^2(\Lambda_x^{-2}-1) + y^2(\Lambda_y^{-2}-1)$$
$$+ z^2(\Lambda_z^{-2}-1)]/2\sigma^2 \, dxdydz$$

$$I_2 = (\Lambda_x\Lambda_y\Lambda_z)^{-1} (2\pi\sigma^2)^{-3/2} \int\int\int \exp\{-[(x/\Lambda_x)^2 + (y/\Lambda_y)^2$$
$$+ (z/\Lambda_z)^2]/2\sigma^2\} \cdot \ln(\Lambda_x\Lambda_y\Lambda_z) \, dxdydz,$$

and the integration limits run from $-\infty$ to $+\infty$
- break I_1 into three terms of the form (prove this):

$$(1-\Lambda_x^2) \cdot (2\pi\sigma^2)^{-1/2} \int \exp[-(X^2/2\sigma^2)] (X^2/2\sigma^2) \, dX = (1-\Lambda_x^2)/2$$

- show that I_2 has the form $I_2 = \ln(\Lambda_x\Lambda_y\Lambda_z)$
- combine these results to obtain Eq. (4.18):

$$\ln\mathscr{P}_a = -(n/2)[\Lambda_x^2 + \Lambda_y^2 + \Lambda_z^2 - 3 - 2\ln(\Lambda_x\Lambda_y\Lambda_z)].$$

Part II
Membranes

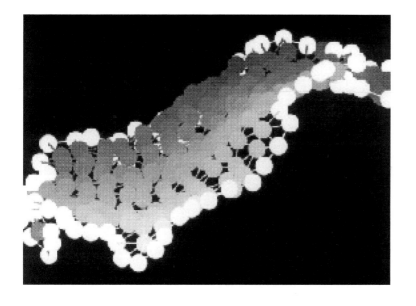

Chapter 5

Biomembranes

The two broad categories of structural components of the cell are filaments, the focus of Part I of this book, and sheets, which we treat in this chapter and its companion. In principle, two-dimensional sheets may display more complex mechanical behavior than one-dimensional filaments, including resistance to both out-of-plane bending and in-plane shear. As examples, the plasma membrane is a two-dimensional fluid having no resistance to in-plane shear, whereas the cell wall possesses shear rigidity as a result of its fixed internal cross-links. In this first chapter of Part II, we introduce the chemical composition of biomembranes and describe their mechanical properties, including elasticity and failure. We focus here on fluid membranes and defer a discussion of polymerized structures, such as the cell wall, to Chapter 6. The softness of biomembranes means that their thermally induced undulations are important, as will be explored in Chapters 6 and 8. More extensive reviews of biomembranes than those provided here can be found in Evans and Skalak (1980), Cevc and Marsh (1987) and Sackmann (1990).

5.1 Composition of biomembranes

The design principles of Chapter 1 argue that the cell's membranes should be very thin, perhaps even just a few molecules in thickness, if their purpose is solely to isolate the cell's contents. The membrane need not contribute to the cell's mechanical strength, as this attribute can be provided by the cytoskeleton or cell wall. Are there materials in nature that meet these design specifications? Evidence for the existence of monolayers just a single molecule thick goes back to Lord Rayleigh, in an experiment that would be environmentally frowned upon today. He poured a known volume of oil onto a calm lake and observed, by reflection, the area covered by the oil. Dividing the volume of the oil by the area over which it spread gave him the thickness of the layer – a good estimate of the size of a single oil molecule.

The oil/water interface in Rayleigh's experiment is flat because the fluids are immiscible and have different densities. In contrast, interfaces with more diverse geometries arise from a class of compounds called surfactants (surface active agents), possessing attractive interactions with a range of fluids whose composition may be so dissimilar as to make them immiscible. For example, sodium stearate (common soap) has a long alkane chain terminating at a carboxylic acid group, as illustrated in Fig. 5.1(a). The hydrocarbon chain is non-polar, and is said to be water-avoiding or hydrophobic, while the (polar) carboxyl group is water-loving or hydrophilic; this dual nature makes the molecule amphiphilic. As a detergent, the hydrophobic end of this amphiphile embeds itself in an oily droplet, leaving the surface of the droplet decorated with polar carboxyl groups that are attractive to water. Fig. 5.2(a) illustrates how a continuous molecular monolayer of amphiphiles can form between oil and water, with the hydrophobic component of the surfactant penetrating the oil phase and the hydrophilic component facing the aqueous phase. Such a structure is close to what we are seeking for the cell boundary, except that we want to have water on both sides of the boundary, not water on one side and oil on the other. By cutting and pasting two copies of Fig. 5.2(a) and omitting the oil, we can construct a bilayer with two leaflets back-to-back, as in Fig. 5.2(b). Are there surfactants capable of preferentially assembling into such bilayers?

A class of molecules broadly referred to as liquid crystals form a variety of condensed phases with properties in between those of solids and isotropic fluids (see Vertogen and de Jeu, 1988). These intermediate phases, called mesophases or mesomorphic phases, are characterized by partial ordering of the molecules over long distances; for instance, the nematic phase of rod-like molecules described in Chapter 4 has long-

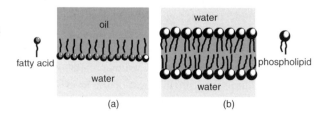

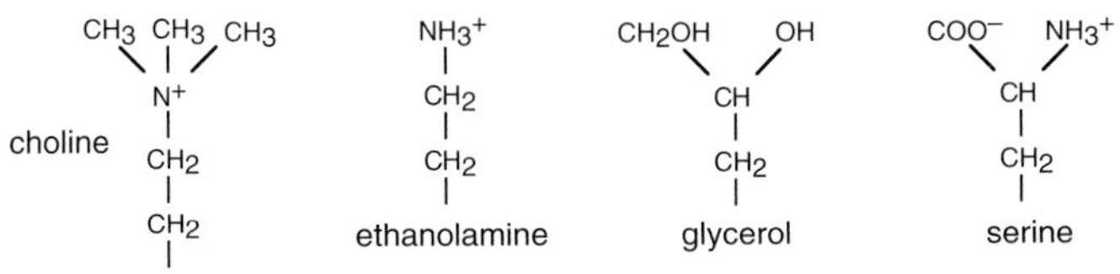

CH3 CH3 CH3
\ | /
N+
|
choline   CH2
|
CH2
|

NH3+
|
CH2
|
CH2
|
ethanolamine

CH2OH   OH
\   /
CH
|
CH2
|
glycerol

COO−   NH3+
\   /
CH
|
CH2
|
serine

Fig. 5.3. Examples of polar head groups commonly found on phospholipid molecules in cellular membranes.

range orientational order without long-range positional order. Molecules capable of forming mesophases are referred to as mesogens, and many amphiphiles of importance in cells are members of this family. Depending on temperature, the amphiphiles present in a bilayer may form a two-dimensional mesophase, although the membranes of most cells operate at sufficiently high temperature, or are sufficiently heterogeneous chemically, that they are isotropic fluids incapable of supporting a shear. More detailed treatments of the formation and characterization of bilayer phases can be found in the reviews by Bloom *et al.* (1991) or Nagle and Tristram-Nagle (2000).

Single-chain fatty acids such as sodium stearate assemble into bilayers only at high concentrations, greater than 50% soap by weight, for instance. This is not what is needed for a cell: the membrane should form at low amphiphile concentration if its materials are to be used efficiently. Consequently, the cell's bilayers are based upon particular varieties of phospholipid molecules having two hydrocarbon chains. Lipids themselves are a broad group of organic compounds that are soluble in organic solvents, but not in water; the generic phospholipid in Fig. 5.1(b) comprises two fatty acids linked to a glycerol, which in turn is linked to a polar head group *via* a phosphate PO_4. The polar head groups of the cell's phospholipids are selected from several organic compounds, none of which is particularly large or complex, as can be seen in Fig. 5.3. In many phospholipids found in biomembranes, one of the fatty acids contains only C–C single bonds while the other contains a C–C double bond. Although a chain may rotate around a C–C single bond, in practice saturated hydrocarbons (single bonds only) tend to straighten out when the phospholipids form a layered structure. In contrast, a double bond generates a kink in the hydrocarbon chain, making the dense packing of phospholipids more difficult and reducing the in-plane viscosity of the bilayer. Other lipids such as cholesterol may fill in the gaps formed by kinks (see Appendix B for a longer discussion of lipid geometry).

Biomembranes are not composed of just one type of lipid, although it is possible to produce and study pure lipid bilayers in the laboratory. Rather, a selection of lipids possessing a range of hydrocarbon chain lengths and polar head groups are found in cell membranes, as illustrated in Table 5.1. The range of chain lengths for membrane lipids is

Table 5.1. *Chain lengths of fatty acids in rat liver membranes, where* n_c *is the number of carbon atoms in the hydrocarbon chain, including the COO group*

Membrane	Percent by weight					
	$n_c = 14$	15	16	17	18	20
mitochondrial (outer)	<1	27	25	14	14	16
mitochondrial (inner)	<1	27	22	16	16	19
plasma membrane	1	37	31	6	13	11
Golgi apparatus	1	35	23	9	18	15

Notes:
The distributions are a percent of the total fatty acid content by weight; not all values of n_c are listed and $n_c = 20$ includes only arachidonic acid. Note that the fatty acids are generally bound as phospholipids and other molecules (from Gennis, 1989).

centered around 15–18 carbon atoms, reflecting the operating temperature and lipid concentration of the cell. If the chains are rather short, the lipids won't form bilayers at low concentrations, while if the chains are overly long, the bilayer is too viscous and lateral diffusion of molecules within the bilayer is restricted. At ~0.1 nm per CH_2 group on the chain, each hydrocarbon chain has a length close to 2 nm in a single layer, for a total bilayer thickness of 4–5 nm. The mean cross sectional area of a single chain is about 0.20 nm², while the average surface area of the bilayer occupied by most membrane lipids is 0.4–0.7 nm² (see Nagle and Tristram-Nagle, 2000). Furthermore, membranes contain more than just phospholipids. For example, rigid cholesterol molecules also may be present, ranging from 0% to 17% by total lipid weight in the plasma membranes of *E. coli* and the red blood cell, respectively. Important for the functioning of the cell, proteins are embedded in membranes to selectively permit and control the passage of material across the bilayer. Such membrane proteins are much larger than lipid molecules, and they enhance the thickness of the membrane.

The bilayer is an attractive arrangement for the binary surfactant/water mixture because the hydrophobic chains of the lipid are not exposed to water, while the hydrophilic head groups are solvated. However, this is not the only way of organizing the hydrophobic chains so that they avoid an aqueous environment: a sphere or cylinder, whose surface is lined by polar groups and whose interior is filled with hydrocarbons, also does the job. Which of these phases is most favored for a given amphiphile? Consider first single-chain fatty acids. At very low

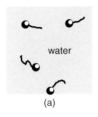
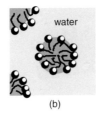
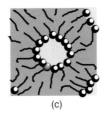

(a)   (b)   (c)

Fig. 5.4. Phases expected for fatty acids in water, where the linear hydrocarbon chain of the acid is not too short. (a) Homogeneous solution at low amphiphile concentration, (b) micelle formation at intermediate concentration, and (c) inverted micelles at high concentration. The roughly circular structures in (b) and (c) are a few nanometers in radius for fatty acids with 10–20 carbon atoms.

concentrations, the molecules disperse in water, as illustrated in Fig. 5.4(a). Yet, owing to their hydrophobic chains, the amphiphiles rapidly reach the saturation point as their concentration is raised, typically in the 10^{-3} molar range (or less for long acyl chains). At higher concentrations, the amphiphiles assemble into collective structures such as the micelles shown in Fig. 5.4(b). Typically formed by fatty acids, the surface of the micelle is covered by polar groups while the hydrocarbon tails gather in the interior. The micelle may have the shape of a spherical droplet, or a long cylinder like a sausage, having dimensions of nanometers (transversely) to microns (longitudinally). The onset of micelle formation occurs at the critical micelle concentration (CMC); raising the amphiphile concentration above the CMC increases the number of micelles, but only weakly affects their average size. At high amphiphile concentrations, micelles fill the volume of the system, adopting regular arrays or forming the inverted micelle phase of Fig. 5.4(c).

The phase favored by a particular amphiphile partly reflects its molecular shape. Consider the three configurations displayed in Fig. 5.5. The spatial region occupied by a molecule in a spherical or cylindrical micelle could be imagined to look like an ice-cream cone or a wedge-shaped slice of pizza, as in Fig. 5.5(a). The interior of the micelle is cramped, and only an amphiphile with either a single hydrocarbon chain, or perhaps two rather short chains, would make a comfortable fit. In contrast, the interior of a bilayer is somewhat roomier, and is favored by dual-chain lipids whose head group and chains have a similar cross sectional area, like the cylinder shown in Fig. 5.5(b). However, if the head group of a dual-chain lipid is too small (say, ethanolamine, which is the smallest of the four common species displayed in Fig. 5.3), the bilayer is less favorable than the inverted micelle phase of Fig. 5.5(c). We see from these examples that the preferred geometry is affected by the

Fig. 5.5. Interface shape is related to molecular geometry. (a) Single-chain fatty acids tend to form micelles, while (b) dual-chain phospholipids with moderate size head groups prefer bilayers. (c) Phospholipids with small head groups may form inverted micelles. The mean molecular shape is indicated by the cones and cylinders drawn beside the amphiphile.

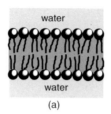
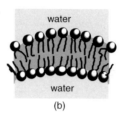

(a)   (b)

ratio of head group area to the cross sectional area of the hydrocarbon region.

A flat bilayer has a mean separation between lipid molecules that minimizes their free energy. When subjected to a deformation such as the bending mode displayed in Fig. 5.6, the mean spacing between head groups in one layer is less than optimal, while the spacing in the other layer is greater than optimal. Thus, a bending mode usually requires an input of energy determined by the bending resistance of the bilayer. Depending upon its structure and environment, the bending resistance of a sheet increases like the square or cube of its thickness, so that very thin sheets are highly flexible. We are all familiar with the unruly behavior of thin sheets of plastic wrap or aluminum foil which, in spite of the rigidity of their parent material, twist and fold in the most exasperating manner. The same is true of bilayers, which, being much thinner than household materials, are very floppy and may exhibit thermal undulations on cellular length scales. In this chapter, we introduce a very simple mathematical description of bending elasticity, exploring its zero-temperature implications here, and finite-temperature effects in Chapters 6 and 8. The role of bilayer elasticity in determining cell shape is investigated in Chapter 7.

Membranes also may be under tension or compression in a cell. For instance, a plant cell wall may bear a mechanical load from the bending of the plant, or a cell in the circulatory system may be subject to shear stress. Internal stresses may arise from an elevated osmotic pressure, or from contractile forces generated by interacting filaments within the cell. Small stresses can be accommodated by small alterations in molecular configurations within the membrane, perhaps including a change in the mean separation between head groups, the magnitude of the change reflecting the membrane's compression modulus. At sufficiently high stress, the membrane may tear or rupture, the amphiphiles rearranging themselves to reduce exposure of their hydrocarbon regions to water. However, configurations such as those shown in Fig. 5.7 are not as energetically favorable as the bilayer itself, resulting in an energy penalty for the formation of holes; this can be characterized by a line tension or edge tension, which is an energy per unit length along the boundary of

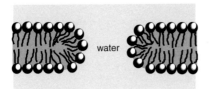

Fig. 5.7. The formation of a hole in a fluid bilayer forces amphiphilic molecules to rearrange themselves to protect the hydrocarbon interior of the bilayer from exposure to water.

the hole. The effective edge tension is temperature-dependent and vanishes at sufficiently high temperature, where the membrane is unstable against hole formation even in the absence of mechanical stress.

We cover a variety of membrane characteristics in this chapter, although the formalism invoked is neither extensive nor challenging. Rather than segregate the formal development of a topic from its application to cell mechanics, as we do in many chapters of this text, here we treat theory and experiment together within each section. We present a mathematically simple model for self-assembly in Section 5.2 that avoids the use of chemical potentials, at the loss of some predictive power available in a completely rigorous treatment. This section also demonstrates the relationship between aggregate structure and molecular shape. Of importance in cell function is not only the self-assembly of membranes, but also their behavior under mechanical stress, and Sections 5.3 and 5.4 begin a discussion of the compression and bending resistance (to be continued throughout Chapters 6 and 7) of bilayers. Lastly, Section 5.5 examines the energetics of hole formation and its relation to membrane topology and structural failure under stress.

5.2 Self-assembly of amphiphiles

Amphiphiles, such as the phospholipids of cell membranes, can form aggregates in aqueous solution if their concentration is above a threshold value commonly called the critical micelle concentration (CMC, although the threshold is more accurately the critical aggregation threshold). The driving force for aggregation is the aversion of the amphiphile's hydrophobic regions to contact with water: the interior of a micelle provides a safe haven for hydrocarbon chains. However, forming a cluster lowers the number of objects in a system, and hence reduces its entropy. Thus, there is a competition between energy, which favors the formation of clusters, and entropy, which favors the distribution of molecules throughout the solution.

Although most treatments of aggregation use the language of chemical potentials (for example, Chapters 2–7 of Tanford, 1980, and Chapter 16 of Israelachvili, 1991), we present here a description based upon the free energies of two idealized phases, invoking just the thermodynamic

Fig. 5.8. A simple two-phase
model for the formation of
aggregates in a
water/amphiphile system. In
(a), the amphiphiles are
condensed into a single blob,
while in (b) they are dispersed
throughout the aqueous
phase.

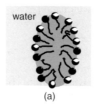

water

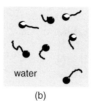

water

(a)   (b)

Fig. 5.8. A simple two-phase model for the formation of aggregates in a water/amphiphile system. In (a), the amphiphiles are condensed into a single blob, while in (b) they are dispersed throughout the aqueous phase.

relations developed in Appendix C. As our reference state, we consider a single blob of amphiphiles in a bath of polar water molecules, as in Fig. 5.8(a). We compare this condensed phase with the solution in Fig. 5.8(b), which we view as a gas of amphiphiles in an aqueous medium. Clearly this picture is an oversimplification: there is a distribution of cluster sizes in any physical system and individual amphiphiles are present in solution even once an aggregate has formed.

We define the reference state in Fig. 5.8(a) to have zero free energy: the entropy of the blob is neglected, and the potential energy between molecules within the blob is defined to be zero. With these definitions, each amphiphile pays an energy penalty E_{bind} to escape from the cluster; we approximate E_{bind} as the energy required to create the new water/hydrocarbon interface once its hydrophobic region is exposed to water in the solution phase. This interfacial energy is crudely equal to the product of the water/hydrocarbon surface tension γ with the effective area of the hydrocarbon chains. We take the hydrocarbon region of the amphiphile to have the geometry of a cylinder of radius R_{hc} and length $n_c \ell_{\text{cc}}$, where n_c is the number of carbon atoms along the hydrocarbon chain and $\ell_{\text{cc}} = 0.126$ nm is the average C–C bond length projected on the chain (see problem set, and Chapter 16 of Israelachvili, 1991). Hence, the area of the hydrophobic region is approximately $2\pi n_c R_{\text{hc}} \ell_{\text{cc}}$, yielding a removal energy per molecule of

$$E_{\text{bind}} = 2\pi n_c R_{\text{hc}} \ell_{\text{cc}} \gamma, \tag{5.1}$$

where we have omitted the end cap area of the cylinder.

In entering the aqueous phase, an amphiphilic molecule pays an energy penalty, but acquires a much larger configuration space to explore, hence increasing the entropy of the system. We assume the solution phase to be sufficiently dilute that dissolved amphiphiles behave like an ideal gas. As established in Appendix C (see also Section 9.10 of Reif, 1965), the entropy per molecule S_{gas} of an ideal gas at number density ρ is

$$S_{\text{gas}} = k_B \{ 5/2 - \ln(\rho \cdot [h/\{2\pi m k_B T\}^{1/2}]^3) \}, \tag{5.2}$$

where m is the mass of each molecule and h is Planck's constant. The length scale $h/\{2\pi m k_B T\}^{1/2}$ is provided by the translational motion of the molecule at $T > 0$.

Given the expressions for E_{bind} and S_{gas}, in our model system, the free energy per molecule F_{sol} in the solution phase at fixed volume is

$$F_{sol} \sim E_{bind} - TS_{gas},\qquad(5.3)$$

where we assume that the entropy of bulk water is largely unchanged by the presence of amphiphiles in solution, and that γ includes changes in entropy arising from the ordering of water molecules near an amphiphile. Whether the energy or entropy of the amphiphiles dominates the free energy F_{sol} depends upon the density ρ. At low densities, the entropy per particle dominates and the solution phase is favored, while at high amphiphile density, E_{bind} favors the condensed phase. With our definition that the free energy of the condensed (reference) state is zero, the cross-over between phases occurs at $F_{sol}=0$, or $E_{bind}=TS_{gas}$, corresponding to an aggregation density ρ_{agg} of

$$\rho_{agg} \cdot [h/\{2\pi m k_B T\}^{1/2}]^3 = \exp(5/2 - E_{bind}/k_B T),\qquad(5.4)$$

from Eqs. (5.1) and (5.2). This expression confirms our intuition that the threshold for aggregation *decreases* as the binding energy of an amphiphile *increases*.

To evaluate the accuracy of Eq. (5.4), we calculate ρ_{agg} for two generic amphiphiles – single chain and double chain phospholipids. Both amphiphiles are taken to have 10 carbon atoms per hydrocarbon chain, corresponding to a length $n_c \ell_{cc}$ of 1.26 nm. With molecular masses in the range of 400 Da (single) and 570 Da (double), the phospholipids have length scales $h/\{2\pi m k_B T\}^{1/2}$ of 5.1×10^{-12} m (single) and 4.3×10^{-12} m (double) at $T=300$ K. To estimate E_{bind}, we use a surface tension of $\gamma = 0.05$ J/m^2 (this is γ for short alkanes and water; Weast, 1970), although γ may be as low as 0.02 J/m^2 for some lipids (Parsegian, 1966). The effective hydrocarbon radius of a single chain is 0.2 nm; the effective radius of a double chain lipid is less than twice that of a single chain, and we take $\sqrt{2} \cdot 0.2 = 0.3$ nm as a reasonable approximation. These values for γ, n_c and R_{hc} give $E_{bind}/k_B T = 20$ (single) and 30 (double), such that Eq. (5.4) yields

$$\rho_{agg} \text{ (single)} \sim 2 \times 10^{26}/\text{m}^3 \sim 0.3 \text{ molar}\qquad(5.5)$$
$$\rho_{agg} \text{ (double)} \sim 1.4 \times 10^{22}/\text{m}^3 \sim 2 \times 10^{-5} \text{ molar}.$$

We see immediately that the aggregation threshold for double-chain lipids is much less than that of single-chain lipids, and is low in an absolute sense; we return to the importance of this finding to properties of the cell at the end of this section. How does Eq. (5.5) compare to experiment? The observed values of the CMC of single- and double-chain lipids depend strongly on the head group and lie between 10^{-3} to a few times 10^{-2} molar for single-chain lipids and 10^{-5}–10^{-3}

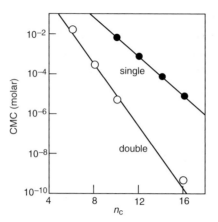

Fig. 5.9. Semilogarithmic plot of critical micelle concentration (molar) as a function of the number of carbon atoms n_c in an individual chain. Lyso-phosphatidylcholines (single chain) are indicated by the solid disks and di-acyl phosphatidylcholines (double chain) are indicated by open circles. The straight lines are the fitted functions 690 exp($-1.15n_c$) molar and 570 exp($-1.8n_c$) molar for the single- and double-chain lipids, respectively. Data compilations are from Marsh (1990) and Chapter 16 of Israelachvili (1991).

molar for dual-chain lipids, all with 10 carbon atoms per hydrocarbon chain. The values calculated in Eq. (5.5) are not in strong disagreement with the range observed experimentally.

A better test of the physics underlying Eq. (5.4) can be made by examining the chain length dependence of the CMC for a given type of lipid. Examples of both single- and double-chain lipids with the chemical composition R_nPC and R_nR_nPC are displayed in Fig. 5.9, where R_n is a linear hydrocarbon with n_c carbon atoms and PC refers to the phosphatidylcholine head group. The first thing to note about the figure is that the single-chain lipids have uniformly higher CMCs than double-chain lipids for the same number of carbon atoms per chain. This is as expected, because the symmetric dual-chain lipids in Fig. 5.9 have twice as many carbon atoms as their single chain counterparts, resulting in a larger hydrophobic area and hence a greater tendency to form aggregates at low concentration. The second feature is the exponential dependence of the CMC on the number of carbon atoms, namely 690 exp($-1.15n_c$) molar and 570 exp($-1.8n_c$) molar for the specific R_nPC and R_nR_nPC sequences in the figure. For each additional CH_2 group added to one of the chains, the CMC drops approximately by factors of 3 and 2.5 for the single and dual chain phospholipids in the figure, respectively. This exponential behavior of the CMC can be seen in Eq. (5.4) when it is written as

$$\rho_{agg} = [\{2\pi m k_B T\}^{1/2}/h]^3 \; e^{5/2} \exp(-2\pi n_c R_{hc}\ell_{cc}\gamma/k_B T), \qquad (5.6)$$

after substituting Eq. (5.1). The exponential factors 1.15 and 1.8 from the fits to the data in Fig. 5.9 can be compared directly to $2\pi R_{hc}\ell_{cc}\gamma/k_B T$ in Eq. (5.6). Taking the values for R_{hc}, ℓ_{cc} and γ used in the calculation of Eq. (5.5), the exponential slopes of Fig. 5.9 are predicted to be 2.0 and 3.0 respectively, about a factor of two larger than observed in the fit. Considering the approximations involved in obtaining Eq. (5.6), and the assumed values of the parameters, this prediction is not unimpressive. In particular, if the surface tension at the molecular scale is closer to 0.030 J/m², the predicted exponents are close to the measured values ($\gamma = 0.02$ J/m² has been obtained in an analysis of phase transitions in water + fatty-acid salt systems by Parsegian, 1966).

A more rigorous treatment of aggregation, which yields a result similar to Eq. (5.6) is performed in Chapter 16 of Israelachvili (1991), where the two-phase assumption (solution or aggregate) is dropped and a distribution of aggregate sizes is permitted. This approach allows one to calculate the average cluster size as a function of amphiphile concentration, and to show how the free amphiphile concentration in solution approaches its asymptotic value near the CMC. Readers familiar with the concepts of chemical equilibrium will not find this approach difficult. The phase behavior of multicomponent systems, say with more than one type of amphiphile, is more complex than the two-component system explored here; introductions to this topic can be found in Mouritsen (1984) and Gompper and Schick (1994).

Our model for aggregation considers only two states – a single cluster *vs.* an ideal solution – without regard to the shape of the cluster. However, we know experimentally that aggregates can have different geometries, including spherical or cylindrical micelles and bilayers. Is there a way of understanding which of these geometries is favored by a given amphiphile without having to calculate the free energy of each possible phase? Israelachvili *et al.* (1976) demonstrated the relationship between amphiphile shape and aggregate geometry by comparing the mean shape of the amphiphile with the optimal molecular packing arrangement in a cluster.

Using known carbon–carbon bond lengths and angles, a single saturated hydrocarbon chain with n_c carbon atoms has a volume of $v_{hc} = 27.4 + 26.9 n_c \times 10^{-3}$ nm³ (consistent with data from Lewis and Engelman, 1983) and a maximum chain length of $\ell_{cc} = 0.154 + 0.126 n_c$ nm (see Tanford, 1980). Note that the actual thickness of a single bilayer leaflet will be less than the length of a fully extended chain because of the presence of kinks (Lewis and Engelman, 1983; Bloom *et al.*, 1991, review bond orientation; Nagle and Tristram-Nagle, 2000, review lipid sizes). As n_c increases, the constant terms in the expressions for v_{hc} and ℓ_{hc} become less important, and the effective cross sectional area of the chain becomes $a_{hc} = v_{hc}/\ell_{hc} = 0.21$ nm². In a micelle or bilayer, the average surface area a_0 occupied by the head group of an amphiphile is greater than a_{hc} of a single chain, being near 0.5 nm² for many phospholipids. The optimal packing arrangement of molecules in a cluster reflects the difference between a_0 and the average cross sectional area of

Fig. 5.10. Packing constraints experienced by a typical amphiphile in four aggregates: (a) spherical micelle, (b) cylindrical micelle, (c) bilayer, and (d) inverted micelle. The cross sectional area of a head group is denoted by a_o, while the radius and thickness of the molecular shape are R and t, as applicable.

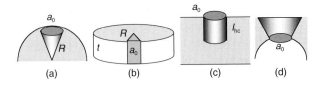

the hydrocarbon chain or chains, taking into account any kinks or coils in the chains. We now examine the molecular packing in several aggregate shapes.

Spherical micelles

As shown in Fig. 5.10(a), a spherical micelle of radius R has a surface area $4\pi R^2$ and volume $4\pi R^3/3$. Knowing the average surface area a_o occupied by a molecule, as well as the volume of its hydrocarbon chain(s) v_{hc}, one can deduce the number of molecules in the micelle as either $4\pi R^2/a_o$ or $(4\pi R^3/3)/v_{hc}$ from surface or volume considerations, respectively. For both of these expressions to give the same number of molecules, R must satisfy

$$R = 3v_{hc}/a_o. \tag{5.7}$$

Now, the molecules of a micelle are arranged so that their polar head groups reside on the exterior of the cluster, while the chains reside inside. The interior of the cluster cannot contain a void, so the distance from the centre to the surface cannot exceed the length of one amphiphile, as shown in Fig. 5.4(b). Thus, the radius of the micelle must be less than or equal to the projected length of the hydrocarbon chain ℓ_{hc} (shown in Fig. 5.1 for a saturated chain; note that a saturated chain would have to be placed under considerable tension to produce a completely linear chain from the zig-zag configuration displayed). Since $R \le \ell_{hc}$, Eq. (5.7) becomes

$$v_{hc}/a_o\ell_{hc} \le 1/3 \text{ (spherical micelles).} \tag{5.8}$$

According to this expression, spherical micelles are favored by large values of a_o compared to v_{hc}/ℓ_{hc}, corresponding to the conical shape shown in Fig. 5.10(a). The combination $v_{hc}/a_o\ell_{hc}$ is called the shape factor.

Cylindrical micelles

As the volume of the hydrocarbon region of an amphiphile increases for a given head group, perhaps because the chains are longer or because there are two chains instead of one, the micelle must distort from its spherical shape to a prolate ellipsoid (cigar-shaped). The length of the ellipsoid can increase to accommodate a larger internal volume until the micelle becomes a cylinder, of which Fig. 5.4(b) could represent a section perpendicular to the axis of cylindrical symmetry. To obtain the upper bound on the shape factor in the micelle domain, consider a section of a cylindrical micelle of thickness t as shown in Fig. 5.10(b), where the axis of cylindrical symmetry is vertical. The surface area of this section around the ring is $2\pi R t$, and the volume is $\pi R^2 t$. Repeating the same argument as was used for the number of molecules in a spherical micelle, we determine that the number of molecules in the cylindrical section is $2\pi R t / a_o$ according to area and $\pi R^2 t / v_{hc}$ according to volume. The radius must satisfy

$$R = 2 v_{hc} / a_o, \tag{5.9}$$

if these two expressions for the number of molecules are to be consistent. Substituting $R \le \ell_{hc}$, as before, leads to

$$1/3 < v_{hc} / a_o \ell_{hc} \le 1/2 \text{ (cylindrical micelles)} \tag{5.10}$$

for the range of shape factor favoring cylindrical micelles.

Bilayers

At even larger values of the internal volume than permitted by Eq. (5.10), cylindrical micelles distort towards bilayers by spreading so that they become elliptical in cross section rather than circular. The packing geometry of the bilayer is best satisfied by cylindrical molecules, whose hydrocarbon volume is given by $v_{hc} = a_o \ell_{hc}$, as shown in Fig. 5.10(c). Hence, the "ideal" bilayer satisfies $v_{hc} / a_o \ell_{hc} = 1$, and the range of the shape factor favored for bilayers is

$$1/2 < v_{hc} / a_o \ell_{hc} \le 1 \text{ (bilayers).} \tag{5.11}$$

Given that the typical value of a_o for phospholipids is 0.5 nm^2, about double the average cross sectional area of a single hydrocarbon chain, it is no surprise that dual chain phospholipids preferentially form bilayers compared with single-chain lipids.

Inverted micelles

Lastly, we consider what happens when the hydrocarbon volume v_{hc} is larger than the product of the head group area a_o with the chain length ℓ_{hc}, such that the shape factor $v_{hc} / a_o \ell_{hc}$ exceeds unity. Here, the head group is relatively small, and lies near the apex of a truncated cone as shown in Fig. 5.10(d); this situation is the inverse of the molecular shape favored by micelles. Such molecules form inverted micelles, with the head groups on the "inside" of the micelle, and the hydrocarbon regions radiating away from the aqueous core. Thus, the range of shape factor favoring inverted micelles is

$$v_{hc}/a_o\ell_{hc} > 1 \text{ (inverted micelles)}. \tag{5.12}$$

Amphiphiles in the cell

From Eqs. (5.8) to (5.12) we have established a correspondence between the value of the shape factor $v_{hc}/a_o\ell_{hc}$ and aggregate geometry preferred by an amphiphile:

0 to 1/3	spherical micelles
1/3 to 1/2	cylindrical micelles
1/2 to 1	bilayers
greater than 1	inverted micelles.

What does this imply for aggregates in the cell? Let's consider amphiphiles having sufficiently long hydrocarbon chains that their length and volume are linear functions with zero intercept, with $\ell_{hc} = 0.126n_c$ nm for chains in a linear zig-zag configuration (less for chains in a bilayer; see Lewis and Engelman, 1983) and $v_{hc} = 26.9n_c \times 10^{-3}$ nm³, where n_c is the number of carbon atoms in a single chain. These values imply a cross sectional area of 0.21 nm² for a single chain, consistent with observations from X-ray scattering on single layers of single-chain surfactants. The shape factor is then $0.21/a_o$ for a single chain amphiphile, and $0.42/a_o$ for a double chain, where a_o is quoted in nm². Head groups in the cell commonly have cross sectional areas near 0.50 nm², ranging from ~0.40 nm² for phosphatidylethanolamines to 0.5–0.7 nm² for phosphatidylcholines, depending upon their liquid crystalline phase (Huang and Mason, 1978; Lewis and Engelman, 1983; Nagle and Tristram-Nagle, 2000). From these estimates of ℓ_{hc}, a_o, and v_{hc}, the shape factor should be roughly 0.4 for single-chain lipids, and 0.8 for double-chain lipids. Thus, molecular packing arguments predict that single-chain lipids in the cell are most likely to be found in spherical or distorted micelles, and double-chain lipids probably form bilayers.

The fact that dual-chain lipids form bilayers has another implication for cells: we have established that such lipids also have a very low CMC, less than 10^{-5} molar for many lipids of interest. This characteristic is consistent with our design principles for the cell: a low CMC means that the concentration of lipids need not be very high before a bilayer will condense from solution, and that the concentration of lipids remaining in solution once the bilayer forms is fairly low. In other words, the building materials of the cell, once produced by metabolic or other means, are used very efficiently to build membranes.

5.3 Bilayer compression resistance

The molecular shape of an amphiphile influences the preferred geometry of its mesoscopic aggregate, as described in Section 5.2. The conformations of amphiphiles in an aggregate can be investigated by experimental techniques such as NMR (Marcelja, 1974; for a review, see Bloom *et al.*, 1991) and by computer simulations (for example, Pink *et al.*, 1980; van der Ploeg and Berendsen, 1983; Egberts and Berendsen, 1988; Cardini *et al.*, 1988; Heller *et al.*, 1993; for a review, see Pastor, 1994), from which the elastic behavior of membranes can be obtained (see Goetz and Lipowsky, 1998, Goetz *et al.*, 1999, and references therein). In this and the following section, we explore how the elasticity of a bilayer reflects its molecular constituents. Here, we present two models for the area compression resistance of a bilayer and evaluate them in light of recent experiment. We do not treat the thermal expansion of bilayers, but refer to Evans and Waugh (1977b), Evans and Needham (1987) and Marsh (1990) for further reading.

As a first and perhaps naïve approach, we view the bilayer as a homogeneous rigid sheet, much like a thin metallic plate in air. To find its area compression modulus K_A, consider what happens when a tensile stress is applied to the square plate of thickness d_p shown in Fig. 5.11. We impose two non-vanishing components on the stress tensor, $\sigma_{xx} = \sigma_{yy} = S$, which generate the in-plane elements of the strain tensor

$$u_{xx} = u_{yy} = S\,(2/9K_V + 1/6\mu), \tag{5.13}$$

where K_V and μ are the volume compression and shear moduli in three dimensions, respectively (see Eq. (D.29)). Now, the two-dimensional description of a deformation invokes an applied tension $\tau = Sd_p$ (force per unit length) which gives rise to a relative change in area $u_{xx} + u_{yy}$ according to $\tau = K_A(u_{xx} + u_{yy})$. Substituting Eq. (5.13), we find

$$K_A = d_p K_V / (4/9 + K_V/3\mu) \text{ (uniform rigid plate).} \tag{5.14}$$

For many materials, $K_V \sim 3\mu$, and the denominator is of order unity. The important result from this equation is that K_A increases linearly with plate thickness.

Fig. 5.11. Application of an in-plane tension $\tau = Sd_p$ to a uniform plate of thickness d_p.

Fig. 5.12. Dependence of the energy per molecule E at a water + amphiphile interface as a function of the mean interface area a occupied by a surfactant molecule. The equilibrium value of a, defined as a_o, occurs where dE/da vanishes (after Chapter 17 of Israelachvili, 1991).

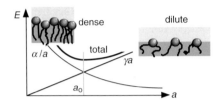

Useful as this approach might be, it ignores an important feature of amphiphilic condensates that their constituents rearrange easily in response to a stress. Consider the interface formed by water and a monolayer of amphiphiles, as in the insets of Fig. 5.12. The mean interfacial area occupied by an amphiphile reflects a competition between the steric repulsion among the molecules and the surface tension of the interface. At low density, as on the right-hand side of Fig. 5.12, the energy penalty per molecule for exposing the amphiphile's hydrocarbon chains is equal to γa, where a is the mean interface area occupied by an amphiphile and γ is the surface tension of the water + amphiphile interface. At high density, as on the left-hand side of Fig. 5.12, the molecules are crowded together, and their repulsive energy rises with density as a power of $1/a$. The simplest dependence of the interface energy per molecule on a is then $E = \alpha/a + \gamma a$, although higher order terms in a^{-1} become more important as a decreases (Chapter 17 of Israelachvili, 1991). As evident in Fig. 5.12, E has a minimum at the mean value of a at equilibrium, namely a_o. Setting $dE/da = 0$ gives $a_o = \sqrt{(\alpha/\gamma)}$, permitting E to be written as

$$E = 2\gamma a_o + (\gamma/a) \cdot (a - a_o)^2. \tag{5.15}$$

Only the second term changes as a departs from a_o.

Near equilibrium, the energy per amphiphile changes by $(\gamma/a_o) \cdot (a - a_o)^2$ relative to E at $a = a_o$, giving an elastic energy density of $\gamma[(a - a_o)/a_o]^2$ when divided by the area per molecule a_o. Now, this energy density also can be written as $(K_A/2) \cdot (u_{xx} + u_{yy})^2$, where $u_{xx} + u_{yy}$ equals the relative change in area, $(a - a_o)/a_o$. Comparing these expressions for the energy at small deformation, we find that K_A is 2γ for a monolayer, and 4γ for a bilayer:

$$K_A = 2\gamma \text{ (monolayer)}$$
$$K_A = 4\gamma \text{ (bilayer)}, \tag{5.16}$$

where the difference between the monolayer and bilayer arises because there are two interfaces to the latter. A polymer brush model for the bilayer gives the closely related expression for $K_A = 6\gamma$ (see Rawicz et al., 2000). Experimentally, the surface tension of the water + amphiphile interface is in the range 0.02–0.05 J/m² (see discussion in Section 5.2), so Eq. (5.16) predicts that K_A of a lipid bilayer should lie in the range 0.08–0.2 J/m^2, and should be independent of bilayer thickness.

Experimental measurements of K_A and K_V

The area and volume compression moduli have been measured for a number of lipid bilayers using variations of a technique in which membrane strain is recorded as a function of applied stress. In most cases, the membrane is a component of a cell or a pure bilayer vesicle, with linear dimension of the order of microns, and the deformation is observed microscopically; for example, the membrane may be placed under stress through aspiration by a micropipette with a diameter of a few microns (Evans and Waugh, 1977a; Kwok and Evans, 1981; Evans and Needham, 1987; Needham and Nunn, 1990; Evans and Rawicz, 1990; Zhelev, 1998). The stress-induced change in membrane area is extracted from microscope images, and K_A is found from a fit to the stress–strain curve. Images and schematic representations of aspirated cells appear in Sections 1.3 and 5.5. In a rather different approach first suggested by Parsegian *et al.* (1979), a stack of bilayers is placed under osmotic pressure causing the interface area per lipid to decrease, rather than increase as it does in the aspiration technique. A recent version of this technique (Koenig *et al.*, 1997) uses X-ray scattering and the ordering of hydrocarbon chains to determine independently the average interface area per lipid; parenthetically, interface areas per lipid of 0.59, 0.61 and 0.69 nm^2 are found for diMPC, SOPC and SDPC, respectively (definitions follow Table 5.2).

Apparent values of K_A from the slope of the stress–strain curve are displayed in Table 5.2 for a number of pure lipid bilayers; many of the measurements can be seen to lie in the range of 0.1–0.2 J/m^2, as expected from the surface tension argument of Eq. (5.16). However, there are two contributions to the apparent area compression resistance for soft membranes, one arising from a change in intermolecular separation (K_A) and one arising from membrane undulations. At low tension, the membrane may fluctuate transversely such that its average in-plane area is less than its true area, an effect with the same entropic roots as the reduction in the end-to-end displacement of a floppy chain compared to its contour length (see Chapter 2). Thus, when the membrane is placed under *small* tensions, its in-plane area increases because out-of-plane undulations are suppressed, not because the interfacial area increases (Helfrich and Servuss, 1984; Evans and Rawicz, 1990; Marsh, 1997). At larger tensions, once the undulations have been ironed out, the membrane area does increase, as governed by K_A (shown analytically in Chapter 6). When bending motion is accounted for, K_A of many lipids in Table 5.1 rises to about 0.24 J/m^2 even though the bilayer thickness varies by up

Table 5.2. *Selected measurements of the apparent area compression modulus K_A of lipid bilayers and cell membranes*

Membrane	T (C)	Apparent K_A (J/m²)	Reference
diAPC	15	0.057±0.014	Needham and Nunn, 1990
	18	0.135±0.020	Evans and Rawicz, 1990
	21	0.183±0.008	Rawicz *et al.*, 2000
diGDG	23	0.160±0.007	Evans and Rawicz, 1990
diMPC	21	0.150±0.014	Rawicz *et al.*, 2000
	29	0.145±0.010	Evans and Rawicz, 1990
	30	0.14	Koenig *et al.*, 1997
diOPC	21	0.237±0.016	Rawicz *et al.*, 2000
SDPC	30	0.12	Koenig *et al.*, 1997
SOPC	15	0.19±0.02	Needham and Nunn, 1990
	30	0.22	Koenig *et al.*, 1997
egg PC		0.14	Kwok and Evans, 1981
		0.17	Zhelev, 1998
red cell plasma membrane		0.45	Evans and Waugh, 1977

Notes:
For a summary of measurements up to 1990, see Table II.10.2 of Marsh (1990). Correcting for bending motion, the extracted value of K_A rises to about 0.24 J/m² for many phospholipids with 13–22 carbons atoms per hydrocarbon (Rawicz *et al.*, 2000).
Abbreviations: diAPC = diarachidonyl-phosphatidylcholine; diGDG = digalactosyl-diacylglycerol;
diMPC = dimyristoyl-phosphatidylcholine; diOPC = dioleoyl-phosphatidylcholine; SDPC = 1-stearoyl-2-docosahexaenoyl-phosphatidylcholine; SOPC = 1-stearoyl-2-oleoyl-phosphatidylcholine; egg PC = egg phosphatidylcholine.

to 25% as measured by X-ray scattering. These data favor the surface tension approach of Eq. (5.16), rather than the rigid plate model of Eq. (5.14); the former model predicts K_A is independent of membrane thickness d_p, while the latter predicts K_A increases linearly with d_p.

More chemically heterogeneous than a pure lipid bilayer is the plasma membrane of the red blood cell, which contains membrane-embedded proteins as well as ~40 mol % cholesterol. As can be seen from Table 5.2, this membrane has a significantly higher compression modulus than the pure lipid bilayers. In general, it is found that the presence of cholesterol increases the membrane's resistance both to

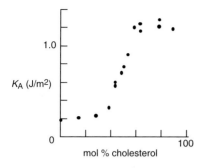

Fig. 5.13. Measured values of the area compression modulus K_A for SOPC bilayers containing cholesterol, as a function of cholesterol concentration expressed in mol % (redrawn with permission, after Needham and Nunn, 1990; © 1990 by the Biophysical Society).

compression and to bending (see Section 5.4), although the change in K_A as a function of the cholesterol fraction is not at all linear, as illustrated in Fig. 5.13 for an SOPC bilayer (Needham and Nunn, 1990). There is a gentle linear rise at low cholesterol concentrations, followed by a dramatic jump as the concentration passes through 50 mol %, reaching a plateau value that is about a factor of six larger than K_A of pure SOPC. Similar behavior is observed for mixtures of diMPC and cholesterol (Evans and Needham, 1987).

Lastly, we note that the volume compression modulus K_V has been measured for some phospholipids commonly used in the study of bilayers (for a summary, see Table II.10.1 of Marsh, 1990). For example, the compression modulus of diPPC (dipalmitoyl-phosphatidylcholine) is observed to be 2–3×10^9 J/m^3 at physiologically relevant temperatures (Mitaku *et al.*, 1978). These values of the bulk modulus are in the same range as everyday liquids such as water ($K_V = 1.9 \times 10^9$ J/m^3) and ethyl alcohol ($K_V = 1.1 \times 10^9$ J/m^3) at 0 °C and atmospheric pressure.

Other estimates of K_A

Other simple models for estimating the area compression modulus prove less accurate than the surface tension argument leading to Eq. (5.16). For example, K_A of a two-dimensional ideal gas is equal to the pressure of the gas, which in turn equals $\rho k_B T$, where ρ is the molecular number density. Viewing the interface as a two-dimensional gas, we would assign $\rho = a_o^{-1}$ for a monolayer or $2a_o^{-1}$ for a bilayer. Taking a typical value of a_o to be 0.5 nm^2 for phospholipids gives $\rho k_B T = 0.016$ J/m^2 for a bilayer, about an order of magnitude low compared with experiment. Alternatively, the thin plate model of Eq. (5.14) predicts $K_A \sim K_V d_p$, where d_p is the thickness of the plate. A typical value of K_V is 2–3×10^9 J/m^3 and a typical bilayer thickness is 4 nm, yielding ~ 10 J/m^3 for $K_V d_p$. This estimate is more than an order of magnitude larger than the measured values of K_A, indicating that the sliding of amphiphiles reduces the bilayer's compression resistance. In summary, the surface tension approach best captures both the bilayer thickness dependence and the absolute magnitude of K_A.

Rupture tension

As the bilayer is stretched, it becomes thinner and its hydrophobic core is increasingly exposed to water. At what point is the bilayer stretched so thin that it ruptures, releasing tension through the formation of a hole that permits the bilayer to return to near its optimal density? The answer to this question is time-dependent: it depends on how long you wait to see the membrane fail. On laboratory time scales, a membrane fails once its area is stretched just a few percent, often 2–5%, beyond its equilibrium value. That is, applying a small tension to the bilayer causes it to stretch, but applying a sufficiently large tension causes it to fail once its area has increased by a few percent. The corresponding tension at which failure occurs on laboratory time-scales, referred to as the lysis tension, is of the order 0.01 J/m², with significant variation. In general, the lysis tension increases with K_A of the membrane, and its magnitude can be estimated by noting that the tension is equal to the product of K_A and the critical strain at lysis, typically a few percent. We discuss membrane failure again in Section 5.5; the reader is referred to Needham and Nunn (1990) for further discussion.

5.4 Bilayer bending resistance

Displacing the amphiphiles of a bilayer from their mean or equilibrium positions and configurations requires an input of energy. One example of this resistance to deformation is encountered in the compression or stretch of a bilayer, as described in Section 5.3, where compression is opposed by steric interactions between amphiphiles and stretch is opposed by the water + amphiphile surface tension. Another example is the bending of a bilayer, which involves both a stretching of one leaflet, and a compression of the other, as seen in Fig. 5.6(b). In this section, we introduce the energetics of bending deformations, and both summarize and interpret experimental measurements of bending elastic moduli. We defer to Chapter 7 a discussion of membrane structure and its influence on elasticity and cell shape.

Fig. 5.14. Uniformly curved shells: (a) is a spherical shell of radius R, while (b) is a cylindrical shell of radius R and length L. Sets of vectors tangent to the surface are displayed for both configurations.

Radius of curvature

Consider the two uniformly curved surfaces in Fig. 5.14: a spherical shell and a hollow cylinder, both of radius R. For the time being, we define the

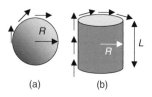

(a)   (b)

curvature C of a surface observed along a particular direction as the rate of change of a unit vector tangent to the surface, similar to our treatment of polymers in Section 2.2. For the sphere in Fig. 5.14(a), for instance, the rate of change of the unit tangent vector along a ring of radius R is equal to $1/R$; hence $C = 1/R$, where R must also be the radius of curvature. The sphere is a special case; in general, two curvature parameters are needed to describe a surface, as can be demonstrated from the cylinder in Fig. 5.14(b). Around the circumference of the cylinder, the tangent vectors change at the same rate as they do along an equator of the sphere, so the curvature is $1/R$. Yet parallel to the symmetry axis, the tangent vectors are parallel and the curvature vanishes, or alternatively, the radius of curvature is infinite. Thus, the cylinder is described by two inequivalent curvatures: $C_1 \equiv 1/R_1 = 1/R$ and $C_2 \equiv 1/R_2 = 0$. The curvature formalism for surfaces of arbitrary shape is developed in Chapter 6; in this section, we consider only simple shapes whose curvature is obvious from inspection.

For a given molecular composition, the energy per unit surface area to bend a bilayer increases with the curvature. The simplest form for this energy density involves the squared mean curvature $(C_1/2 + C_2/2)^2 = (1/R_1 + 1/R_2)^2/4$ and the Gaussian curvature $C_1 C_2 = 1/R_1 R_2$, where R_1 and R_2 are the two radii of curvature (Canham, 1970; Helfrich, 1973; Evans, 1974). Thus, we write the energy density $\mathscr{F}$ as

$$\mathscr{F} = (\kappa_b/2) \cdot (1/R_1 + 1/R_2)^2 + \kappa_G/(R_1 R_2), \tag{5.17}$$

where the material-specific parameters κ_b and κ_G are the bending rigidity and the Gaussian bending rigidity, respectively; both parameters have units of energy. The Gaussian term has the interesting property that its integral is invariant under shape deformation for a given topology; e.g., if a spherical shell is deformed into an ellipsoid, $\int (R_1 R_2)^{-1} dA$ doesn't change. As will be explained in Chapter 7, Eq. (5.17) can be modified to include bilayers whose native curvature is not zero. For the two shells in Fig. 5.14, the bending energy compared to a flat surface is

$$E = 4\pi(2\kappa_b + \kappa_G) \quad \text{(sphere)} \tag{5.18a}$$

$$E = \pi\kappa_b L/R \quad \text{(cylinder)}, \tag{5.18b}$$

according to Eq. (5.17). Note that the bending energy of a spherical shell is independent of its radius. Extensive measurements have been reported for κ_b, as we now summarize, but rather few for κ_G; in fact, some experiments find κ_G is negative (Section 4.5.9 of Petrov, 1999).

Experimental measurements of κ_b

As will be shown later, the bending resistance of a thin sheet is about a factor of ten less than the product of its area compression modulus K_A

with the *square* of its thickness. With $K_A \sim 0.2$ J/m^2 from Table 5.2, the very thinness of bilayers at 4×10^{-9} m means that their bending rigidity must be of the order 10^{-19} J, or some tens of $k_B T$. As a consequence, an isolated membrane readily undulates at room temperature unless its geometry is constrained by some means such as osmotic swelling. Analysis of a bilayer's thermal undulations is the heart of many measurements of the bending modulus, as pioneered by Brochard and Lennon (1975) and Servuss *et al.* (1976). A commonly employed technique examines the shape fluctuations of synthetic vesicles (pure bilayer membranes: Schneider *et al.*, 1984; Faucon *et al.*, 1989; Duwe *et al.*, 1990; Méléard *et al.*, 1997) or flat bilayer sheets (Mutz and Helfrich, 1990). Many of these experiments require the development of extensive theoretical machinery to analyse the amplitudes of the shape fluctuations and obtain a bending modulus. An alternate approach is based on the observation that the effective in-plane area of a membrane is reduced by thermal fluctuations, as described in Section 5.3, such that a tension applied to a membrane irons out these fluctuations first before increasing the true area of the bilayer. The apparent area compression modulus $K_{A, app}$ is given by (Evans and Rawicz, 1990; see Section 6.5)

$$K_{A, app} = K_A / [1 + K_A k_B T / (8\pi \kappa_b \tau)], \tag{5.19}$$

where τ is the applied tension. Eq. (5.19) becomes $K_{A, app} \sim K_A$ at low temperature or large tension, but is $K_{A, app} \sim 8\pi\kappa_b\tau/k_B T$ at small tension. Thus, one can extract κ_b from $K_{A, app}$ obtained from a stress/strain curve at small stress.

A selection of bending moduli obtained experimentally is displayed in Table 5.3, where it is seen that many of the measurements cluster in the range $\kappa_b \sim 10$–20 $k_B T$. The table also includes the red cell plasma membrane, whose bending modulus is similar to pure bilayers. However, this similarity probably doesn't hold for membranes in general: a recent study of *Dictyostelium discoideum* cells found bending moduli in the hundreds of $k_B T$ range, more than an order of magnitude larger than that of pure bilayers (Simson *et al.*, 1998). Similar to the behavior of the area compression modulus, κ_b is found to rise with cholesterol concentration. For example, both Duwe *et al.* (1990) and Méléard *et al.* (1997) find that κ_b of diMPC increases by a factor of three as the cholesterol concentration is raised from 0 to 30 mol%.

Interpretation of κ_b

Several models for the molecular arrangements in a bilayer permit the prediction of how κ_b depends on the bilayer thickness d_{bl}. As displayed in Fig. 5.6, the bending of a bilayer, or any plate for that matter, involves

Table 5.3. *Bending rigidity κ_b for selected lipid bilayers and cell membranes. Most measurements are made at 20–30 °C; definitions follow Table 5.2*

Membrane	κ_b $(\times 10^{-19} \text{ J})$	$(k_B T)$	Reference
diAPC	0.44±0.05	11	Evans and Rawicz, 1990
diGDG	0.44±0.03	11	Evans and Rawicz, 1990
	0.15–0.4		Duwe et al., 1990
	0.2±0.07	5	Mutz and Helfrich, 1990
diMPC	0.56±0.06	14	Evans and Rawicz, 1990
	1.15±0.15	29	Duwe et al., 1990
	0.46±0.15	11	Zilker et al., 1992
	1.30±0.08	33	Méléard et al., 1997
diMPE	0.7±0.1	18	Mutz and Helfrich, 1990
diOPC	0.85±0.10	21	Rawicz et al., 2000
SOPC	0.90±0.06	23	Evans and Rawicz, 1990
egg PC	1–2		Schneider et al., 1984
	0.4–0.5	11	Faucon et al., 1989
	1.15±0.15	29	Duwe et al., 1990
	0.8	20	Mutz and Helfrich, 1990
red blood cell	0.13–0.3	3–8	Brochard and Lennon, 1975
plasma membrane	1.3	32	Evans, 1983
	0.3–0.7	8–18	Duwe et al., 1990
	1.4–4.3	35–108	Peterson et al., 1992
	0.2±0.05	5	Zilker et al., 1992
	2.0	50	Hwang and Waugh, 1997

a compression of one surface and an extension of the other, the magnitude of the strain being governed by the area compression modulus K_A. We show in the problem set that the bending rigidity of a uniform plate obeys $\kappa_b = K_A d_{bl}^2/12$, which drops to $\kappa_b = K_A d_{bl}^2/48$ if the plate is sliced into two leaflets free to slide past each other. A polymer brush model for the bilayer predicts the intermediate result $\kappa_b = K_A d_{bl}^2/24$ (Rawicz et al., 2000). These models can be summarized by

$$\kappa_b = K_A d_{bl}^2/\alpha, \tag{5.20}$$

where the numerical constant $\alpha = 12$, 24 or 48 (see Section 17.9 of Israelachvili, 1991, for a slightly different functional dependence). If K_A is independent of d_{bl} for the bilayers of interest, as indicated in Section

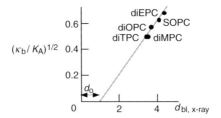

5.3, then Eq. (5.20) predicts $\kappa_b \propto d_{bl}^2$; a rigid plate, however, obeys $K_A \propto d_{bl}$, which leads to $\kappa_b \propto d_{bl}^3$.

The bilayer thickness dependence of κ_b has been tested by Rawicz *et al.* (2000) as displayed in Fig. 5.15. For a selection of lipids, the bilayer thickness $d_{bl, X\text{-ray}}$ was determined by X-ray scattering, although it is possible that there is an offset d_o between $d_{bl, X\text{-ray}}$ and the mechanically relevant d_{bl}. The straight line drawn through the data has $d_o = 1$ nm, and $\alpha = 24$ in Eq. (5.20); the data favor the d_{bl}^2 dependence of κ_b of Eq. (5.20) as well as the value of α from the polymer brush model. Just as the constancy of K_A argues against the rigid-plate model, so too is there little experimental support for the rigid-plate prediction $\kappa_b \propto d_{bl}^3$.

5.5 Edge energy

Self-assembly of micelles and bilayers in a water + amphiphile system is driven by the differential attraction of an amphiphilic molecule to both polar and non-polar environments. Further, the favored geometry of the condensed phase reflects, in part, a match between the mean shape of the amphiphilic molecule and the packing space within the aggregate. However, while we now understand why some amphiphiles condense into bilayers, we have not yet explained why a bilayer sheet with a free boundary closes up into the spherical topology of the cell's plasma membrane, an otherwise seamless surface punctuated by protein pores. Is the presence of a free edge on a bilayer unfavorable for some reason?

Two possible edge configurations that a bilayer might adopt are shown in Fig. 5.16. Configuration (a) is energetically costly because the hydrocarbon core is exposed to water at the boundary (Deryagin and Gutop, 1962; Helfrich, 1974a; Litster, 1975), in contrast to configuration

Fig. 5.16. Possible configurations that a bilayer may adopt at a free edge. In (a), the hydrocarbon chains are exposed to water, while in (b), the molecular packing is not matched to the shape of a cylindrical lipid.

(a)

(b)

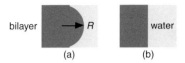

Fig. 5.17. Face view of two
bilayer–water boundaries with
different curvature, where the
darker region represents the
lipids (Fig. 5.16 is a cross
section or side view).

(b) which has the hydrocarbons hidden from water, at the price of introducing a curved surface. Now, we argued in Section 5.2 that bilayer-forming molecules tend to have a cylindrical shape, implying that the curved surface in Fig. 5.16(b) is not their favorite packing geometry. Both configurations in Fig. 5.16 suffer an energy penalty for creating a free edge, and we parametrize this penalty by an edge tension λ, an energy per unit length, analogous to the surface tension γ.

Characterizing the boundary by a fixed value of λ is an oversimplification. Consider the shape of the boundaries illustrated in Fig. 5.17, which are face views of a bilayer, Fig. 5.16 being side views. Even if the bilayer is planar, the surface at the boundary has two principal curvatures, which are reciprocals of the radii of curvature, as explained in Section 5.4. The curvature of the surface defined by the head groups along the cross section in Fig. 5.16(b), has a value of $2/d_{bl}$, where d_{bl} is the thickness of the bilayer. However, the curvature in the bilayer plane is different for the two configurations in Fig. 5.17: R^{-1} and 0 for (a) and (b), respectively, where R is the radius indicated. If the edge tension depends upon the mean curvature of the boundary region, it would have different values for the two configurations (Evans and Wortis, private communication). The curvature-dependence of edge tension has not yet been treated in the literature; here we assume that λ is curvature-independent.

In this section, we explore the implications of edge tension for several aspects of cell stability. First, we examine the competition between bending resistance and edge tension that compels a large, flat membrane to close into a sphere. Next, we calculate the energetics of hole formation in a planar membrane under tension. The analytical treatment of these topics is restricted to zero temperature, and we show by computer simulation what effects membrane and boundary fluctuations have at finite temperature. In closing Section 5.5, the measured values of the edge tension are summarized and interpreted.

Vesicle formation: edge energy vs. bending energy

To illustrate the energetics driving the closure of fluid membranes into vesicles and cells, we consider two sample shapes shown in Fig. 5.18: a closed spherical shell of radius R_V and a flat disk with a circular boundary (Helfrich, 1974a; Fromhertz, 1983). If the sphere and disk have the same area, the radius of the disk must be $2R_V$. Describing the bending

Fig. 5.18. Two idealized membrane shapes: (a) a spherical shell of radius R_V and (b) a flat disk with a circular boundary. The radius of the disk must be $2R_V$ if the disk and sphere have the same area.

energy of the membrane by the simple curvature model presented in Section 5.4, the energy E_{sphere} required to bend a flat membrane into the shape of a sphere is independent of the sphere radius, and is given by

$$E_{\text{sphere}} = 4\pi(2\kappa_b + \kappa_G), \qquad (5.21)$$

where κ_b and κ_G are the bending rigidities. This energy may be compared to that of a flat disk,

$$E_{\text{disk}} = 4\pi R_V \lambda, \qquad (5.22)$$

to which the only contribution is from the edge energy along the perimeter $4\pi R_V$. At zero temperature, the shape adopted by the membrane is the one with the lowest energy. If R_V is small, the disk is favored. However, the disk energy increases with perimeter, and the sphere becomes the preferred shape for radii above

$$R_V^* = (2\kappa_b + \kappa_G)/\lambda, \qquad (5.23)$$

although there may be an energy barrier between the disk and sphere configurations (see problem set).

At non-zero temperature, the curvature of the surface and the boundary fluctuate locally. Two snapshots from a computer simulation of a membrane with bending resistance at $T > 0$ are displayed in Fig. 5.19, and the undulations of the membrane are clearly visible. Because of their entropy, these shape fluctuations favor the "magic carpet" configuration of Fig. 5.19(a) over the handbag in Fig. 5.19(b); hence, λ must be larger to seal the free boundary at $T > 0$. Simulations show that the sheet closes only if λ exceeds a threshold value of

$$\lambda^* = 1.36 \, k_B T / b \text{ (three dimensions)} \qquad (5.24)$$

Fig. 5.19. Simulation of membranes with open (a) and closed (b) topologies. The membrane boundary is indicated by the white spheres (from Boal and Rao, 1992b).

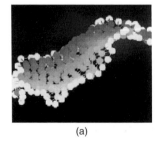 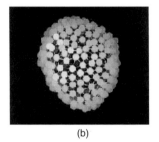

(a)   (b)

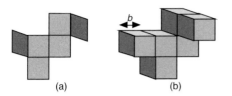

Fig. 5.20. Sample configurations of an open (a) and closed (b) self-avoiding surface on a cubic lattice. Each plaquette has length b to the side.

where b is a length scale from the simulation (Boal and Rao, 1992b). For the simulation of Fig. 5.19, b is set by the mean separation between vertices.

The minimum edge tension needed to close the boundary can be calculated within a lattice model where the surface is represented by a connected set of identical square plaquettes. Two sample configurations of a membrane on a lattice are displayed in Fig. 5.20: configuration (a) has a free boundary while configuration (b) is closed. For a fixed number of plaquettes N, Glaus (1988) has explicitly evaluated by computer the number of simply connected configurations C_N (where "connected" means no internal voids or gaps) for both ensembles, finding

$$C_N \cong 12.8^N \cdot N^{-1.48} \text{ (open)} \qquad (5.25)$$
$$C_N \cong 1.73^N \cdot N^{-1.51} \text{ (closed)}.$$

The mean perimeter Γ of the open configurations is close to $2Nb$, where b is the bond length, so that the free energy, $F = \Gamma\lambda - TS$, of the two ensembles is

$$F \cong 2N\lambda b - k_B T \ln C_N \cong 2N\lambda b - k_B TN \ln(12.8) \text{ (open)} \qquad (5.26a)$$

$$F \cong -k_B TN \ln(1.73) \text{ (closed)}. \qquad (5.26b)$$

Eqs. (5.26a) and (5.26b) are equal at the transition point, yielding $\lambda^* b = -0.5 k_B T \cdot \ln(1.73/12.8)$ or

$$\lambda^* = 1.0 \, k_B T / b, \qquad (5.27)$$

which is near the value observed in the fluid membrane simulation. We describe below how to relate b to a physical length scale of the membrane, and hence extract a physical bound on the edge tension.

Membrane rupture: edge energy vs. tension

Subjected to a small tensile stress compared to the area compression modulus, a bilayer will expand and sustain a strain of a few percent, as described in Section 5.3. However, if the stress is too large, the membrane will rupture. A simple model for membrane failure employs the creation of a circular hole in an incompressible fluid sheet (Deryagin and Gutop, 1962; Litster, 1975; for more general configurations, see

Fig. 5.21. (a) Enthalpy difference ΔH of a membrane +hole system compared to an unbroken membrane as a function of hole radius R at zero temperature. (b) Reduction in the free energy of a membrane +hole as the temperature increases.

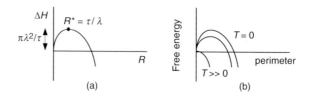

Netz and Schick, 1996; for a comparison with solid membranes, see Zhou and Joós, 1997). At zero temperature, the system acts to minimize its enthalpy H,

$$H = E - \tau A, \tag{5.28}$$

where τ is the two-dimensional tension ($\tau > 0$ is tension, $\tau < 0$ is compression). The energy of a circular hole in the sheet is $E = 2\pi R\lambda$, and the area difference of the sheet + hole system with respect to the intact sheet is just πR^2, where R is the radius of the hole. Hence, the difference in enthalpy ΔH of the membrane + hole system compared to the unbroken membrane is

$$\Delta H = 2\pi R\lambda - \tau \, \pi R^2. \tag{5.29}$$

For small holes, the term linear in R dominates Eq. (5.29) and ΔH grows with R for $\lambda > 0$. However, the term quadratic in R dominates for large holes, and ΔH becomes increasingly negative for membranes under tension, as shown in Fig. 5.21. The maximum value of ΔH occurs at a hole radius of

$$R^* = \tau/\lambda. \tag{5.30}$$

The physical meaning of R^* is that at zero temperature, holes with $R < R^*$ shrink, while those with $R > R^*$ expand without bound.

For a membrane under tension, Fig. 5.21(b) tells us that small holes are metastable in the sense that they do not represent configurations with the minimum free energy. Rather, there is an energy barrier of height $\pi\lambda^2/\tau$ (at zero temperature) which a hole must cross before it can expand without bound, as favored thermodynamically. At finite temperature, energy exchange with its environment permits a hole to form and cross the energy barrier (for discussion of the dynamics of hole formation, see Deryagin and Gutop, 1962; Wolfe *et al.*, 1985; Wilhelm *et al.*, 1993). Further, the entropy of a membrane + hole system at $T > 0$ lowers its free energy as shown in Fig. 5.21(b). Simulations not only confirm that holes form more easily as the temperature rises (or the edge tension decreases), they also show that, even at $\tau = 0$, there is a minimum λ required to keep the membrane from fragmenting (Shillcock and Boal,

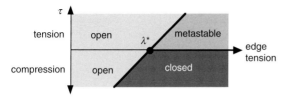

Fig. 5.22. Schematic phase diagram illustrating the stability of a two-dimensional membrane + hole system. Holes may arise from thermal fluctuations anywhere in the parameter space of the diagram, but are thermodynamically favored only in the "open" and "metastable" regions (after Shillcock and Boal, 1996).

1996). For planar membranes in two dimensions, this minimum edge tension λ^* is

$$\lambda^* = 1.66 \, k_B T / b \quad \text{(two dimensions)}, \tag{5.31}$$

where b is the mean separation between network points in the simulation (the points define the discrete membrane of the simulation; they appear as spheres in Fig. 5.19). This threshold λ is somewhat larger than what is obtained in three dimensions, Eq. (5.24). A schematic phase diagram found in the simulation of membrane rupture in two dimensions is shown in Fig. 5.22. Membranes are unstable against hole formation, even under compression, in the regions labelled "open", and have a finite lifetime in the "metastable" region. The threshold λ^* is clearly visible on the $\tau = 0$ axis. Systems with multiple holes have been investigated by simulation as well, although a consensus on the effects of multiple holes has not yet emerged (Nielsen, 1999; Shillcock and Seifert, 1998).

Measured edge tensions

Physical values of the edge tension have been obtained from several different techniques, which have in common the measurement of a stress and, directly or indirectly, a hole size. One approach uses the competition between bending resistance and edge tension in the formation of vesicles, as described earlier in this section. Fromhertz *et al.* (1986) measured the maximum size of disk-like membranes containing egg lecithin and taurochenodesoxycholate. The mixture was sonicated to destroy large-scale aggregates, then stained and imaged by electron microscopy. The images showed uniform disk-like membranes arranged in stacks, like columns of coins, with average radii of ~20 nm, and a distribution of small vesicles corresponding to disk radii of 30 nm. From these and related observations, the authors conclude that the edge tension of pure lecithin bilayers is 4×10^{-11} J/m, based upon an analysis similar to Eq. (5.23) to relate λ to the maximum disk size.

A second approach places a bilayer in a vesicle under large enough tension to cause pore formation. The earliest version of this technique

Fig. 5.23. In the experiments
of Zhelev and Needham
(1993), a pure bilayer vesicle
is placed under pressure by
aspirating it with a
micropipette. Applying an
electric field across the vesicle
thins the bilayer, creating a
hole.

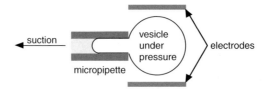

used osmotic pressure to swell a small vesicle (Taupin *et al.*, 1975), while a more recent refinement uses micromechanical manipulation of a vesicle (Zhelev and Needham, 1993). As illustrated in Fig. 5.23, a giant bilayer vesicle (25–50 μm in diameter) can be aspirated into a micropipette, and a hole created in the membrane by applying an electric field (see Chernomordik and Chizmadzhev, 1989). The hole is long-lived, an attribute discussed further in Moroz and Nelson (1997), and its radius R can be determined by measuring the rate of fluid loss through the opening. The tension τ experienced by the membrane can be determined from the known aspiration pressure. Balancing the force from the edge tension against that of the tensile stress (where the force $F = -dV/dx$ is obtained from the derivative of Eq. (5.29)) implies $\lambda = R/\tau$. The experiment yields $\lambda = 0.9 \times 10^{-11}$ J/m for pure stearoyl-oleoyl phosphatidylcholine (SOPC) and 3.0×10^{-11} J/m for SOPC with 50 mol% cholesterol. These results are similar to the osmotic swelling experiments of Taupin *et al.* (1975), who find $\lambda \sim 0.7 \times 10^{-11}$ J/m for dipalmitoyl phosphatidylcholine (diPPC).

How do these measured values of λ compare with the bound $\lambda^* > k_B T/b$ found in simulations of rupture at $T > 0$? The length scale b of the simulations is an effective length scale associated with the bending of the boundary, similar to the persistence length of polymer chains in Chapter 2, and is not necessarily a molecular dimension. The persistence length of the boundary is a measure of the distance over which the direction of the membrane boundary is decorrelated and must be at least equal to the membrane thickness. If the membrane has the cross section of Fig. 5.16(b), the length scale of the curved region in the bilayer plane must be greater than or equal to the inverse curvature perpendicular to the plane, namely $d_{bl}/2$. Taking $b \sim d_{bl} \sim 4 \times 10^{-9}$ m, the simulations predict that the edge tension must exceed $\lambda^* \sim 10^{-12}$ J/m at ambient temperatures for membrane stability ($k_B T = 4 \times 10^{-21}$ J at $T = 300$ K). The measured values of λ for bilayers in the cell exceed this estimate of λ^* by an order of magnitude, easily sufficient to make the bilayer resistant against rupture under moderate tensions, but not sufficient to enable it to operate at an elevated osmotic pressure. To withstand such pressures, a cell must augment its plasma membrane with a cell wall.

5.6 Summary

Membranes in the cell are based upon a lipid bilayer structure with thickness d_{bl} of about 4 nm and vanishing in-plane shear resistance, like a two-dimensional fluid. Belonging to a class of molecules called amphiphiles, membrane lipids have a relatively small electrically polar region called the head group, to which are indirectly attached one or more non-polar hydrocarbon chains. In aqueous solution, amphiphiles condense into aggregates if their concentration exceeds the so-called critical micelle concentration (CMC). In some models of micelle formation, the CMC is proportional to the product of a volume term $[\{2\pi m k_B T\}^{1/2}/h]^3$ and a probability factor $\exp(-E_{bind}/k_B T)$, where m and E_{bind} are the molecular mass and the energy required to remove the amphiphile from an aggregate, respectively (h and k_B are the usual Planck and Boltzmann constants). The more energy required to extract an amphiphile, the smaller the CMC, which, for the types of lipid found in membranes, may be as low as 10^{-5} molar (single hydrocarbon chain) and 10^{-10} molar (two chains). The aggregate geometry preferred by a lipid depends upon its molecular geometry as characterized by a shape factor $v_{hc}/a_o\ell_{hc}$, where a_o is the interface area occupied by the polar head group and v_{hc} and ℓ_{hc} are the volume and length of the hydrocarbon region, respectively. Geometrical packing arrangements argue that $v_{hc}/a_o\ell_{hc} \leq 1/3$ favors spherical micelles, while $1/3 < v_{hc}/a_o\ell_{hc} \leq 1/2$ favors cylindrical micelles. Bilayers are preferred in the range $1/2 < v_{hc}/a_o\ell_{hc} \leq 1$, and inverted micelles are the choice if $v_{hc}/a_o\ell_{hc}$ exceeds 1.

With their symmetry axis normal to the bilayer interface, amphiphiles occupy an interfacial area per molecule a_o that represents a competition between steric repulsion (at high density) and water/amphiphile surface tension γ (at low density). Viewed as an isotropic plate, a membrane is expected to obey $K_A \sim K_V d_{bl}$, where K_A and K_V are the area and volume compression moduli, respectively, and d_{bl} is the bilayer thickness. This prediction is not supported by experiment, however, and K_A is relatively constant at 0.2 J/m² for many bilayers, a behavior consistent with $K_A \sim (4\text{–}6)\gamma$ from models based on the surface tension (γ) of bilayers. Bending deformations of the membrane alter the area per molecule at an energy cost characterized by two material parameters: the bending rigidity κ_b and the Gaussian rigidity κ_G. The simplest energy density for bending has the form $(\kappa_b/2)\cdot(1/R_1 + 1/R_2)^2 + \kappa_G/R_1 R_2$, where the local shape of the membrane is described by the two radii of curvature R_1 and R_2. Typical values found experimentally for lipid bilayers are $\kappa_b \sim 0.2\text{–}1.0$ J $\sim 5\text{–}25\, k_B T$. Several models for bending deformations lead to the general expectation $\kappa_b = K_A d_{bl}^2/\alpha$ where α may be 12, 24 or 48 according to the model. Current experiment supports

$\kappa_b \propto d_{bl}^2$ (not d_{bl}^3 of rigid plates) and favors $\alpha = 24$ of a polymer brush model.

Under moderate tension, membrane strain is governed by the area compression modulus K_A. However, if the strain exceeds 2–4%, most membranes rupture, at a corresponding stress usually less than $K_A/20 \sim 10^{-2}$ J/m². Rupture is accompanied by the opening of a hole in the bilayer which, at its simplest level, can be characterized by an edge tension λ along its boundary. The bilayer boundary probably involves the formation of a smooth interface with the headgroups arranged in cross section like the letter U, leading to the crude estimate that $\lambda \sim \kappa_b/d_{bl}$, or $\lambda \sim 2 \times 10^{-11}$ J/m; in fact, many measurements of λ cluster around $1-4 \times 10^{-11}$ J/m. The edge energy is one of the forces driving an open membrane with a free boundary to close up into a sphere if its perimeter exceeds $(2\kappa_b + \kappa_G)/\lambda$, which is in the range of a few tens of nanometers for typical values of λ and κ_b. The simplest free energy density $\mathcal{F}$ describing the formation of a circular hole in a planar membrane under a tension τ at zero temperature is $\mathcal{F} = 2\pi R\lambda - \pi R^2 \tau$, which predicts that the thermodynamically favored state for $\tau > 0$ (tension) is a hole, although to reach this state, the system faces an energy barrier of $\pi \lambda^2/\tau$, often $10-100 \, k_B T$ or more. However, at finite temperature, the barrier is reduced, and vanishes for λ less than a critical value $\lambda^* \sim k_A T/b$, where b is a length scale characterizing the curvature of the boundary. As b is greater than d_{bl}, then the threshold λ^* must be less than 10^{-12} J/m, which is an order of magnitude below the physical values observed for the bilayers in a cell.

5.7 Problems

Biological applications

5.1

(a) Find the length of a saturated hydrocarbon chain as a function of the number of carbon atoms $n_c \geq 3$, assuming that all bond angles are 109.5° and the bond lengths and radii are:

$$C-C = 0.154 \text{ nm} \quad C-H = 0.107 \text{ nm}.$$

Take the chain to have a linear zig-zag configuration.

(b) Evaluate this length for $n_c = 14$ and 20, the range of importance for membrane lipids.

5.2 Evaluate S_{gas}/k_B per molecule for an ideal gas of hydrogen molecules at STP ($T = 273$ K and $P = 1$ atmosphere). Compare this contribution to the free energy (i.e. $TS_{gas}/k_B T$) with $PV/k_B T$ per molecule.

5.3 The CMCs of several single-chain lipids are

lauroyl PC	7.0×10^{-4} molar
myristoyl PC	7.0×10^{-5} molar
palmitoyl PC	7.0×10^{-6} molar

After determining the number of carbons for each hydrocarbon chain, use Eq. (5.6) to find the nominal surface tension γ of the hydrocarbon region at $T = 300$ K (see Table B.1). Take the lipid mass to be 500 Da.

5.4 Find the range of the head group area a_o expected for bilayer formation for lipids with 1, 2, or 3 chains. Assume that $v_{hc} = 0.2 \, \ell_{hc}$ for a single hydrocarbon chain, where v_{hc} and ℓ_{hc} are quoted in cubic nanometers and nanometers, respectively. Do the results of Koenig *et al.* (1997) agree with your predictions?

5.5 Suppose that the plasma membrane of a bacterium were constructed from an initially flat fluid membrane. How much energy (in units of $k_B T$) is required to bend this membrane into the shape of a spherocylinder 3 μm long and 1 μm in diameter? Assume $\kappa_b = \kappa_G = 10 k_B T$.

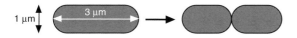

1 μm   3 μm

Now let this cell divide, keeping the volume and diameter the same. What is the total bending energy of the daughter cells?

5.6 Estimate the bending rigidity of the red blood cell cytoskeleton using $\kappa_b = K_A t^2 / 12$. Take the thickness t of the cytoskeleton to be 25 nm. Obtain K_A from the shear resistance $\mu = 10^{-5}$ J/m² by modelling the cytoskeleton as a triangular network of springs at zero temperature and zero stress (see Section 3.3). Quote your answer in $k_B T$ (at $T = 300$ K). Compare this value of κ_b with that of a typical lipid bilayer such as diMPC in Table 5.3 to determine the relative importance of the cytoskeleton to the bending resistance of the red cell.

5.7 Find the edge tension λ for the two bilayer configurations in Fig. 5.16 as a function of the bilayer thickness d_{bl}.

(a) In Fig. 5.16(a), assume the hydrocarbons form a planar surface with $\gamma = 0.050$ J/m² as the hydrocarbon/water surface tension.

(b) In Fig. 5.16(b), assign a curvature resistance $\kappa_b/2$ to the curved monolayer at the bilayer edge to find the bending energy of a cylindrical surface of radii of curvature $R_1 = d_{bl}/2$ and $R_2 = \infty$. Take $\kappa_b = 10\,k_BT$ (at $T = 300$ K).

(c) If the bilayer thickness is approximately $d_{bl} = 0.26 n_c$ nm, calculate λ from parts (a) and (b) for $n_c = 10$ and 20. Which configuration in Fig. 5.16 more closely represents experiment?

5.8 The disk and sphere configurations of a fluid membrane have the same energy at the special disk radius $2(2\kappa_b + \kappa_G)/\lambda$, but there is an energy barrier between them of $\Delta E = \pi(2\kappa_b + \kappa_G)$ (see Problem 5.13). The probability of a system to hop over an energy barrier is proportional to $\exp(-\Delta E/k_BT)$, among other terms. Evaluate this contribution to the hopping probability for $\kappa_b = \kappa_G = 10k_BT$, typical of a lipid bilayer. What does your result tell you about the likelihood of a disk-like bilayer closing up into a spherical cell or vesicle at this special radius?

5.9 Evaluate the enthalpy barrier against hole formation for a planar membrane under tension with $\lambda = 2 \times 10^{-11}$ J/m and $\tau = 10^{-3}$ J/m² and compare the result to k_BT. The probability that a hole could hop over the barrier at $T > 0$ is proportional to $\exp(-\Delta H/k_BT)$, among other factors. Calculate this contribution to the hopping probability for the value of ΔH you just obtained. At what value of τ is the barrier height equal to k_BT if $\lambda = 2 \times 10^{-11}$ J/m?

Formal development and extensions

5.10 Consider a thin solid sheet of thickness t characterized by a Young's modulus Y and a Poisson ratio σ_p. When subjected to a pressure P along its edges, it deforms according to a strain tensor whose non-vanishing components are (Section 5 of Landau and Lifshitz, 1986):

$$u_{xx} = u_{yy} = -(P/Y)(1 - \sigma_p) \quad u_{zz} = +2P\sigma_p/Y.$$

(a) Prove that the three-dimensional energy density of this deformation is $P^2(1 - \sigma_p)/Y$.

(b) Find the corresponding two-dimensional energy density in terms of K_A, Y and σ_p.

(c) Compare your results from (a) and (b) to establish

$$K_A = K_V t \cdot \alpha,$$

where K_V is the bulk modulus and $\alpha = 3(1-2\sigma_p)/2(1-\sigma_p)$.
(d) Evaluate α for $\sigma_p = 1/3$, which is a common value for many solids. At what value of σ_p does $\alpha = 1$?

You will need the (three-dimensional) relations $\mu = Y/2(1+\sigma_p)$ and $K_V = Y/3(1-2\sigma_p)$.

5.11. Obtain a relationship between the bending rigidity κ_b and the two-dimensional area compression modulus K_A by considering the gentle bending of a sheet into a section of a cylinder with principal curvatures $1/R$ and 0. Take the surface of zero strain (the neutral surface) to run through the middle of the sheet, and assume that the thickness is unchanged by the deformation.

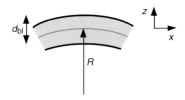

(a) Show that the energy density of the cylindrical deformation is $\mathcal{F} = \kappa_b/2R^2$.
(b) The strain u_{xx} varies linearly with z from 0 at the neutral surface to $\pm u_m$ at the boundary. Determine the average value of u_{xx}^2 across the membrane, and find the corresponding energy density if $\mathcal{F} \sim \langle u_{xx}^2 \rangle K_A/2$.
(c) For small deformations, demonstrate that $u_m = d_{bl}/2R$.
(d) Combine your results from (a) to (c) to obtain $\kappa_b = K_A d_{bl}^2/12$.

This problem is treated more thoroughly in Section 11 of Landau and Lifshitz (1986).

5.12 Use the result $\kappa_b = K_A d_{bl}^2/12$ from Problem 5.11 for a single uniform sheet to show that $\kappa_b = K_A d_{bl}^2/48$ if the sheet is sliced in two along the neutral surface, with each slice permitted to slide smoothly past the other.

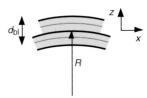

5.13 Consider an open membrane with the shape of a bowl of constant radius of curvature R. Defining the area of the surface to be πr_0^2, the perimeter of the bowl is equal to $2\pi r_0 (1 - r_0^2/4R^2)^{1/2}$ (a result which enthusiasts may wish to prove!).

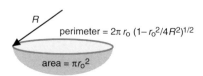

R

perimeter = $2\pi r_0 (1 - r_0^2/4R^2)^{1/2}$

area = πr_0^2

(a) Show that the contributions of the edge energy and bending energy give a total energy of

$$E = 2\pi \lambda r_0 (1 - r_0^2 C^2/4)^{1/2} + (2\kappa_b + \kappa_G)(\pi r_0^2 C^2),$$

where the curvature C is equal to $1/R$.

(b) The disk and sphere have the same energy at $r_0 = 2(2\kappa_b + \kappa_G)/\lambda$. Plot the reduced energy E/E_{sphere} at this value of r_0 from $C = 0$ (disk) to $C = 2/r_0$ (sphere), where $E_{\text{sphere}} = 4\pi(2\kappa_b + \kappa_G)$.

(c) At what value of C is the energy in (b) a maximum? Establish that $E_{\text{max}}/E_{\text{sphere}} = 5/4$ at this value. What does the fact that $E_{\text{max}} > E_{\text{sphere}}$ and $E_{\text{max}} > E_{\text{disk}}$ imply about the stability of the sphere and disk configurations?

5.14 Show that the barrier between open and closed configurations for the energy in Problem 5.13(a) disappears when the radius of the flat disk r_0 exceeds $4(2\kappa_b + \kappa_G)/\lambda$. Are disks with radii above this value stable, metastable or unstable? [*Hint: evaluate* dE/dC *and determine when* $dE/dC < 0$ *for very small* C (*Helfrich, 1974a, b*)].

5.15 Consider a closed shell in the shape of a pancake with rounded edges and dimensions as shown:

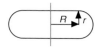

R r

(a) Prove that the area of the curved region is

$$2\pi r^2 (\pi \rho + 2),$$

where $\rho = R/r$.

(b) Find the energy density at the equator using Eq. (5.17).

(c) If your result from (b) applied to all curved surfaces (is this an underestimate or overestimate?) show that the corresponding bending energy of the shell would be

$$2\pi\,(\pi\rho+2)\cdot[(\kappa_b/2){\cdot}(\rho+2)^2+\kappa_G(\rho+1)]/(\rho+1)^2.$$

(The energy density is treated correctly in Chapter 7.)

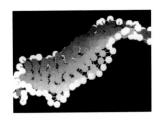

Chapter 6

Membrane undulations

Membranes of the cell are characterized by several elastic parameters, such as the area compression modulus, that reflect the membrane's quasi-two-dimensional structure. As described in Chapter 5, these parameters have small values for a lipid bilayer just 4–5 nm thick, yet they properly describe the energetics of membrane deformation at zero temperature where thermal fluctuations in shape are unimportant. But what happens at finite temperature? In the discussion of polymer and network elasticity in Part I, we saw that the entropic contribution to the elasticity of very flexible filaments is significant at ambient temperatures, owing to the large configuration space that these filaments can explore. Do we expect similar behavior for flexible sheets? In this chapter, we develop a mathematical description of surfaces, and explore the characteristics of membrane undulations. Membranes are treated in isolation here, and in interaction with other surfaces in Chapter 8. For a general review of membrane fluctuations, see Leibler (1989).

6.1 Thermal fluctuations in membrane shape

The bending rigidity κ_b of a phospholipid bilayer lies close to 10^{-19} J, or $10–20\ k_BT$ at ambient temperatures, as summarized in Table 5.3. What does such a small value of κ_b imply about the undulations of a membrane with the dimensions of a cell? For illustration, we calculate the change in energy of the flat, disk-shaped membrane in Fig. 6.1(a) as it is deformed into the surface of constant curvature in Fig. 6.1(b). Configuration (b) has radius of curvature R and energy density $2\kappa_b/R^2$,

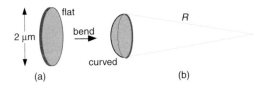

Fig. 6.1. Bending of a disk with diameter 2 μm (a) into a surface with a constant radius of curvature R (b). The surface normal vectors in (b) meet at an angle θ.

175

Fig. 6.2. Just as the
persistence length of a
polymer characterizes the
change in orientation of its
tangent vectors (a), so too
the persistence length of a
membrane characterizes the
change in orientation of the
surface normal vectors (b).

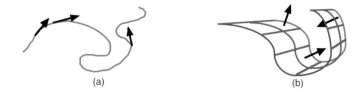

(a)   (b)

Fig. 6.2. Just as the persistence length of a polymer characterizes the change in orientation of its tangent vectors (a), so too the persistence length of a membrane characterizes the change in orientation of the surface normal vectors (b).

where the Gaussian rigidity is neglected (see Sections 5.4 and 6.2). Taking the disk diameter as 2 μm and $\kappa_b \sim 10\ k_B T$, the deformation energy is $20\pi k_B T/R^2$, where R must be quoted in microns. What is the typical value of R when the disk flexes at non-zero temperature? The thermal energy scale is set by $k_B T$ and the disk energy rises to this value (from zero for a flat disk with $R=\infty$) when R declines to ~ 8 μm. This radius of curvature corresponds to $\theta \sim 15°$ in Fig. 6.1(b), meaning that undulations may have measurable effects in cells of dimension 10 μm.

As introduced in Chapter 2, thermal fluctuations of flexible filaments are characterized by a temperature-dependent length scale called the persistence length ξ_p, equal to $\kappa_f/k_B T$, where κ_f is the flexural rigidity. Can membrane undulations be characterized by a persistence length in the same way as a filament? Inspection of Fig. 6.2(a) reminds us that the persistence length of a filament is a measure of the contour distance over which tangent vectors become decorrelated, as indicated by the directions of the arrows. For membranes, there is more than one tangent vector to a given surface element, and a better indicator of the surface orientation is provided by the normal vector. As shown in Fig. 6.2(b), normal vectors wobble and sway across a wavy surface, and their relative orientation over large length scales also can be described by a persistence length. However, for sheets, ξ_p depends exponentially on the bending rigidity as $\xi_p \sim b \exp(-4\pi\kappa_b/3k_B T)$, where b is an elementary length scale of the membrane. For example, if b is of the order of nanometers, and κ_b is $10k_B T$, then ξ_p is of the order of kilometers, implying that the undulations of cellular membranes are smooth. The exponential dependence of ξ_p on κ_b in membranes, vs. a linear dependence on κ_f in polymers, is not unexpected: a change in direction of a filament involves just a few neighboring segments, whereas a change in direction of a surface affects many adjacent elements in the membrane plane.

Flexible polymers exhibit a universal scaling behavior with contour length; for ideal polymers, the mean linear dimension (e.g., the end-to-end displacement) scales like $N^{1/2}$, where N is the number of segments on the chain (see Section 2.3). Is there analogous scaling behavior for membranes? The answer is a subtle "yes", with the scaling exponent depending on the self-avoidance and connectivity properties of the membrane. As exemplified in Fig. 6.3(a), most bilayers in the cell at

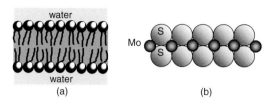

Fig. 6.3. Schematic cross section through two membranes: (a) a fluid bilayer with no shear resistance, (b) a polymerized sheet that resists in-plane shear (e.g., MoS$_2$). The thickness of the membrane is nanometers in (a) and ångstroms in (b).

physiologically relevant temperatures are fluid in the sense that neighboring lipid molecules are not covalently bound, and slide past each other when the membrane is subject to an in-plane shear. These fluid membranes can be contrasted with what we will refer to as polymerized membranes, such as the MoS$_2$ monolayer illustrated in Fig. 6.3(b), which acquires in-plane shear resistance from ionic or covalent bonds between neighboring atoms or molecules. Although either type of membrane may be viewed as a two-dimensional generalization of a one-dimensional polymer chain, in fact fluid and polymerized membranes each have distinctive geometrical and elastic properties.

As evidenced in many simulation studies, polymerized membranes are commonly flat over long length scales, meaning that transverse undulations decrease relative to the longitudinal dimension of the membrane (e.g., the square root of the area). In contrast, a fluid membrane behaves like a branched polymer at distances long compared to the membrane persistence length, developing wavy arms and protrusions. Such structures cannot be made from an initially flat polymerized membrane, in the same way as a piece of paper cannot be used to wrap an apple without developing creases and folds. Two sample configurations of polymerized and fluid membranes are displayed in Figs. 6.9 and 6.11; we caution that the complete geometry of membranes is more complicated than just these two phases (flat and branched polymer) as will be described later.

This chapter is somewhat more mathematical than others in this book, even though the formal proofs are kept to a minimum. In Section 6.2, some aspects of the mathematical definition and properties of surfaces are presented, including a simplified representation of a gently undulating surface. This formalism is applied immediately in Section 6.3 to obtain the persistence length of a fluid membrane. The mathematical machinery required to predict the geometrical scaling of membranes is too intricate for inclusion here, and so, in Section 6.4, we sketch out some simple expectations for scaling, which are then confirmed by computer simulation. Lastly, several experimental analyses of membrane undulations are reported in Section 6.5; the values of the elastic moduli deduced from these experiments can be found in Section 5.4.

6.2 Surface curvature

For a moment, let's return to the properties of a one-dimensional curve embedded in two dimensions. To give the curve a physical meaning, suppose that it is the trajectory of a ball thrown from one person to another, executing an arc as it travels. Measuring the height h of the ball as a function of the distance x along the ground allows us to construct the function $z = h(x)$ of the trajectory. Alternatively, if we know the time evolution of the trajectory, we can describe the motion by two parametric functions $x(t)$ and $h(t)$, where t is the time. The utility of each of these descriptions varies with the application. When we introduced polymers in Chapter 2, the parametric approach based on the arc length s provided simple expressions for geometrical quantities such as the unit tangent vector $\mathbf{t} = \partial\mathbf{r}/\partial s$ and the curvature $C = \mathbf{n}\cdot(\partial\mathbf{t}/\partial s) = \mathbf{n}\cdot(\partial^2\mathbf{r}/\partial s^2)$, where $\mathbf{n}$ is the normal to the curve at position $\mathbf{r}$.

The analytical description of a surface involves a choice of coordinates and representation (e.g., parametric). One coordinate convention uses a set of basis vectors embedded in the surface itself, much like the arc length along a curve, as indicated by the pair $\mathbf{u,v}$ in Fig. 6.4(a). However, a common approach in membrane studies employs Cartesian coordinates to express a point $\mathbf{r}$ on the surface by

$$\mathbf{r} = [x,\ y,\ h(x,y)], \tag{6.1}$$

where h is the "height" away from the xy plane, as illustrated in Fig. 6.4(a). Unfortunately, $h(x,y)$ may not be single-valued for general surfaces: for instance, $h(x,y)$ has three values in the overhang region of Fig. 6.4(b). In this text, we adopt the Monge representation in which $h(x,y)$ of Eq. (6.1) is constrained to be single-valued; i.e. overhangs are forbidden. This formalism allows us to establish results applicable at long length scales, so long as the surfaces are not too large compared with their persistence length.

Our first task is to express tangent vectors and normals within the Monge representation. Suppose that we take a slice of a surface along the x-direction (i.e. at constant y). We can construct a tangent vector $\mathbf{r}_x$ in this direction by making a unit step in the x-direction, and a step

Fig. 6.4. (a) A surface can be defined by a function $h(x,y)$ of a set of external Cartesian coordinates x,y or a set u,v resident on the surface itself. (b) The function $h(x,y)$ is not single-valued if overhangs are allowed.

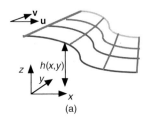

(a)

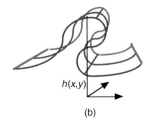

(b)

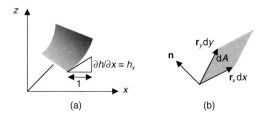

(a)   (b)

Fig. 6.5. (a) For a line tangent
to the surface and lying in the
xz-plane, a step $\Delta x = 1$
corresponds to $\Delta z = \partial h / \partial x$,
resulting in a tangent vector
$\mathbf{r}_x = (1, 0, \partial h / \partial x)$. (b) An area
element $dx\,dy$ on the xy-plane
corresponds to an element dA
$= |\mathbf{r}_x \times \mathbf{r}_y|\,dx\,dy$ on the surface.
The unit normal to the surface
is $\mathbf{n} = \mathbf{r}_x \times \mathbf{r}_y / |\mathbf{r}_x \times \mathbf{r}_y|$.

$\partial h / \partial x \equiv h_x$ in the z-direction, as shown in Fig. 6.5(a). Repeating this procedure for the y-direction gives the pair of tangent vectors

$$\mathbf{r}_x = (1, 0, h_x) \quad \mathbf{r}_y = (0, 1, h_y), \tag{6.2}$$

where $\mathbf{r}_x$ represents a vector, not the x-component of the position. These vectors are *not* unit vectors, and are *not* generally orthogonal; however, they do define the plane tangent to the surface at point $[x, y, h(x,y)]$, and can generate the unit normal vector $\mathbf{n}$ to the surface *via* the cross product

$$\mathbf{n} \equiv (\mathbf{r}_x \times \mathbf{r}_y)/|\mathbf{r}_x \times \mathbf{r}_y| = (-h_x, -h_y, 1)/(1 + h_x{}^2 + h_y{}^2)^{1/2}. \tag{6.3}$$

As shown in Fig. 6.5(b), a segment of length dx along the x-axis corresponds to a vector $\mathbf{r}_x dx$ along the surface. The cross product of the vectors $\mathbf{r}_x dx$ and $\mathbf{r}_y dy$ gives the area element dA on the surface, which is, from Eq. (6.2),

$$dA = |\mathbf{r}_x \times \mathbf{r}_y|\,dx\,dy = (1 + h_x{}^2 + h_y{}^2)^{1/2}\,dx\,dy. \tag{6.4}$$

The quantity $(1 + h_x{}^2 + h_y{}^2)$ is called the metric g of the surface

$$g \equiv (1 + h_x{}^2 + h_y{}^2), \tag{6.5}$$

such that $dA = \sqrt{g}\,dx\,dy$. A more thorough description of the metric and parallel transport on curved surfaces can be found in texts such as Goetz (1970) or Kreyszig (1959); we return to the metric momentarily.

The curvature of a surface can be obtained from the same definition as that for a line, namely $C = \mathbf{n} \cdot (\partial^2 \mathbf{r} / \partial s^2)$, where s is the arc length. However, the curvature of a surface is not unique, and a direction must be specified for the path on a surface. For example, the curvature of a cylindrical shell of radius R is zero in a direction parallel to the cylindrical axis, and $1/R$ along a circle perpendicular to the axis. One could choose other directions in this example as well, each having a different curvature. The extremal values of the curvature as a function of direction are called the principal curvatures, denoted here by C_1 and C_2; the combinations $(C_1 + C_2)/2$ and $C_1 \cdot C_2$ are the mean and Gaussian curvatures, respectively. Unfortunately, it takes some mathematical effort to obtain expressions for the curvatures; here, we forgo some elegant results from differential geometry and work strictly within the Monge representation.

With Eq. (6.3) for $\mathbf{n}$ now in hand, we need an expression for the vector $\mathbf{r}''$, where primes denote differentiation with respect to s: $\mathbf{r}' = \partial\mathbf{r}/\partial s$ and $\mathbf{r}'' = \partial^2\mathbf{r}/\partial s^2$. Starting with

$$\mathbf{r}' = \mathbf{r}_x x' + \mathbf{r}_y y', \tag{6.6}$$

where $x' = \partial x /\partial s$ and $\mathbf{r}_x = \partial\mathbf{r}/\partial x$ as in Eq. (6.2), the second derivative must be

$$\mathbf{r}'' = \mathbf{r}_{xx}(x')^2 + \mathbf{r}_{yy}(y')^2 + 2\mathbf{r}_{xy}x'y' + \mathbf{r}_x x'' + \mathbf{r}_y y'', \tag{6.7}$$

where $\mathbf{r}_{xx} \equiv \partial\mathbf{r}_x/\partial x$, etc. The vector character of Eq. (6.7) should be clear from the bold symbols. Our goal is to evaluate $C = \mathbf{n}\cdot\mathbf{r}''$, to which the last two terms in Eq. (6.7) do not contribute, owing to the orthogonality of $\mathbf{r}_x$ and $\mathbf{r}_y$ with $\mathbf{n}$. Thus

$$\mathbf{n}\cdot\mathbf{r}'' = b_{xx}(x')^2 + b_{yy}(y')^2 + 2b_{xy}x'y'. \tag{6.8}$$

The scalar coefficients $b_{\alpha\beta}$ are

$$b_{\alpha\beta} \equiv \mathbf{n}\cdot\mathbf{r}_{\alpha\beta} \quad (\alpha, \beta = x, y), \tag{6.9}$$

where we note that the subscripts on $b_{\alpha\beta}$ are indices, whereas the subscripts on $\mathbf{r}_\alpha$ refer to derivatives. In the literature, the coefficients b_{uu} are sometimes replaced by $b_{uu} = L$, $b_{uv} = M$, $b_{vv} = N$ in notation introduced by Gauss (see Chapter IV of Kreyszig, 1959), where u and v are a general coordinate set, as in Fig. 6.4(a). Differentiating the orthogonality condition $\mathbf{n}\cdot\mathbf{r}_\alpha = 0$ leads to the relation $\mathbf{n}\cdot\mathbf{r}_{\alpha\beta} + \mathbf{n}_\beta\cdot\mathbf{r}_\alpha = 0$, or, from the definition (6.9),

$$b_{\alpha\beta} = -\mathbf{r}_\alpha\cdot\mathbf{n}_\beta, \quad (\alpha, \beta = x, y), \tag{6.10}$$

where $\mathbf{n}_x \equiv \partial\mathbf{n}/\partial x$, etc. To summarize, Eqs. (6.8) and (6.10) provide a formal way of obtaining the curvature $C = \mathbf{n}\cdot\mathbf{r}''$ from a knowledge of $\mathbf{r}$.

The direction-dependence of the curvature is implicit in Eq. (6.8) through $\mathbf{r}''$, the second derivative of the position with respect to arc length along a particular direction. The maximum and minimum values of $\mathbf{n}\cdot\mathbf{r}''$ are the principal curvatures, and can be obtained from Eq. (6.8) using the method of Lagrange multipliers (see Section 1.5 of Safran, 1994) or other techniques (see David, 1989, or Chapter IV of Kreyszig, 1959). The mathematics is something of a digression, so we simply quote the results for the mean and Gaussian curvatures:

$$(C_1 + C_2)/2 = (g_{xx}b_{yy} + g_{yy}b_{xx} - 2g_{xy}b_{xy})/2g \tag{6.11}$$

$$C_1 C_2 = (b_{xx}b_{yy} - b_{xy}{}^2)/g, \tag{6.12}$$

where g is the metric familiar from Eq. (6.5). The metric is the determinant of the metric tensor $g_{\alpha\beta}$, whose components here are

$$g_{xx} = \mathbf{r}_x\cdot\mathbf{r}_x \qquad g_{xy} = g_{yx} = \mathbf{r}_x\cdot\mathbf{r}_y \qquad g_{yy} = \mathbf{r}_y\cdot\mathbf{r}_y. \tag{6.13}$$

As with the tensor $b_{\alpha\beta}$, the subscripts on $g_{\alpha\beta}$ are indices; they do not refer to derivatives. It is straightforward to verify Eq. (6.5) by substituting Eq. (6.2) into (6.13). In the general coordinate set $(\mathbf{u},\mathbf{v})$ the element of arc length squared is written in terms of the metric tensor as

$$ds^2 = g_{uu}(du)^2 + g_{vv}(dv)^2 + 2g_{uv}du\,dv. \tag{6.14}$$

The tensors $g_{\alpha\beta}$ and $b_{\alpha\beta}$ are referred to as the first and second fundamental forms, respectively, in some texts on differential geometry.

We now determine the mean and Gaussian curvatures in the Monge representation. First, we evaluate $\mathbf{n}_\alpha$ using Eq. (6.3), finding

$$\mathbf{n}_x = -\{ \ ([1+h_y^2]h_{xx} - h_x h_y h_{xy}), \ ([1+h_x^2]h_{xy} - h_x h_y h_{xx}),$$
$$(h_x h_{xx} + h_y h_{xy}) \ \} \ / \ (1+h_x^2+h_y^2)^{3/2}. \tag{6.15}$$

$$\mathbf{n}_y = -\{ \ ([1+h_y^2]h_{xx} - h_x h_y h_{yy}), \ ([1+h_x^2]h_{yy} - h_x h_y h_{xy}),$$
$$(h_x h_{xy} + h_y h_{yy}) \ \} \ / \ (1+h_x^2+h_y^2)^{3/2}.$$

These somewhat intimidating relations can be combined with Eqs. (6.2) and (6.10) to yield the simple

$$b_{xx} = h_{xx}/(1+h_x^2+h_y^2)^{1/2}$$
$$b_{xy} = b_{yx} = h_{xy}/(1+h_x^2+h_y^2)^{1/2} \tag{6.16}$$
$$b_{yy} = h_{yy}/(1+h_x^2+h_y^2)^{1/2}.$$

Lastly, Eqs. (6.13) and (6.16) can be substituted into (6.11) and (6.12) to find the mean and Gaussian curvatures:

$$(C_1+C_2)/2 = \{(1+h_x^2)h_{yy} + (1+h_y^2)h_{xx} - 2h_x h_y h_{xy}\}/2(1+h_x^2+h_y^2)^{3/2} \tag{6.17}$$

$$C_1 C_2 = (h_{xx}h_{yy} - h_{xy}^2)/(1+h_x^2+h_y^2)^2. \tag{6.18}$$

For many situations of interest, $h(x,y)$ is a slowly varying function corresponding to gentle undulations, such that the curvatures can be approximated by

$$(C_1+C_2)/2 \cong (h_{xx}+h_{yy})/2 \tag{6.19}$$

$$C_1 C_2 \cong h_{xx}h_{yy} - h_{xy}^2, \tag{6.20}$$

if we retain just the leading-order terms from Eqs. (6.17) and (6.18).

To confirm the geometrical meaning of Eqs. (6.19) and (6.20) consider the shape of a surface in the region of a minimum or inflection point, as illustrated in Fig. 6.6. For notational simplicity, place the minimum directly over the coordinate origin at $x=y=0$. A Taylor series expansion describes the height of the surface for small regions near the minimum, namely

$$h(x,\ y) = h_o + h_x x + h_y y + h_{xx}x^2/2 + h_{yy}y^2/2 + h_{xy}xy + \dots, \tag{6.21}$$

Fig. 6.6. The derivatives h_x
and h_y vanish at the minimum
or inflection points in these
two shapes. Part (a) is a
section of a sphere where the
principal curvatures have the
same sign; (b) is a saddle
region where the curvatures
have opposite signs: $R_u \cdot R_v < 0$.

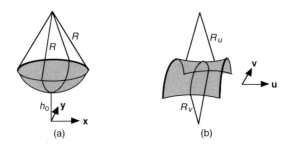

Fig. 6.6. The derivatives h_x and h_y vanish at the minimum or inflection points in these two shapes. Part (a) is a section of a sphere where the principal curvatures have the same sign; (b) is a saddle region where the curvatures have opposite signs: $R_u \cdot R_v < 0$.

where the height at the origin is h_o. For surfaces like Fig. 6.6(a), the derivatives h_x and h_y vanish at the local minimum in height, showing that the surface is quadratic in x and y with coefficients proportional to the local curvature. Eqs. (6.19) and (6.20) also establish that principal curvatures need not have the same sign. If Fig. 6.6(a) is a section of a sphere, for instance, the principal curvatures have the same sign and are both equal to $1/R$ in magnitude, where R is the radius of the sphere. In contrast, the curvatures at the saddle point in Fig. 6.6(b) have opposite sign: the normals are canted towards each other in the **u**-direction, but away from each other in the **v**-direction; that is $R_u \cdot R_v < 0$, where R_u and R_v are the radii of curvature in the **u** and **v** directions, respectively.

The simplest form for the energy density of bending deformations involves the mean and Gaussian curvatures, usually written in the combination (Canham, 1970; Helfrich, 1973; Evans, 1974)

$$\mathcal{F} = (\kappa_b/2)(C_1 + C_2 - C_0)^2 + \kappa_G C_1 C_2, \tag{6.22}$$

where κ_b and κ_G are the bending rigidity (or bending modulus) and Gaussian bending rigidity (or saddle-splay modulus) respectively. Introducing the constant C_0, known as the spontaneous curvature, permits Eq. (6.22) to describe bilayers that are curved in their equilibrium state because their two (monolayer) leaflets are compositionally inequivalent. In most of our discussions, C_0 will be set equal to zero. Now, it is easy to show that Eq. (6.22) has a particularly appealing form when $\kappa_b = -\kappa_G$ (at $C_0 = 0$), namely

$$\mathcal{F} = (\kappa_b/2)(\mathbf{n}_x^2 + \mathbf{n}_y^2), \tag{6.23}$$

in the limit of gentle undulations (Kantor, 1989; Nelson, 1989); we remind the reader that $\mathbf{n}_x$ and $\mathbf{n}_y$ have units of inverse length. This form of the energy density is a natural generalization of the bending resistance of a flexible filament, introduced in Section 2.2. Whether physical bilayers are described by $\kappa_b = -\kappa_G$ is not resolved; the existing measurements are consistent with the rather broad range $0 > \kappa_b/\kappa_G > -2$ (see Section 4.5.9 of Petrov, 1999). Fortunately, the integral over the surface of the Gaussian curvature does not change with the shape of the surface

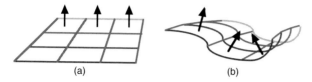

Fig. 6.7. The orientation of a surface normal is independent of position on a planar surface (a) but changes with position on a rough surface (b).

as long as the topology of the surface is fixed, courtesy of the Gauss–Bonnet theorem, and so our ignorance of κ_G does not impede our describing shape changes of cells with fixed topology.

6.3 Membrane bending and persistence length

At zero temperature, a membrane minimizes its bending energy by adopting a shape that is flat or uniformly curved, unless factors like compositional inhomogeneity force it to do otherwise. However, at finite temperature, the membrane exchanges energy with its environment, allowing it to explore a configuration space that includes energetically expensive bending deformations. The presence of wrinkles and ripples cause the orientation of the surface to change with position, as illustrated in Fig. 6.7. Mathematically, we represent the local orientation of a surface at position $\mathbf{r}$ by a normal vector $\mathbf{n}(\mathbf{r})$. For the flat configuration in Fig. 6.7(a), the normals are parallel such that the scalar product $\mathbf{n}(\mathbf{r}_1){\cdot}\mathbf{n}(\mathbf{r}_2)$ is independent of positions $\mathbf{r}_1$ and $\mathbf{r}_2$. For the wavy configuration in Fig. 6.7(b), the normals change with position and $\mathbf{n}(\mathbf{r}_1){\cdot}\mathbf{n}(\mathbf{r}_2)$ follows suit.

The scalar product of the normals provides a measure of the surface fluctuations. For nearby positions on the surface, $\mathbf{n}(\mathbf{r}_1)$ and $\mathbf{n}(\mathbf{r}_2)$ are roughly parallel and $\mathbf{n}(\mathbf{r}_1){\cdot}\mathbf{n}(\mathbf{r}_2)$ should not be too much less than unity. In contrast, if the positions are far apart, $\mathbf{n}(\mathbf{r}_1)$ and $\mathbf{n}(\mathbf{r}_2)$ may point in such different directions that $\mathbf{n}(\mathbf{r}_1){\cdot}\mathbf{n}(\mathbf{r}_2)$ may not even be positive. Thus, the average of the normal-normal product, $\langle\mathbf{n}(\mathbf{r}_1){\cdot}\mathbf{n}(\mathbf{r}_2)\rangle$, taken over the entire surface, should decrease from a value near unity at small separations to near zero at large separations. The spatial decorrelation of the normals is characterized by the persistence length ξ_p through

$$\langle\mathbf{n}(\mathbf{r}_1){\cdot}\mathbf{n}(\mathbf{r}_2)\rangle = \exp(-\Delta r/\xi_p), \tag{6.24}$$

where $\Delta\mathbf{r} \equiv \mathbf{r}_1 - \mathbf{r}_2$. This formalism is the same as the persistence length of a polymer, as described in Chapter 2, except that tangent vectors $\mathbf{t}$ are used for polymers whereas normals $\mathbf{n}$ are used for surfaces.

The persistence length of a membrane depends upon its bending resistence κ_b, just as ξ_p of a polymer depends upon the filament's flexural rigidity. Obtaining a mathematical formula for ξ_p for a surface is rather involved; the steps we follow are:

- transform $\langle \mathbf{n}(\mathbf{r}_1) \cdot \mathbf{n}(\mathbf{r}_2) \rangle$ to the Fourier representation $h(\mathbf{q})$ of the height function
- find $h(\mathbf{q})$ in terms of κ_b and temperature
- solve for ξ_p.

The proof is presented in detail because we draw intermediate results from it when discussing membrane undulations elsewhere in the text.

Fourier components of $\langle \mathbf{n}(\mathbf{r}_1) \cdot \mathbf{n}(\mathbf{r}_2) \rangle$

As demonstrated in Eq. (6.3), the surface normal is $(-h_x, -h_y, 1) \cdot (1 + h_x^2 + h_y^2)^{-1/2}$ in the Monge representation. Using the binomial approximation $(1 + x)^\alpha \sim 1 + \alpha x$ at small x permits $\mathbf{n}$ to be expressed to order h^2 as

$$\mathbf{n}(\mathbf{r}) = (-h_x, -h_y, 1 - [h_x^2 + h_y^2]/2). \tag{6.25}$$

Accordingly, the product $\mathbf{n}(\mathbf{r}_1) \cdot \mathbf{n}(\mathbf{r}_2)$ must be

$$\mathbf{n}(\mathbf{r}_1) \cdot \mathbf{n}(\mathbf{r}_2) = h_x(\mathbf{X}_1) h_x(\mathbf{X}_2) + h_y(\mathbf{X}_1) h_y(\mathbf{X}_2)$$
$$+ 1 - [h_x(\mathbf{X}_1)^2 + h_y(\mathbf{X}_1)^2]/2 - [h_x(\mathbf{X}_2)^2 + h_y(\mathbf{X}_2)^2]/2 + O(h^3)$$

$$= 1 - (\partial/\partial x[h(\mathbf{X}_1) - h(\mathbf{X}_2)])^2/2 - (\partial/\partial y[h(\mathbf{X}_1) -$$
$$h(\mathbf{X}_2)])^2/2 + O(h^3), \tag{6.26}$$

also to order h^2, where $\mathbf{X}$ is a two-dimensional vector comprising the xy coordinates of the position $\mathbf{r}$. It is convenient to express $\mathbf{n} \cdot \mathbf{n}$ in terms of the Fourier representation of the height function. For readers who have not seen the Fourier representation before, the idea is to expand an arbitrary function $f(x)$ in terms of a series of standard functions such as trigonometric functions. For example, the linear function $f(x) = x$ on the domain $-1 < x < 1$ can be written as a sum of sine functions multiplied by coefficients b_j

$$f(x) = \Sigma_j b_j \sin(j\pi x) \quad b_j = 2(-1)^{j+1}/j\pi, \tag{6.27}$$

where the sum runs from $j = 1$ to infinity (see Chapter 1 of Powers, 1999). For most numerical problems, the series is truncated at the accuracy appropriate to the situation. Functions other than sines can be used in Fourier representations, although something like $\cos(j\pi x)$ obviously won't work in Eq. (6.27) because our function $f(x)$ vanishes at $x = 0$ whereas $\cos(j\pi x)$ does not.

The height function $h(\mathbf{X})$ can be written in terms of a set of complex functions $\exp(i\mathbf{q} \cdot \mathbf{X}) = \cos(\mathbf{q} \cdot \mathbf{X}) + i \sin(\mathbf{q} \cdot \mathbf{X})$ as

$$h(\mathbf{X}) = (A / 4\pi^2) \int d\mathbf{q} \, \exp(i\mathbf{q} \cdot \mathbf{X}) h(\mathbf{q}), \tag{6.28}$$

where $\mathbf{q}$ is a wavevector with units of inverse length and A is the membrane area. Having units of length, the function $h(\mathbf{q})$ is the continuous analog of the discrete coefficients b_j in Eq. (6.27). Both $\mathbf{X}$ and $\mathbf{q}$ are two-dimensional quantities, as must be $d\mathbf{q}$ in the integral. We first evaluate $\partial/\partial x[h(\mathbf{X}_1)-h(\mathbf{X}_2)]$ and its square in the Fourier representation before attempting to transform Eq. (6.26):

$$\partial/\partial x[h(\mathbf{X}_1)-h(\mathbf{X}_2)]=(A/4\pi^2)\,\partial/\partial x\,\{\int d\mathbf{q}\,\exp(i\mathbf{q}\cdot\mathbf{X}_1)h(\mathbf{q})$$
$$-\int d\mathbf{q}'\exp(i\mathbf{q}'\cdot\mathbf{X}_2)h(\mathbf{q}')$$
$$=(A/4\pi^2)\int iq_x\,d\mathbf{q}\,[\exp(i\mathbf{q}\cdot\mathbf{X}_1)-\exp(i\mathbf{q}\cdot\mathbf{X}_2)]\,h(\mathbf{q}).$$
$$(6.29)$$

We have permitted the functions $h(\mathbf{q})$ in Eq. (6.28) to be complex, so we must take the complex square of Eq. (6.29) to ensure that Eq. (6.26) is real:

$$|\partial/\partial x[h(\mathbf{X}_1)-h(\mathbf{X}_2)]|^2=(A/4\pi^2)^2\int d\mathbf{q}'q_xq_x'\,d\mathbf{q}\,d\mathbf{q}'\,[\exp(i\mathbf{q}\cdot\mathbf{X}_1)$$
$$-\exp(i\mathbf{q}\cdot\mathbf{X}_2)]\cdot[\exp(-i\mathbf{q}'\cdot\mathbf{X}_1) \qquad (6.30)$$
$$-\exp(-i\mathbf{q}'\cdot\mathbf{X}_2)]\,h(\mathbf{q})h^*(\mathbf{q}'),$$

where $h^*(\mathbf{q})$ is the complex conjugate of $h(\mathbf{q})$.

Now, the expectation $\langle\mathbf{n}(\mathbf{r}_1)\cdot\mathbf{n}(\mathbf{r}_2)\rangle$ in Eq. (6.24) involves both an average over thermally allowed configurations, and an average over the positions $\mathbf{r}_1$ and $\mathbf{r}_2$ within each configuration, subject to the constraint $\mathbf{r}_1-\mathbf{r}_2=\Delta\mathbf{r}$. The positional average of Eq. (6.30) can be accomplished by integrating with $A^{-1}\int d\mathbf{X}_1\,d\mathbf{X}_2\,\delta(\mathbf{X}-[\mathbf{X}_1-\mathbf{X}_2])$, leading to

$$\langle|\partial/\partial x[h(\mathbf{X}_1)-h(\mathbf{X}_2)]|^2\rangle_{\text{pos}}=[A/(4\pi^2)^2]\int d\mathbf{X}_1\,d\mathbf{X}_2\,\delta(\mathbf{X}-[\mathbf{X}_1\text{-}\mathbf{X}_2])$$
$$\cdot\int q_xq'_x\,d\mathbf{q}\,d\mathbf{q}'\,[\exp(i\mathbf{q}\cdot\mathbf{X}_1)-\exp(i\mathbf{q}\cdot\mathbf{X}_2)]$$
$$\cdot[\exp(-i\mathbf{q}'\cdot\mathbf{X}_1)-\exp(-i\mathbf{q}'\cdot\mathbf{X}_2)]\,h(\mathbf{q})h^*(\mathbf{q}').$$
$$(6.31)$$

Integrating the delta function $\delta(\mathbf{X}-[\mathbf{X}_1-\mathbf{X}_2])$ over $\mathbf{X}_2$ has the effect of replacing $\mathbf{X}_2$ by $\mathbf{X}_1-\mathbf{X}$; collecting terms in $\mathbf{X}_1$ yields

$$\langle|\partial/\partial x[h(\mathbf{X}_1)-h(\mathbf{X}_2)]|^2\rangle_{\text{pos}}=[A/(4\pi^2)^2]\int d\mathbf{X}_1\int q_xq'_x\,d\mathbf{q}\,d\mathbf{q}'$$
$$\exp(i[\mathbf{q}-\mathbf{q}']\cdot\mathbf{X}_1)\cdot[1-\exp(-i\mathbf{q}\cdot\mathbf{X})]$$
$$[1-\exp(i\mathbf{q}'\cdot\mathbf{X})]\,h(\mathbf{q})h^*(\mathbf{q}'). \qquad (6.32)$$

The integral over $\mathbf{X}_1$ involves just the exponential $\exp(i(\mathbf{q}-\mathbf{q}')\cdot\mathbf{X}_1)$, and can be replaced by the two-dimensional generalization of $(2\pi)^{-1}$ $\int\exp(iqx)\,dx=\delta(q)$ to give

$$\langle|\partial/\partial x[h(\mathbf{X}_1)-h(\mathbf{X}_2)]|^2\rangle_{\text{pos}}=(A/4\pi^2)\int q_xq'_x\,d\mathbf{q}\,d\mathbf{q}'\,\delta(\mathbf{q}-\mathbf{q}')$$
$$\cdot[1-\exp(-i\mathbf{q}\cdot\mathbf{X})][1-\exp(i\mathbf{q}'\cdot\mathbf{X})]\,h(\mathbf{q})h^*(\mathbf{q}'),$$
$$(6.33)$$

or, after a trivial integration over $\mathbf{q}'$

$$\langle|\partial/\partial x[h(\mathbf{X}_1)-h(\mathbf{X}_2)]|^2\rangle_{\text{pos}}=(A/4\pi^2)\int q_x^2\,d\mathbf{q}\,[2-2\cos(\mathbf{q}\cdot\mathbf{X})]$$
$$h(\mathbf{q})h^*(\mathbf{q}), \qquad (6.34)$$

where we have used the trigonometric identity $[\exp(ix)-1]\cdot[\exp(-ix)-1]=2-2\cos(x)$. Eq. (6.34) and the corresponding expression for $\langle|\partial/\partial y[h(\mathbf{X}_1)-h(\mathbf{X}_2)]|^2\rangle_{pos}$ give the full thermal average of Eq. (6.26):

$$\langle \mathbf{n}(\Delta\mathbf{r})\cdot\mathbf{n}(0)\rangle = 1-(A/4\pi^2)\int q^2\,d\mathbf{q}\,[1-\cos(\mathbf{q}\cdot\mathbf{X})]\,\langle h(\mathbf{q})h^*(\mathbf{q})\rangle, \quad (6.35)$$

which can be evaluated knowing the thermal average of the Fourier components.

Fourier components and the equipartition theorem

Let's consider a membrane under tension such that a given configuration has an energy E governed by the applied tension τ and bending rigidity κ_b

$$E = \tau \int dA + (\kappa_b/2)\int (C_1+C_2)^2 dA, \quad (6.36)$$

where we assume that the topology of the surface is fixed so that the Gaussian curvature term is simply a constant. The tension opposes the creation of new contour area compared to the projected area on the xy plane. In the Monge representation, the surface area element dA is $(1 + h_x^2+h_y^2)^{1/2}d\mathbf{X}$ from Eq. (6.4), and the sum of the curvatures is $C_1+C_2 = h_{xx}+h_{yy}$ from Eq. (6.19). Expanding the square root of the metric, and keeping only terms of order h^2, the energy of a configuration can be written as

$$E = \int d\mathbf{X}\,(\tau/2)(h_x^2+h_y^2)+(\kappa_b/2)(h_{xx}+h_{yy})^2, \quad (6.37)$$

where an overall additive constant of the form $\tau\int d\mathbf{X}$ has been neglected.

As before, we use the Fourier representation $h(\mathbf{X})=(A/4\pi^2)\int d\mathbf{q}$ $\exp(i\mathbf{q}\cdot\mathbf{X})h(\mathbf{q})$ to write the surface tension term as

$$h_x(\mathbf{X})^2 + h_y(\mathbf{X})^2 = [A/(4\pi^2)]^2 \int [q_x q'_x + q_y q'_y]d\mathbf{q}d\mathbf{q}'$$

$$\exp(i[\mathbf{q}-\mathbf{q}']\cdot\mathbf{X})h(\mathbf{q})h^*(\mathbf{q}'), \quad (6.38)$$

where the left-hand side implicitly involves a complex square. This expression is integrated over $d\mathbf{X}$ in Eq. (6.37), allowing us to replace $(2\pi)^{-2}\int d\mathbf{X}$ $\exp(i[\mathbf{q}-\mathbf{q}']\cdot\mathbf{X})$ by $\delta(\mathbf{q}-\mathbf{q}')$. Then integrating over $\mathbf{q}'$ gives the simple

$$\int d\mathbf{X}\,[h_x(\mathbf{X})^2 + h_y(\mathbf{X})^2] = (A^2/4\pi^2)\int q^2 d\mathbf{q}\,h(\mathbf{q})h^*(\mathbf{q}). \quad (6.39)$$

These steps can be repeated for $(h_{xx}+h_{yy})^2$ to give

$$\int d\mathbf{X}\,[h_{xx}(\mathbf{X})+h_{yy}(\mathbf{X})]^2 = (A^2/4\pi^2)\int q^4 d\mathbf{q}\,h(\mathbf{q})h^*(\mathbf{q}). \quad (6.40)$$

Substituting Eqs. (6.39) and (6.40) into (6.37) yields

$$E = (1/2)\cdot(A^2/4\pi^2)\int d\mathbf{q}\,(\tau q^2 + \kappa_b q^4)h(\mathbf{q})h^*(\mathbf{q}), \quad (6.41)$$

up to an overall additive constant, as before.

Our next task is to find the thermal expectation $\langle h(\mathbf{q})h^*(\mathbf{q})\rangle$ given an energy of the form Eq. (6.41). We proceed by recalling the behavior of a one-dimensional harmonic oscillator, as described in Appendix C. This system is governed by the energy $E(x)=k_{sp}x^2/2$, and its thermal average is

$$\langle E\rangle=k_{sp}\langle x^2\rangle/2=(k_{sp}/2)\int x^2\exp(-E/k_BT)dx/\int\exp(-E/k_BT)dx.$$

$$(6.42)$$

Integrating this expression gives $\langle x^2\rangle=k_BT/k_{sp}$, and, consequently, $\langle E\rangle=k_BT/2$. This latter result arises from the equipartition of energy: each oscillation mode has an average energy of $k_BT/2$. Our surface energy Eq. (6.41) can be looked upon as a generalization of the harmonic oscillator. The factor $\int d\mathbf{q}$ is a sum over oscillator modes, with one mode per $(2\pi)^2/A$ in $\mathbf{q}$-space, each with an energy $(A/2)(\tau q^2+\kappa_b q^4)h(\mathbf{q})h^*(\mathbf{q})$. Equating the average energy of an individual mode with $k_BT/2$ gives

$$\langle h(\mathbf{q})h^*(\mathbf{q})\rangle=k_BT/A\cdot(\tau q^2+\kappa_b q^4).\qquad(6.43)$$

which demonstrates that $\langle|h(q)|^2\rangle$ increases with temperature.

Persistence length

We can now obtain the persistence length by substituting Eq. (6.43) into Eq. (6.35). The resulting integral is

$$\langle\mathbf{n}(\Delta\mathbf{r})\cdot\mathbf{n}(0)\rangle=1-(k_BT/4\pi^2)\int q^2\,d\mathbf{q}\,[1-\cos(\mathbf{q}\cdot\mathbf{X})]/(\tau q^2+\kappa_b q^4),\quad(6.44)$$

where the normalization areas have cancelled. We evaluate this in the tensionless limit $\tau=0$, or

$$\langle\mathbf{n}(\Delta\mathbf{r})\cdot\mathbf{n}(0)\rangle=1-(k_BT/4\pi^2\kappa_b)\int d\mathbf{q}\,[1-\cos(\mathbf{q}\cdot\mathbf{X})]/q^2.\qquad(6.45)$$

This is a two-dimensional integral with $d\mathbf{q}=q\,dq\,d\theta$, where θ is the angle between $\mathbf{q}$ and $\mathbf{X}$, or $\mathbf{q}\cdot\mathbf{X}=q|\mathbf{X}|\cos\theta$. Defining the θ part of the integral to be $I_\theta(z)$, we have

$$\langle\mathbf{n}(\Delta\mathbf{r})\cdot\mathbf{n}(0)\rangle=1-(k_BT/2\pi\kappa_b)\int I_\theta(z)\,dq/q,\qquad(6.46)$$

with $z=q|\mathbf{X}|$ and

$$I_\theta(z)=1-(2\pi)^{-1}\int d\theta\cos(z\cos\theta),\qquad(6.47)$$

where the integral runs from 0 to 2π. This integral appears frequently in physical problems, and is one of a class of functions called Bessel functions; this particular one is defined as $J_0(z)\equiv(2\pi)^{-1}\int d\theta\cos(z\cos\theta)$. Bessel functions do not have an analytical form, but they do have approximate representations (see Chapter 9 of Abramowitz and Stegun, 1970) which are used to construct the graph of $I_\theta(z)$ in Fig. 6.8. One sees

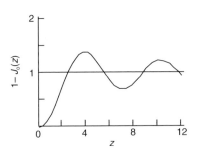

Fig. 6.8. Behavior of the integral $I_\theta(z)=1 - J_0(z)$ as a function of the argument z, where J_0 is a Bessel function.

that $I_\theta(z)$ rises from 0 at $z=0$, passes by unity at $z\sim 2$–3 after which it oscillates ever closer to 1 as z increases. Hence,

- at large z, $I_\theta(z)\sim 1$ and can be ignored in Eq. (6.46); at large q, the integral is truncated by the intermolecular spacing b
- at small z, $I_\theta(z)\sim 0$, and the integral is truncated at $q\sim 2$–$3/\Delta r$.

With q ranging from approximately $\pi/\Delta r$ to π/b, Eq. (6.46) becomes

$$\langle \mathbf{n}(\Delta\mathbf{r})\cdot\mathbf{n}(0)\rangle \sim 1 -(k_BT/2\pi\kappa_b) \int_{\pi/\Delta r}^{\pi/b} dq/q \sim 1-(k_BT/2\pi\kappa_b)\ln(\Delta r/b).$$
(6.48)

How do we relate this to the persistence length? The correlation function approaches zero (or $1/e$, strictly speaking) at distances close to the persistence length ξ_p. The condition $\langle \mathbf{n}(\Delta\mathbf{r})\cdot\mathbf{n}(0)\rangle\sim 0$ in Eq. (6.48) is satisfied by

$$\xi_p \sim b \exp(2\pi\kappa_b/k_BT) \quad \text{(de Gennes and Taupin),}$$
(6.49)

a result first obtained by de Gennes and Taupin (1982). A more detailed treatment by Peliti and Leibler (1985), based on a formalism called the renormalization group, leads to a similar expression with a numerical factor in the exponent which is 2/3 of that in Eq. (6.49),

$$\xi_p \sim b \exp(4\pi\kappa_b/3k_BT) \quad \text{(Peliti and Leibler).}$$
(6.50)

Computer simulations (Gompper and Kroll, 1995) are consistent with Eq. (6.50). The exponential dependence of ξ_p on the bending resistance of a surface should be contrasted with the linear dependence of ξ_p on the bending resistance of a polymer.

As demonstrated by Eq. (6.50), the persistence length decreases as the temperature rises, meaning that the undulations have shorter wavelengths. Put another way, it becomes easier to bend a membrane as the temperature increases, an observation that can be quantified by the q-dependent effective bending rigidity

$$\kappa(q) = \kappa_b + (3k_BT/4\pi\kappa_b)\ln(q/b),$$
(6.51)

as introduced by Peliti and Leibler (1985; see also Helfrich, 1986). As q^{-1} is less than b, the logarithm in Eq. (6.51) is negative and $\kappa(q)$ decreases from its zero-temperature value κ_b. This relation also shows that the surface becomes softer when viewed at longer wavelengths. What does Eq. (6.50) tell us about the persistence lengths of lipid bilayers? Numerically, if we set $b \sim 1$ nm, and $\kappa_b \sim 10k_B T$, we expect $\xi_p \sim 10^6$ km. Dramatic as this result is, it does not mean that the membranes of a cell are planar, only that they undulate smoothly on cellular length scales.

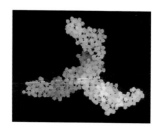

Fig. 6.9. Computer simulation of a fluid membrane with no bending resistance other than self-avoidance; beads and tethers represent the surface. This membrane has the topology of a sphere (no holes or handles) with a surface area that hovers close to a fixed mean and an enclosed volume that fluctuates freely (Boal and Rao, 1992a).

6.4 Scaling of polymers and membranes

If they are sufficiently large, self-avoiding fluid membranes appear like the branched polymers described in Section 2.3, in that they form dangling arms, such as illustrated in Fig. 6.9. In contrast, a fluid sheet without self-avoidance adopts exotic configurations in which the floppy arms can intersect each other, creating a topologically daunting labyrinth of tubes, pockets and chambers. Do membrane shapes have universal characteristics when taken as an ensemble, like the configurations of ideal polymers, branched polymers etc.? We know that polymer shapes obey a scaling law that depends upon the polymer's interaction and structure: for instance, the linear size of an ideal unbranched polymer scales like $(L_c/\xi_p)^{1/2}$, where L_c is the contour length and ξ_p the persistence length. In analogy, how does the linear size of a membrane scale with its area or, in the case of enclosed surfaces, its volume? Not unexpectedly, distinct classes of membranes display different scaling behavior.

Membranes without self-avoidance

We can use the squared radius of gyration R_g^2 to represent the linear size of a square membrane of contour area L^2; for instance, a flat membrane obeys $R_g^2 \sim L^2$. Subject to thermal fluctuations, the size of a fluid membrane without self-avoidance (called a *phantom* membrane) grows very slowly with contour area, obeying (see Nelson, 1989, for a review)

$$R_g^2 \propto \ln L, \tag{6.52}$$

when the reduced bending resistance $\kappa_b/k_B T$ is small. This result can be proven using a Fourier decomposition of the surface (reproduced in Section 8.5 of Plischke and Bergersen, 1994) and is easy to verify by computer simulation (Kantor et al., 1986, 1987). We conclude that the geometry of membranes can be characterized by observables such as R_g, although the power-law exponents, and even the functional form of the scaling, may be very different from polymers.

Flory scaling of polymers and membranes

Self-avoidance tends to increase the linear size of both chains and sheets. The effects are strongest when the embedding dimension is low; for instance, because it cannot double-back on itself, a self-avoiding chain in one dimension looks like a rod and obviously obeys $R_g \sim L_c^{1}$, not the scaling $R_g \sim L_c^{1/2}$ of ideal chains. We summarized the scaling behavior of polymers in Table 2.1, but only derived the scaling exponents of the simplest systems. In 1953, Flory introduced a generic model for self-avoiding polymers that used energy and entropy arguments to successfully predict their scaling behavior (see Chapter X of Flory, 1953). Here, we obtain the Flory scaling exponent for polymers, then quote its extension to membranes. The proof has two components:

- determining the dependence of the free energy on the chain's linear size r and number of segments N
- finding the value of r that minimizes the free energy as a function of N.

Following the presentation of de Gennes (1979; see also McKenzie, 1976), the behavior of the free energy is evaluated in two regimes.

(i) *Short distances* Steric repulsion between segments causes a chain to swell compared to an ideal chain. The repulsive energy experienced by one segment through its interaction with others is proportional to the product of the concentration of segments and the volume v_{ex} within which they interact. Approximating the segment concentration by N/r^d for a chain in a d-dimensional space, the total repulsive energy experienced by all N segments is proportional to

[*number of segments*] · [*interaction volume*]
· [*concentration of segments*] $= v_{ex} N^2/r^d$.

If v_{ex} represents a hard-core steric interaction which defines only a length scale $v_{ex}^{1/3}$, then the energy scale of the interaction is $k_B T$. Thus, the steric contribution to the free energy at short distances behaves like

$$F_{short} = k_B T\, v_{ex}\, N^2/r^d, \tag{6.53}$$

where all constants, including factors of two from double-counting the interacting pairs, have been absorbed into v_{ex}.

(ii) *Long distances* As an ideal chain is stretched, the probability of finding a given end-to-end distance r decreases exponentially as $\exp(-dr^2/2Nb^2)$, where b is the elementary segment length (see Section 2.4). Recalling that the entropy S is proportional to the logarithm of the probability, then to within a constant

$$S/k_B = -\, dr^2/2Nb^2. \tag{6.54}$$

Since the free energy is related to the entropy through $F = E - TS$, the entropic contribution to the free energy at long distances is

$$F_{long} = + k_B T dr^2 / 2Nb^2. \qquad (6.55)$$

Combining Eqs. (6.53) and (6.55) and discarding overall normalization constants, the free energy of the self-avoiding chain can be approximated by

$$F \cong k_B T v_{ex} N^2 / r^d + k_B T dr^2 / 2Nb^2, \qquad (6.56)$$

showing that there are energy penalties for pushing the chain elements close together (small r) or for stretching out the chains (large r). The value of r that minimizes F can be found by equating to zero the derivative of Eq. (6.56) with respect to r, holding other quantities fixed. This minimization shows that r scales like

$$r \sim N^{3/(2+d)}. \qquad (6.57)$$

Eq. (6.57) applies to any measure of the linear dimension of the system as a whole, such as the end-to-end distance or the root mean square radius. The exponent on the right hand side of Eq. (6.57) is called the Flory exponent v_{Fl}

$$v_{Fl} = 3/(2+d). \qquad (6.58)$$

As a trivial check, Eq. (6.58) predicts $v_{Fl} = 1$ for $d = 1$, just as we found for the one-dimensional self-avoiding chain.

For two and three dimensions, the Flory exponent predicts that the end-to-end displacement $\langle r_{ee}^2 \rangle^{1/2}$ scales like $N^{3/4}$ and $N^{3/5}$ respectively. The prediction for $d = 2$ has been obtained by analytic means (Nienhuis, 1982), while the $d = 3$ prediction has been verified numerically (see McKenzie, 1976). For $d = 4$, the Flory exponent is the same as the ideal exponent, indicating that chain self-avoidance is irrelevant in more than three dimensions. In other words, self-avoidance causes chains to swell in $d = 1$, 2 and 3 dimensions.

A generalized Flory argument for the scaling behavior of a D-dimensional surface embedded in d-dimensional space reads (see de Gennes, 1979, and Kantor, 1989)

$$v_{Fl} = (2 + D)/(2 + d) \qquad (6.59)$$

for the scaling exponent v in

$$R_g^2 \propto L^{2v}, \qquad (6.60)$$

where L is a contour distance along the surface. For polymers, $D = 1$ and Eq. (6.59) reverts to (6.58). For membranes, $D = 2$ and Eq. (6.59) predicts $v = 1$ for two dimensions (as expected for a membrane confined to a plane) and $\nu=4/5$ for three dimensions, a prediction which is not obeyed, as we now report.

Fluid and polymerized membranes

Membranes are categorized broadly as polymerized or fluid according to whether or not they can resist in-plane shear, although variations of this categorization would include mixed fluid/polymerized systems, and sheets with holes. Shear resistance affects the scaling properties of self-avoiding membranes because the smooth arms and other structures illustrated for fluid membranes in Fig. 6.9 are not accessible to a polymerized membrane that is initially flat. In addition to shear, membranes also may resist out-of-plane bending, as quantified by the bending rigidities κ_b and κ_G. For polymers, the flexural rigidity κ_f only sets the length scale beyond which the asymptotic scaling behavior appears; the specific value of κ_f does not change the scaling exponent. What about membranes? Does κ_b just set the length scale ξ_p but not influence the scaling law?

Extensive computer simulations have been performed on self-avoiding fluid membranes and random surfaces. Some studies explicitly count configurations on a lattice (Glaus, 1988), while others may use an algorithm for fluid sheets developed by Baumgärtner and Ho (1990), as illustrated in Fig. 6.10. Most of the simulation work has been performed on closed fluid sheets – computational vesicles, if you will – to avoid interpretational ambiguities arising from edge fluctuations. Analyzing the radius of gyration of a fluid vesicle according to Eq. (6.60) yields $\nu=0.8$ for systems of modest size, in agreement with the Flory prediction for fluid membranes (Baumgärtner and Ho, 1990). However, more extensive analyses, which include larger systems and examine the appearance of the vesicle as a function of bending rigidity, favor branched polymer scaling (Kroll and Gompper, 1992; Boal and Rao, 1992a). The exponents of branched polymer scaling are dimension-dependent; for an enclosed shape in three dimensions, the scaling laws are

$$R_g^2 \propto A^1 \text{ and } \langle V \rangle \propto A^1 \quad \text{(branched polymer)}, \tag{6.61}$$

where A is the surface area and V is the enclosed volume. Analytical treatment (Gross, 1984) and lattice-based simulations (Glaus, 1988) of

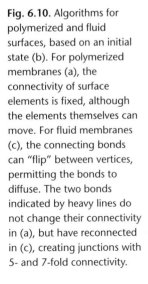

Fig. 6.10. Algorithms for polymerized and fluid surfaces, based on an initial state (b). For polymerized membranes (a), the connectivity of surface elements is fixed, although the elements themselves can move. For fluid membranes (c), the connecting bonds can "flip" between vertices, permitting the bonds to diffuse. The two bonds indicated by heavy lines do not change their connectivity in (a), but have reconnected in (c), creating junctions with 5- and 7-fold connectivity.

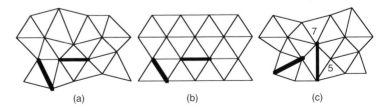

(a)   (b)   (c)

random surfaces also favor branched polymer scaling for fluid membranes. The branched appearance of a fluid membrane is very clear in the simulation of Fig. 6.9; at large length scales, the membrane gains entropy by forming arms and fingers even if there is a modest cost in energy.

Both phantom and self-avoiding polymerized membranes display different behavior than fluid membranes at long distances. Computer simulations show that phantom polymerized membranes are convoluted at low rigidity $\kappa_b/k_B T$ as in Eq. (6.52) but become asymptotically flat as $\kappa_b/k_B T \gg 0$ (Kantor *et al.*, 1986; Kantor and Nelson, 1987). Further, the simulations exhibit a phase transition separating the crumpled phase at small $\kappa_b/k_B T$ from the flat phase at large $\kappa_b/k_B T$, as expected analytically (Peliti and Leibler, 1985; Nelson, 1989; Guitter *et al.*, 1989). This behavior is different from phantom (or ideal) chains, which always obey $R_g^2 \propto L_c$ at any flexural rigidity as long as the viewing scale is much larger than the persistence length.

What happens when self-avoidance is included: does the sharp transition between flat and crumpled phases persist? The earliest simulations to address this question supported the idea of a phase transition; however, most simulations find that polymerized membranes are flat, but rough, even in the absence of bending rigidity (Plischke and Boal, 1988, Lipowsky and Girardet, 1990, Gompper and Kroll, 1991, and many others; see also Baumgärtner, 1991, and Guitter and Palmeri, 1992). Because of its in-plane shear rigidity, a polymerized membrane cannot form entropy-generating arms, as in Fig. 6.9, without introducing entropy- and energy-consuming folds and creases. As a result, a polymerized membrane at finite temperature looks like Fig. 6.11, with only gentle undulations at long length scales. Choosing a reference plane parallel to the membrane plane, the height fluctuations $\langle h^2 \rangle$ of a polymerized membrane in the Monge representation scale with the membrane area A like

$$\langle h^2 \rangle \sim A^\zeta, \tag{6.62}$$

where the roughness exponent ζ is found analytically to be 0.59 (Le Doussal and Radzihovsky, 1992), in the range anticipated by early

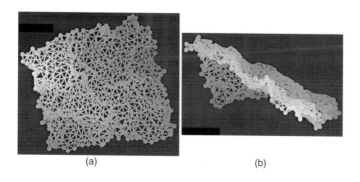

(a)   (b)

Fig. 6.11. Top (a) and side (b) view of a sample configuration taken from a simulation of a polymerized membrane. The ratio of the membrane's transverse thickness to its overall length declines with increasing membrane size.

simulation studies (Plischke and Boal, 1988; Lipowsky and Girardet, 1990; Gompper and Kroll, 1991). Allowing a membrane to change topology through the creation of voids or handles does not change the scaling behavior of the systems studied thus far (Levinson, 1991; Plischke and Fourcade, 1991; Jeppesen and Ipsen, 1993). Lastly, we mention that a phase intermediate between polymerized and fluid membranes was predicted some time ago (see Nelson, 1989) and has been seen in simulation studies (Gompper and Kroll, 1997).

The elastic properties of flat polymerized membranes at finite temperature have been investigated for at least a decade (Aronovitz and Lubensky, 1988; Nelson and Peliti, 1987). One of the interesting features emerging from such studies is the prediction that the Poisson ratio of polymerized membranes is $-1/3$ universally (Le Doussal and Radzihovsky, 1992). As we discussed in our treatment of networks in Chapter 3, a negative Poisson ratio means that the membrane expands transversely when stretched longitudinally. One can imagine how this behavior arises by considering the configuration in Fig. 6.11, or experimenting with a crumpled sheet of aluminum foil: as the folds are pulled out by applying a stress in one direction, the sheet unfolds in the transverse direction as well (D. Nelson, private communication). Simulation studies have confirmed the negative Poisson ratio (Zhang *et al.*, 1996; Falcioni *et al.*, 1997).

Effects of pressure and attraction

Beyond bending resistance, polymers and membranes may experience other interactions, like long-range attraction and repulsion, and may be subject to external stresses. Further, a presure difference may occur across the boundary of closed shapes like rings in two dimensions and vesicles in three dimensions. These factors span a potentially large parameter space; here, we consider only a slice of that space, namely pressure and attraction, to illustrate qualitatively the general scaling behavior of a system. We include both rings and fluid vesicles, the former because of its possible relevance to DNA (Baumgärtner, 1982).

The various scaling laws applicable to closed systems in two and three dimensions are summarized in Table 6.1, showing that the scaling exponents are unique to each system as long as two independent geometrical characteristics are taken into account. These families of configurations are illustrated pictorially in Fig. 6.12, which displays the phase behavior in the pressure/attraction plane at zero bending resistance (Boal, 1991). By *attraction*, we mean short-range attractive interactions among all elements of a chain. Two trends are apparent in the figure:

Table 6.1. *Scaling behavior of closed shapes in two and three dimensions (rings of contour length L_c and fluid bags of area A, respectively)*

Configuration	Scaling law		Reference
Two dimensions	$\langle R_g^2 \rangle \propto L_c^{2\nu}$	$\langle A \rangle \propto L_c^{2\eta}$	
inflated	$\nu = 1$	$\eta = 1$	
self-avoiding walk	$\nu = 3/4$	$\eta = 3/4$	Duplantier, 1990, Nienhuis, 1982
branched polymer	$\nu = 0.64$	$\eta = 1/2$	Derrida and Stauffer, 1985
dense	$\nu = 1/2$	$\eta = 1/2$	
Three dimensions	$\langle R_g^2 \rangle \propto A^{\nu}$	$\langle V \rangle \propto A^{\eta}$	
inflated	$\nu = 1$	$\eta = 3/2$	
Flory-type argument	$\nu = 4/5$		
branched polymer	$\nu = 1$	$\eta = 1$	
dense	$\nu = 2/3$	$\eta = 1$	

- the enclosed area increases as the pressure outside the ring drops below the pressure inside, at fixed attraction
- the linear size of the configuration decreases with increasing attraction at fixed pressure.

Under compression, for example, the gangly arms of a branched polymer stick to each other at high attraction, thus reducing the linear size of the ring. Similar dependence on pressure is found for rings subject to bending rigidity without attraction (Leibler *et al.*, 1987). Note that a finite-size system may not adopt the phase displayed in the figure, particularly for systems under pressure (Maggs *et al.*, 1990).

Studies of closed rings have been extended to both fluid (see Gompper and Kroll, 1995, and references therein) and polymerized (Komura and Baumgärtner, 1991) membranes in three dimensions. Subject to a pressure difference at $\kappa_b = 0$, fluid membranes exhibit a size-dependent transition between the branched polymer phase at zero pressure difference to

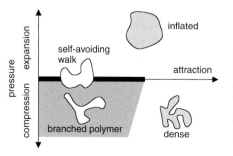

Fig. 6.12. Phase diagram for two-dimensional rings subject to pressure and attraction but very small bending rigidity. There are considerable finite size scaling effects near the phase boundaries (from Boal, 1991).

the inflated phase when the inside pressure greatly exceeds the outside pressure. Inclusion of attractive interactions among elements of a polymerized membrane also leads to a collapsed or dense phase, although the nature of the transition is not as simple as it is for rings (Abraham and Kardar, 1991; Liu and Plischke, 1992).

6.5 Measurement of membrane undulations

Where not suppressed by the presence of a cell wall or other constraints, the thermal undulations of biomembranes may be large enough to have observable consequences in a cell. For example, flickering can be seen on the plasma membrane of the mature human red blood cell, which is supported only by a membrane-associated cytoskeleton. Undulations are even more pronounced in pure lipid bilayers, as illustrated by the oscillations of a bilayer vesicle in Fig. 1.13. Before we interpret such experimental observations, let us briefly recap some formal results for fluid membranes from Sections 6.2–6.4.

In the Monge representation of a surface, a position $\mathbf{r}$ has Cartesian coordinates $[x, y, h(x,y)]$, the displacement from the xy-plane [or height function $h(x,y)$] being single-valued if the surface has no overhangs (unlike a cresting wave). An area element dA on the surface is related to its projected area $dx\,dy$ on the xy-plane by the metric g via $dA = \sqrt{g}\,dx\,dy$, where $g = 1 + h_x^2 + h_y^2$ and $h_x = \partial h/\partial x$. The curvature C characterizes the rate at which the orientation of the surface, as measured by the normal $\mathbf{n}$, changes with arc length s according to $\mathbf{n}\cdot\partial^2\mathbf{r}/\partial s^2$ in a given direction. The largest and smallest values of the curvature at a point are called the principal curvatures C_1 and C_2. For gently undulating surfaces, the mean and Gaussian curvatures, which are combinations of C_1 and C_2, are given by $(C_1 + C_2)/2 = (h_{xx} + h_{yy})/2$ and $C_1 C_2 = h_{xx}h_{yy} - h_{xy}^2$, respectively, where $h_{xx} = \partial^2 h/\partial x^2$, etc. These combinations enter the simplest form for the bending energy density as $\mathcal{F} = (\kappa_b/2)(C_1 + C_2)^2 + \kappa_G C_1 C_2$, where κ_b and κ_G are material constants called the bending rigidity and Gaussian bending rigidity, respectively. The mean direction of the surface normal $\mathbf{n}$ changes with separation $\Delta\mathbf{r} = \mathbf{r}_1 - \mathbf{r}_2$ between positions $\mathbf{r}_1$ and $\mathbf{r}_2$ as $\langle \mathbf{n}(\Delta\mathbf{r})\cdot\mathbf{n}(0)\rangle = \exp(-\Delta r/\xi_p)$, where ξ_p is the persistence length. For fluid membranes, $\xi_p = b\,\exp(4\pi\kappa_b/3k_B T)$, demonstrating that the persistence length decreases with increasing temperature as surface undulations become more pronounced.

One manifestation of thermally induced undulations is the response of a membrane to an applied tension τ. As illustrated in Fig. 6.13, the application of a small tension irons out the wrinkles in a bilayer without significantly stretching it. As the applied tension is raised further, the membrane area expands according to the area compression modulus

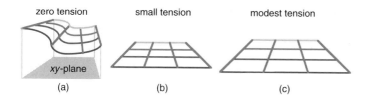

zero tension	small tension	modest tension

xy-plane

(a)   (b)   (c)

Fig. 6.13. (a) At zero tension, a membrane undergoes thermal fluctuations, reducing its projected area. (b) Applying a small tensile stress suppresses these undulations. (c) At larger tension, the intermolecular separation increases according to K_A.

K_A. For the model membrane presented in Section 6.3, the energy E of a particular configuration is described by

$$E = \tau \int dA + (\kappa_b/2) \int (C_1 + C_2)^2 dA, \tag{6.63}$$

where $\tau > 0$ corresponds to tensile stress and where the Gaussian curvature term has been omitted (Helfrich, 1975; see also Helfrich and Servuss, 1984). The first term expresses how the tension opposes growth in the contour area $\int dA$ compared to the area in the xy-plane.

The contour area is $\int dA = \int \sqrt{g}\, d\mathbf{X}$, where the two-dimensional integral $d\mathbf{X} = dx\, dy$ runs over the xy-plane and g is the metric. In the Monge representation, $\sqrt{g}$ is equal to $(1 + h_x^2 + h_y^2)^{1/2}$, which is approximately $1 + (h_x^2 + h_y^2)/2$ for gentle undulations corresponding to small derivatives $h_x = \partial h/\partial x$. Thus, the contour area A of a given configuration is

$$A \cong \int d\mathbf{X} + (1/2) \int (h_x^2 + h_y^2) d\mathbf{X}. \tag{6.64}$$

In other words, $(1/2) \int (h_x^2 + h_y^2) d\mathbf{X}$ is the amount by which the projected area $\int d\mathbf{X}$ is reduced below its contour area A by thermal fluctuations; the reduction must be obtained by averaging over all configurations with the appropriate Boltzmann weight. From Eq. (6.39), the thermal average of $\int (h_x^2 + h_y^2) d\mathbf{X}$ is

$$\left\langle \int (h_x^2 + h_y^2) d\mathbf{X} \right\rangle = (A^2/4\pi^2) \int q^2 d\mathbf{q}\, \langle h(\mathbf{q})h^*(\mathbf{q}) \rangle, \tag{6.65}$$

where $\langle \ldots \rangle$ indicates an ensemble average and $\mathbf{q}$ is a two-dimensional wavevector. As established in Eq. (6.43) for a configurational energy of the form in Eq. (6.63), the Fourier transform $h(\mathbf{q})$ of the height function obeys

$$\langle h(\mathbf{q})h^*(\mathbf{q}) \rangle = k_B T / (\tau q^2 + \kappa_b q^4) \cdot A, \tag{6.66}$$

where $h(\mathbf{X}) = (A/4\pi^2) \int d\mathbf{q}\, \exp(i\mathbf{q}\cdot\mathbf{X}) h(\mathbf{q})$.

Combining Eqs. (6.64) to (6.66), the reduction A_{red} of the in-plane area with respect to the contour area becomes

$$A_{red}(\tau) = (1/2) \left\langle \int (h_x^2 + h_y^2) d\mathbf{X} \right\rangle = A(k_B T/8\pi^2) \int d\mathbf{q}/(\tau + \kappa_b q^2), \tag{6.67}$$

or, after making the replacement $d\mathbf{q} = (1/2)\, dq^2\, d\theta$ and integrating $d\theta$ over 2π,

$$A_{red}(\tau)/A = (k_B T/8\pi) \int dq^2/(\tau + \kappa_b q^2), \tag{6.68}$$

As discussed in Section 6.3, the domain of the wavevector q is set by the largest and smallest length scales of the system, and ranges from a low of $\pi/A^{1/2}$, as dictated by the size of the membrane, to π/b, as dictated by the molecular spacing b. Changing integration variables to $z = q^2 + \tau/\kappa_b$ gives

$$A_{red}(\tau)/A = (k_B T/8\pi\kappa_b) \int dz/z, \tag{6.69}$$

where the integral runs from $\pi^2/A + \tau/\kappa_b$ to $\pi^2/b^2 + \tau/\kappa_b$, yielding a logarithm

$$A_{red}(\tau)/A = (k_B T/8\pi\kappa_b) \ln([\pi^2/b^2 + \tau/\kappa_b]/[\pi^2/A + \tau/\kappa_b]). \tag{6.70}$$

This expression, obtained by Helfrich (1975), differs somewhat from that found by Milner and Safran (1987) for quasi-spherical vesicles, although both approaches yield the same effective compression modulus. Eq. (6.70) predicts that $A_{red}(\tau)/A$ vanishes in two limits: $k_B T/\kappa_b \rightarrow 0$ or $\tau/\kappa_b \rightarrow \infty$; ripples are suppressed at low temperature or high tension, and the projected area is the same as the contour area.

We have shown that the projected area of a membrane is less than its contour area because of thermal fluctuations. The change in contour area arising from increased separation between molecules at higher tensions is governed by the (zero temperature) area compression modulus K_A, which can be superimposed on the area change from fluctuations (Helfrich and Servuss, 1984; see also Helfrich, 1975). To determine the effective area compression modulus, Evans and Rawicz (1990) evaluate the area change with respect to the tensionless reference state, which has the largest area reduction $A_{red}(0)/A = (k_B T/8\pi\kappa_b) \ln(A/b^2)$. Then, applying a tension increases the projected area by an amount $\Delta A/A$ given by

$$\Delta A/A = [A_{red}(0) - A_{red}(\tau)]/A$$
$$= (k_B T/8\pi\kappa_b) \ln([1 + \tau A/\pi^2\kappa_b]/[1 + \tau b^2/\pi^2\kappa_b]). \tag{6.71}$$

Given that the membrane area A is very much larger than the squared distance between molecules, Eq. (6.71) is well approximated by

$$\Delta A/A = (k_B T/8\pi\kappa_b) \ln(1 + \tau A/\pi^2\kappa_b). \tag{6.72}$$

Adding τ/K_A to the change in the relative area in the low-tension regime, Eq. (6.72), gives an overall area change of

$$\Delta A/A = (k_B T/8\pi\kappa_b) \ln(1 + \tau A/\pi^2\kappa_b) + \tau/K_A. \tag{6.73}$$

Eq. (6.73) permits determination of both κ_b and K_A from a single experiment by recording the membrane area change as a function of tension, as has been done via micropipette aspiration (Evans and Rawicz, 1990; Rawicz et al., 2000). The equation simplifies in two obvious limits.

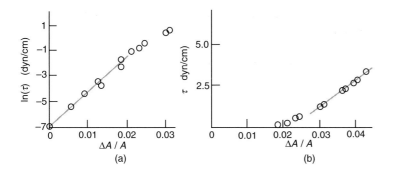

Fig. 6.14. The area change of a lipid bilayer as a function of tension applied to a pure lipid vesicle by aspiration. (a) The measured change in relative area $\Delta A / A$ is proportional to $\ln(\tau)$ at small tension. (b) At higher tension, the area is linear in tension. Data from the application of ascending and descending tension to diGDG bilayers are combined here (redrawn with permission from Evans and Rawicz, 1990; © 1990 by the American Physical Society).

(i) *Low tension* Undulations dominate and the logarithmic term is the most important. Using typical values of $\tau \sim 10^{-4}$ J/m^2, $A \sim 10^{-12}$ m^2, and $\kappa_b \sim 10^{-19}$ J confirms that $\tau A / \pi^2 \kappa_b \gg 1$ for most experimental situations of interest, such that

$$\Delta A / A = (k_B T / 8\pi\kappa_b) \ln(\tau A / \pi^2 \kappa_b) \quad \text{(low tension)}. \tag{6.74}$$

Accordingly, a plot of $\Delta A / A$ *vs.* $\ln(\tau)$ should be linear with a slope of $k_B T / 8\pi\kappa_b$ at low tension, as illustrated in Fig. 6.14(a).

(ii) *High tension* Expansion of the contour area dominates this tension regime, and

$$\Delta A / A = \tau / K_A \quad \text{(high tension)}, \tag{6.75}$$

showing that the area increases linearly with tension, as seen in Fig. 6.14(b).

The data in Fig. 6.14 demonstrate that the apparent compression modulus $K_{A,app}^{-1} = A^{-1}dA/d\tau$ varies with the applied tension; taking the derivative of Eq. (6.73) quickly leads to

$$K_{A,app} = K_A / [1 + K_A k_B T / (8\pi\kappa_b \tau)], \tag{6.76}$$

(Evans and Rawicz, 1990). The cross-over between the two regimes is smooth and occurs over a range in tension estimated in the problem set.

What is the amplitude of the thermal undulations? A few mathematical manipulations are needed to answer this question; readers not wishing to make this detour may skip to Eq. (6.82). Using the Fourier transform of $h(\mathbf{X})$ from Eq. (6.28), we have

$$h(\mathbf{X})^2 = (A/4\pi^2)^2 \int d\mathbf{q} d\mathbf{q}' \exp(i\mathbf{q}\cdot\mathbf{X}) \exp(-i\mathbf{q}'\cdot\mathbf{X}) h(\mathbf{q})h^*(\mathbf{q}'), \tag{6.77}$$

over which we perform a positional average by integrating with $A^{-1}\int d\mathbf{X}$ to obtain

$$\langle h(\mathbf{X})^2 \rangle_{pos} = (A/4\pi^2)^2 A^{-1} \int d\mathbf{X} \int d\mathbf{q} d\mathbf{q}' \exp(i(\mathbf{q}-\mathbf{q}')\cdot\mathbf{X}) h(\mathbf{q})h^*(\mathbf{q}'). \tag{6.78}$$

Grouping the **X**-dependent terms together and integrating just yields a two-dimensional delta function from $\int d\mathbf{X} \exp(i[\mathbf{q}-\mathbf{q}']\cdot\mathbf{X}) = (4\pi^2)\delta(\mathbf{q}-\mathbf{q}')$, which can be removed immediately by performing the $d\mathbf{q}'$ integral. Thus, the full ensemble average is

$$\langle h^2 \rangle = (A/4\pi^2) \int d\mathbf{q} d\mathbf{q}' \, \delta(\mathbf{q}-\mathbf{q}') \, \langle h(\mathbf{q}) h^*(\mathbf{q}') \rangle$$

$$= (A/4\pi^2) \int d\mathbf{q} \, \langle h(\mathbf{q}) h^*(\mathbf{q}) \rangle. \tag{6.79}$$

Eq. (6.66) provides us with an expression for $\langle h(\mathbf{q}) h^*(\mathbf{q}) \rangle$ as a function of tension τ and bending rigidity. We set $\tau = 0$ here (the general case is solved in the problem set) and write

$$\langle h^2 \rangle = (A/4\pi^2) \int d\mathbf{q} \, k_B T / (\kappa_b q^4 A). \tag{6.80}$$

The angular dependence of $d\mathbf{q}$ can be removed immediately, leaving $2\pi q \, dq$, and

$$\langle h^2 \rangle = (k_B T / 2\pi\kappa_b) \int dq / q^3. \tag{6.81}$$

As discussed with Eq. (6.69), the q-integral runs from $\pi/A^{1/2}$ to π/b, to give

$$\langle h^2 \rangle = (k_B T / 4\pi\kappa_b) (A/\pi^2 - b^2/\pi^2)$$
$$\cong (k_B T A / 4\pi^3 \kappa_b), \tag{6.82}$$

where the second equation follows because $A \gg b^2$ (Helfrich and Servuss, 1984). As expected, the amplitude of the undulations grows with temperature as $k_B T / \kappa_b$; the amplitude is also proportional to the square root of the contour area as $\langle h^2 \rangle^{1/2} \sim A^{1/2}$.

In addition to the compression and bending resistance of fluid membranes, polymerized membranes possess non-vanishing shear resistance which modifies the q-dependence of the height function $\langle h(\mathbf{q}) h^*(\mathbf{q}) \rangle$. Simulating a polymerized membrane under pressure against a hard wall, Lipowsky and Girardet (1990) found that their computational membrane was described by

$$\langle h(\mathbf{q}) h^*(\mathbf{q}) \rangle \cong k_B T / [1.3(k_B T Y)^{1/2} q^3 + \kappa_b q^4] \cdot A. \tag{6.83}$$

The two-dimensional Young's modulus Y of the membrane has units of energy per unit area, and is related to the two-dimensional K_A and μ as described in Chapter 3:

$$Y = 4 K_A \mu / (K_A + \mu). \tag{6.84}$$

The apparent elastic behavior of the membrane depends on the magnitude of q: at short wavelengths, q is large and the membrane is fluid-like, while at long wavelengths, q is small and the membrane is solid-like. The crossover between these regimes occurs broadly near $q = 1.3(k_B T Y)^{1/2}/\kappa_b$, corresponding to a distance scale

$$L^* = 2\pi\kappa_b / (k_B T Y)^{1/2}, \tag{6.85}$$

whose magnitude is estimated for the red cell in the problem set.

How do we expect membrane undulations to behave in the solid regime? Neglecting the bending resistance in Eq. (6.83), and substituting the result into Eq. (6.79), we have

$$\langle h^2 \rangle = k_B T / [1.3 \cdot 4\pi^2 (k_B T Y)^{1/2}] \int d\mathbf{q} \, / q^3. \tag{6.86}$$

Performing the usual angular integration, $d\mathbf{q}$ can be replaced by $2\pi q \, dq$ to give

$$\langle h^2 \rangle = (k_B T / Y)^{1/2} / (1.3 \cdot 2\pi) \int dq \, / q^2, \tag{6.87}$$

where q runs from $\pi / A^{1/2}$ to π / b. As usual, $A^{1/2} \gg b$, which simplifies the result of the integral to read

$$\langle h^2 \rangle = (k_B T / Y)^{1/2} A^{1/2} / 2.6. \tag{6.88}$$

Clearly, the undulations of a solid membrane grow more slowly with A than do those of a fluid membrane ($\langle h^2 \rangle \sim A^1$ from Eq. (6.82) for fluids). Comparing with the scaling form Eq. (6.62), namely $\langle h^2 \rangle \sim A^\zeta$, we see that Eq. (6.88) is consistent with a roughness exponent $\zeta = 1/2$. We note that this value is only slightly less than $\zeta = 0.59$ found analytically (Le Doussal and Radzihovsky, 1992) or the range found in other simulations (Plischke and Boal, 1988; Abraham and Nelson, 1990; Gompper and Kroll, 1991), confirming that polymerized membranes are flat, but rough.

The undulations of the combined membrane plus cytoskeleton of the human red blood cell have been interpreted using Eq. (6.83) (Zilker *et al.*, 1992; Peterson *et al.*, 1992). In the experiment, a red cell is attached to a flat substrate at just enough locations on its plasma membrane to prevent lateral motion without suppressing its undulations. Thermal oscillations of the attached membrane are then observed microscopically by interference, much like Newton's rings. As shown in the problem set, the crossover length L^* for the erythrocyte plasma membrane is of the order half a micron, approximately the cell's thickness, but much less than its diameter. The amplitude of the undulations $\langle h^2 \rangle$ was measured over the wavevector range $0.3 \leq q \leq 4 \, \mu m^{-1}$ (or $1.6 \leq 2\pi/q \leq 20 \, \mu m$) and was found to obey q^{-4} over much of that range (Zilker *et al.*, 1992). In other words, the undulations of the composite membrane are dominated by bending resistance in this range of q.

The scaling properties of the red cell cytoskeleton have been measured directly by removing the plasma membrane and determining the structure factor of the network by X-ray or light scattering. In a scattering experiment, the intensity of the light scattered into a given angle is proportional to the structure factor $\mathbf{S(q)}$ (see Chapter 2 of Kittel, 1966), where $\mathbf{q}$ is the difference between the momenta of the incident and scattered photons, $\mathbf{p}_{in}$ and $\mathbf{p}_{scat}$, respectively (see Fig. 6.15)

$$\mathbf{q} = \mathbf{p}_{scat} - \mathbf{p}_{in}. \tag{6.89}$$

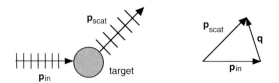

The experimentally measured structure factor is an average over directions of $\mathbf{q}$ because the cytoskeletons are not constrained to a particular orientation. For a theoretical membrane or network with N vertices, such as the computational membrane displayed in Fig. 6.11, the direction-averaged structure factor has the form (see discussion in Kantor *et al.*, 1986)

$$S(q) = N^{-2} \langle \Sigma_{m,n} \exp[i\mathbf{q}\cdot(\mathbf{r}_m - \mathbf{r}_n)] \rangle, \tag{6.90}$$

where m and n are labels attached to the vertices, permitting $S(q)$ to be predicted for model membranes. The observed q-dependence of the averaged structure factor obeys $q^{-2.35}$, which, when compared to the $q^{-3+\zeta}$ expected for gentle undulations of a polymerized membrane at intermediate q (Goulian *et al.*, 1992), leads to a measured value for ζ of 0.65 ± 0.10 (Schmidt *et al.*, 1993). As indicated in the discussion following Eq. (6.88), this value is in the range expected theoretically.

6.6 Summary

In the Monge representation, the position $\mathbf{r}$ on a surface has Cartesian coordinates $[x, y, h(x,y)]$, where h is the displacement of the surface from the xy-plane. An area element dA on the surface is related to its projected area $dx\,dy$ on the xy-plane by the metric g in $dA = \sqrt{g}\,dx\,dy$, where $g = 1 + h_x^2 + h_y^2$ and $h_x = \partial h/\partial x$. The local geometry of a surface or interface can be described by two principal curvatures C_1 and C_2, representing the extremal values of $\mathbf{n}\cdot(\partial^2\mathbf{r}/\partial s^2)$, where $\mathbf{n}$ is the normal to the surface at position $\mathbf{r}$ and s is the arc length along the surface in the direction of interest. For gentle undulations, the combinations $(C_1 + C_2)/2$ and $C_1 C_2$, defined as the mean and Gaussian curvatures, are given in terms of the height functions by $(C_1 + C_2)/2 = (h_{xx} + h_{yy})/2$ and $C_1 C_2 = h_{xx} h_{yy} - h_{xy}^2$, where $h_{xx} = \partial^2 h/\partial x^2$, etc. These combinations enter the simplest form of the bending energy density as $\mathcal{F} = (\kappa_b/2)(C_1 + C_2)^2 + \kappa_G C_1 C_2$, where κ_b and κ_G are the bending rigidity and Gaussian bending rigidity, respectively, and carry units of energy. At finite temperature, a surface may experience thermal undulations, causing its normals to vary in direction from one location to another. The mean orientation of the normals at two positions separated by a distance Δr decays exponentially as $\langle \mathbf{n}(\Delta r)\cdot\mathbf{n}(0) \rangle = \exp(-\Delta r/\xi_p)$, where $\xi_p \sim b \exp(4\pi\kappa_b/3k_B T)$ is the persistence length, b being the elementary length scale of the surface.

It is often convenient to measure the wavelength dependence of the height function by utilizing the Fourier transformation $h(\mathbf{X}) = (A/4\pi^2) \int d\mathbf{q} \exp(i\mathbf{q}\cdot\mathbf{X})h(\mathbf{q})$, where $\mathbf{q}$ is a two-dimensional wavevector. Averaged over an ensemble of configurations at temperature T, the Fourier components $h(\mathbf{q})$ of a fluid membrane subject to a surface tension τ and bending rigidity κ_b are given by

$$\langle h(\mathbf{q})h^*(\mathbf{q})\rangle = k_B T/(\tau q^2 + \kappa_b q^4)\cdot A \qquad \text{(fluid),}$$

where h^* is the complex conjugate of h. In simulation studies, terms of order q^3 must appear in the denominator to describe the undulations of a computational polymerized membrane.

The presence of undulations and ripples reduces the in-plane area of a fluid membrane. As tension is applied, the relative area $\Delta A/A$ increases

* according to $(k_B T/8\pi\kappa_b)\ln(1 + \tau A/\pi^2\kappa_b)$ as the ripples are reduced at low tension
* according to τ/K_A as the intermolecular separation increases at high tension.

From this behavior, the apparent area compression modulus $K_{A,\text{app}}$ is found to be $K_{A,\text{app}} = K_A/[1 + K_A k_B T/(8\pi\kappa_b\tau)]$. Polymerized membranes also have unusual elastic properties; in particular, their Poisson ratio is predicted to be $-1/3$, meaning that, in the membrane plane, they expand transversely when stretched longitudinally.

Although membranes do not adopt a unique configuration at $T > 0$, their mean shape can be characterized by sets of scaling exponents. Considering first self-avoiding membranes not subject to internal interactions or external forces, it is found that fluid membranes behave like branched polymers, while polymerized membranes are rough, but flat. As applied to a closed shape like a fluid vesicle, branched polymer scaling means that the enclosed volume scales like the surface area, $V \sim A^1$; in other words, the vesicle reduces its volume so as to form high-entropy arms and fingers. In contrast, a flat polymerized membrane has an in-plane linear size that grows like the square root of the membrane area, $R^2 \sim A^1$, while its transverse fluctuations grow more slowly like $\langle h^2 \rangle \sim A^{0.59}$. With the inclusion of attractive interactions, both fluid and polymerized membranes collapse to dense configurations. When subject to a tensile stress, the ripples on the surface of a fluid vesicle are suppressed as the vesicle inflates, ultimately scaling like $V \sim A^{3/2}$, although the vesicle attains its asymptotic scaling behavior in a size-dependent way. Polymer rings confined to a plane display similar, but not completely identical, scaling behavior to fluid vesicles.

Most of the predicted elastic behavior of fluid and polymerized membranes has been verified experimentally and used in the analysis of data, some of which are presented in Chapter 5. The behavior $\langle |h(\mathbf{q})|^2 \rangle \sim q^{-4}$ of fluid membranes has been confirmed for bilayers, and provides the foundation for an analysis that obtains both κ_b and K_A from the in-plane membrane area as a function of applied tension. The transverse fluctuations $\langle h^2 \rangle \sim A^{0.59}$ predicted for polymerized membranes also have been observed in two-dimensional cytoskeletons isolated from the red cell.

6.7 Problems

Biological applications

6.1 Consider a dimple on a membrane described by the Gaussian height function $h(x,y) = h_o \exp(-[x^2 + y^2]/ 2w^2)$. The approximate deformation energy of this shape is determined to be $E = \pi \kappa_b (h_o/w)^2$ in Problem 6.11(d).

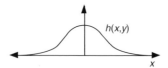

(a) If $w = 1$ μm and $\kappa_b = 10 k_B T$, at what value of h_o does the deformation energy of the surface equal $k_B T$?
(b) Under certain conditions, dimples appear on the surface of a red blood cell, although their occurence is driven by more factors than bending and their shape may not be Gaussian. What is the bending contribution to the deformation energy if $w = 1$ μm, $h_o = 0.1$ μm, and $\kappa_b = 10 k_B T$? Quote your answer in units of $k_B T$.

6.2 We establish in Problem 6.12 that the deformation energy of a membrane undulation of the form $h(x,y) = h_o \cdot \cos(2\pi x/\lambda) \cdot \cos(2\pi y/\lambda)$ on a $\lambda \times \lambda$ square patch is approximately $E = 2\pi^2 \kappa_b (2\pi h_o/\lambda)^2$ where κ_b is the bending rigidity.

(a) What is the value of h_o if $E = k_B T$, assuming $\lambda = 1$ μm and $\kappa_b = 10 k_B T$? Note that the peak to trough height difference is $2h_o$.
(b) Plot $E/k_B T$ as a function of h_o (in μm) up to 0.2 μm to gauge the magnitude of thermally allowed fluctuations on this membrane patch (use λ and κ_b from part (a)).

6.3 A long persistence length does not suppress all undulations on the cellular length scale. Suppose we observe that the root mean square angle between two normals, $\langle \theta^2 \rangle^{1/2}$, is 1/10 of a radian, or about

6 degrees, when the normals are separated by 10 μm. Using the small angle expansion $\cos\theta \sim 1 - \theta^2/2$, find the corresponding ξ_p and $\kappa_b/k_B T$ of the membrane (assume $b = 1$ nm for the molecular length scale). Is this value of $\kappa_b/k_B T$ extraordinarily different from the values observed for cellular bilayers?

6.4 Suppose that a hypothetical bilayer could be made from short chain lipids, giving a bilayer thickness of 3 nm, instead of the conventional 4 nm. By what factor would κ_b change, regarding the bilayer as a uniform sheet of constant bulk modulus K_V? By what factor would the persistence length change if $\kappa_b/k_B T = 10$ for the 4 nm bilayer?

6.5 Most pure lipid bilayers of relevance to the cell have a bending modulus κ_b in the range 10–30 $k_B T$. Plot the persistence length and $\langle h^2 \rangle^{1/2}$ for a thin circular disk of radius 4 μm having this range of κ_b. Use $b \sim 1$ nm for the molecular length scale.

6.6 Plot the difference between a bilayer's contour area and its projected area at zero tension for $\kappa_b = 10k_B T$. Consider the area range 10^{-12}–10^{-10} m^2 and assume the intermolecular separation is $b = 1$ nm.

6.7 Find the approximate tension at the crossover between the logarithmic and linear behavior of membrane area defined by $K_{A.app}/K_A = 1/2$. What is the magnitude of this tension if $\kappa_b = 10k_B T$ and $K_A = 0.15$ J/m^2. How does your answer compare with the rupture tension if the membrane in question fails if its area is 3% higher than its unstressed value?

6.8 Plot the amplitude of thermal undulations $\langle h^2 \rangle$ as a function of tension for a bilayer with $\kappa_b = 10k_B T$. Take the membrane area to be 10^{-11} m^2 and assume the intermolecular separation is $b = 1$ nm. Cover the tension range $10^{-4} \leq \tau \leq 10^{-2}$ J/m^2. Use the expression from Problem 6.14.

6.9
(a) Find the length scale for the crossover between fluid and solid behavior for a polymerized network with $\kappa_b = 0.25\, k_B T$ and $K_A = 2\mu = 2\times10^{-5}$ J/m^2, typical of the erythrocyte cytoskeleton in isolation.
(b) Repeat this calculation for the combined cytoskeleton + bilayer system, with $\kappa_b = 10\, k_B T$, $K_A = 0.2$ J/m^2 (from a bilayer) and μ as in part (a).

For both (a) and (b), determine Y from the two-dimensional relation $Y_{2D} = 4K_A\mu/(K_A + \mu)$ (see Chapter 3 problems).

Formal development and extensions

6.10 A hemispherical surface is governed by the equation $x^2 + y^2 + z^2 - R^2 = 0$ with $z > 0$.

(a) Find $h(x,y)$ and $\mathbf{n}(x,y)$ for $x^2 + y^2 < R^2$ using Eq. (6.3).
(b) Find the mean and Gaussian curvatures from Eqs. (6.11) and (6.12) by explicitly determining $b_{\alpha\beta}$ and $g_{\alpha\beta}$.

Do your results for the curvatures and surface normals agree with your intuition for the properties of a sphere?

6.11 Suppose that a dimple on a membrane is described by the Gaussian form

$$h(x,y) = h_0 \exp(-[x^2 + y^2]/ 2w^2).$$

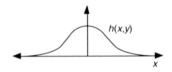

(a) What aspects of the shape do h_0 and w represent?
(b) What are the mean and Gaussian curvatures as a function of (x,y) if h_0 is small?
(c) Use your results from (b) to determine the explicit form of the energy density of the surface based on Eq. (6.22).
(d) Assuming that h_0 is sufficiently small that the metric is close to unity across the surface, integrate your result from (c) to establish that the energy of the dimple is
$$E = \pi\kappa_b(h_0/w)^2,$$
where κ_b is the usual bending rigidity.

6.12 Consider an "egg-carton" surface governed by the equation

$$h(x,y) = h_0 \cdot \cos(2\pi x/\lambda) \cdot \cos(2\pi y/\lambda),$$

where λ is the wavelength of the undulations. For this problem, consider only the xy region $-\lambda/2 \le x,y \le \lambda/2$.

(a) Find the normal $\mathbf{n}$ as a function of (x,y).
(b) Find the mean and Gaussian curvatures if h_0 is small. At what points do these curvatures have their largest magnitudes?

(c) Show that the bending energy of a $\lambda \times \lambda$ square of this surface is

$$E = 2\pi^2 \kappa_b \, (2\pi h_o/\lambda)^2$$

if the energy density is described by Eq. (6.22) and we take $g = 1$.

6.13 The mean square height of a membrane of area A is found to obey $\langle h^2 \rangle \sim A^{0.6}$. What power law form of $\langle |h(\mathbf{q})|^2 \rangle$ is consistent with this scaling behavior?

6.14 Derive the general expression for the amplitude of thermal undulations of a fluid membrane by substituting Eq. (6.66) into Eq. (6.79). Show that when the area $A \gg b^2$ (b is the molecular length scale)

$$\langle h^2 \rangle = (k_B T/4\pi\tau) \, \ln(1 + \tau A / \pi^2 \kappa_b) \quad \text{(Helfrich and Servuss, 1984)}.$$

Confirm that this expression reduces to Eq. (6.82) in the limit of small tensions.

Part III

The whole cell

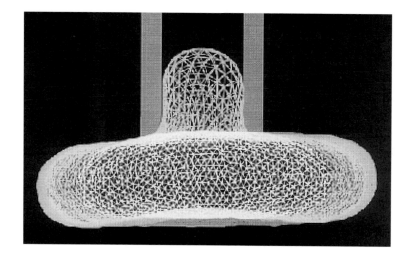

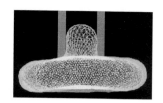

Chapter 7

The simplest cells

Most cells have a complex internal structure of biological rods, ropes and sheets. In terms of the metaphors for the cell in Chapter 1 – a hot air balloon, a sailing ship – our task in Parts I and II of this text was to analyse the individual structural elements of the cell – the masts, rigging and hull – without regard to their interconnection. In Part III, we begin to assemble the units together to form composite systems. As the introduction to Part III, Chapter 7 treats the equilibrium shapes of the simplest systems, namely cells or vesicles without a nucleus or space-filling cytoskeleton. A thorough treatment of complex cells is beyond the reach of our analytical tools; rather, we mention a selection of their properties in Chapters 7–9, before returning to some general attributes of cells in Chapter 10. Many eucaryotic cells live communally as part of a larger organism, and Chapter 8 investigates the interaction between cell membranes, including entropic effects from thermal fluctuations in cell shape. In Chapter 9, we describe mechanisms that cells have developed for changing their shape, as part of cell division or locomotion, for example. The book concludes by returning to the broad perspective established in Chapter 1 and examines some generic features of cell structure, including synthetic cells, and speculates on the mechanical characteristics of the largest and smallest cells.

7.1 Cell shapes

The first figure in Chapter 1 of this book displays examples of the variety of shapes that cells can adopt, ranging from smooth cylindrical bacteria to branched and highly elongated neurons. Yet the primary mechanical elements of all cells are the same: one or more fluid sheets surround the cell and its internal compartments while filaments form a flexible scaffolding primarily, but not exclusively, confined within the cell. What design principles give rise to this wonderful collection of shapes? Let's begin with the lipid bilayer. If the two leaflets of a bilayer have the same

Fig. 7.1. Deformations of the cell boundary can arise from a variety of causes, including: (a) compositional inhomogeneity of the bilayer leaflets, (b) anisotropic network in a cell wall under lateral stress, and (c) pressure from structural elements, such as interacting microtubules during cell division. Note that the length scale of the drawings increases from (a) to (c).

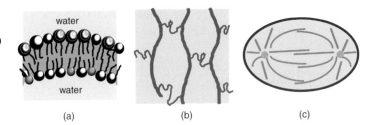

(a)   (b)   (c)

homogeneous composition and are adjacent to equivalent environments, the equilibrium shape of the bilayer is flat at zero temperature. However, compositional differences between leaflets, such as the size or shape of their molecules, may cause a membrane to spontaneously bend; for example, the differing lipid head groups in the schematic bilayer of Fig. 7.1(a) cause its equilibrium shape to have non-vanishing curvature. The rate of migration of lipids between leaflets is sufficiently slow (10^{-7} s^{-1} or less for some phospholipids; see Vance and Vance, 1996) that it may not take much effort for this compositional imbalance to be maintained. Lateral phase separation may occur for leaflets with several molecular components, again opening the door to locally curved regions. Even if the bilayer is symmetric, spontaneous curvature may arise if the external environments of the leaflets are inequivalent.

Anisotropic structural elements, such as the network shown in Fig. 7.1(b), also can influence cell shape. For instance, networks may stretch more easily in one direction than another if the network components have different stiffness. Even when the network in the figure is placed under isotropic tension, it will exhibit a larger strain horizontally than vertically. Further, the magnitude of the lateral stress experienced by a membrane may be direction-dependent according to cell shape. For example, we show in Section 7.2 that the stress around the girth of a cylinder is twice the stress along its symmetry axis.

A third example of shape-changing forces is provided by interacting microtubules during cell division. As described in Chapter 9, microtubules radiate from a center near the cell's nucleus and are dynamic, in the sense that they grow and shrink continuously. During cell division, microtubules grow from two distinct centers, as shown in Fig. 7.1(c); filaments originating from opposite poles of the cell may interact, providing a force that pushes the centers apart and elongates the cell.

In this chapter, we describe several mechanical features of a cell that influence its shape, concentrating on those cells whose elastic elements are assembled at its boundary to form a composite membrane enclosing a structureless interior. In Section 7.2, a number of results from the continuum mechanics of thin shells are derived, and a methodology is presented for determining the shape of a system with axial symmetry.

Only a limited number of structures can be explored analytically with this technique; in general, the equations governing cell shape must be solved numerically.

Biological ropes and sheets can be assembled into cells or cell-like objects with a hierarchy of complexity. Excluding viruses, the simplest structures are cargo-carrying vesicles with just a simple membrane and diameters as little as 50 nm. Without a genetic blueprint, vesicles cannot reproduce; however, model vesicles can be manufactured in the laboratory from lipids and other molecules, allowing their shape to be determined systematically as a function of size and composition. These structures are the subject of Section 7.3. With dimensions of a few hundred nanometers, mycoplasmas have few mechanical elements and are also amenable to theoretical analysis; unfortunately, systematic quantification of their erratic shapes has not yet been performed.

Next to vesicles, the cells whose mechanics have been most thoroughly examined are red blood cells and bacteria, the subjects of Sections 7.4 and 7.5, respectively. The elasticity of these cells is essentially two-dimensional: the cells are bounded by one or two fluid membranes to which is attached a network of flexible links – spectrin for red cells and peptidoglycan for bacteria. Aspects of these networks can be captured, in part, by the harmonic networks of Chapters 3 and 4, as demonstrated in the problem sets of those chapters. Here, we relate the properties of membranes and their associated networks to the shape of cells with axial symmetry; however, we caution that our understanding of these cells is not at all complete.

7.2 Energetics of thin shells

The envelope of the cells of interest in this chapter ranges from about 5 to 50 nm in depth. The thinnest shells are found in small vesicles bounded by a lipid bilayer just 4–5 nm wide, which is about 10% of the vesicle radius. In contrast, the walls of bacteria are much thicker and the cells may be wrapped in a regularly structured S-layer; the two principal designs for the bacterial envelope are as follows (see Fig. 1.7).

- Gram-negative bacteria have two fluid membranes mechanically connected by filaments spanning a periplasmic space often 20 nm wide (with considerable variation among cell types), a region which also contains a thin sheet of peptidoglycan.
- Gram-positive bacteria have a single membrane enclosed by multiple layers of peptidoglycan 20–80 nm across.

Eucaryotic cells such as the red blood cell or auditory outer hair cell have a protein-laden bilayer along with a membrane-associated cytoskeleton

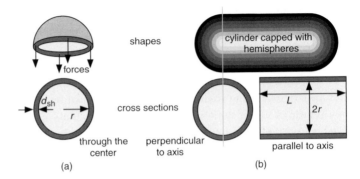

Fig. 7.2. Two shells of finite thickness d_{sh} and inner radius r; the appearance of the shell is shown on the top row, their cross sections are on the bottom. Pressurized shells bear forces perpendicular to the cross sections. (a) Half of a spherical shell, whose cross section is a thick ring. (b) Cylindrical shell, capped by hemispheres; the two cross sections are perpendicular and parallel to the symmetry axis.

perhaps 30–40 nm across. For cells with complex boundaries, then, the thickness of the outer shell (often 40–60 nm) is about 5–10% of the cell radius (~1/2 micron for common bacteria), permitting us to view the cell boundary as essentially two-dimensional for mechanical purposes. In developing a set of analytical tools to describe cell shape, we will first include the thickness of the shell as an explicit parameter before folding it into the stress tensor to obtain formal results for two-dimensional sheets.

Unlike animal cells, bacteria (and plant cells) may operate at an elevated internal pressure which is sustained by their cell wall. The tension created in the wall is proportional to its curvature and the pressure difference it experiences. To see how this arises, we first analyze the forces in a spherical shell defined by inner and outer radii r and $r + d_{sh}$, respectively, where d_{sh} is the shell thickness. Let's take an imaginary slice through the center of the sphere, dividing the shell in two and exposing the ring shown in the bottom row of Fig. 7.2(a). An interior pressure P results in a force of $\pi r^2 P$ across the plane of the ring. Regarding the hemisphere as a free body, this "upward" force is balanced by the "downward" force across the ring itself, as indicated by the arrows in Fig. 7.2(a). Because the force is spread over the cross sectional area of the ring, which is $\pi[(r + d_{sh})^2 - r^2] \cong 2\pi r d_{sh}$ at small d_{sh}/r, the mean stress $\langle \sigma \rangle$ in the shell is

$$\langle \sigma \rangle = [force]/[area]$$
$$\cong rP/2d_{sh} \quad \text{(spherical shell).} \tag{7.1}$$

An exact result can be obtained by using the full expression for the area of the ring. As a technical note, we could have evaluated the force by summing $P \, dA$ over all area elements dA of the hemisphere, rather than considering the force on the midplane. However, we would then have to determine the component of the force in the direction normal to the midplane ($P \, dA \cos\theta$) in calculating the overall force, leaving us to evaluate the integral $2\pi r^2 P \int \sin\theta \cos\theta \, d\theta$, where the polar angle θ runs from 0 to $\pi/2$. Explicit evaluation gives the previous result: the force is $\pi r^2 P$.

Stresses in the cylindrical shells of many bacteria are only slightly more complex than those in spheres. The model bacterium in Fig. 7.2(b) consists of a cylindrical tube capped at each end by hemispherical shells to form what is sometimes referred to as a spherocylinder. The mean stress experienced by the endcaps is the same as that of the sphere in Fig. 7.2(a), namely $rP/2d_{sh}$. However, the stresses in the cylindrical section are not isotropic, but are different for area elements in directions perpendicular and parallel to the symmetry axis, which we will define as σ_z and σ_θ, respectively. A cross section through the cylinder perpendicular to the axis, has the shape of a ring, as in the bottom left corner of Fig. 7.2(b). Determination of the stress across an area element of the ring follows the same steps as the sphere and leads to the same conclusion, $\langle\sigma_z\rangle = rP/2d_{sh}$. However, a slice through the tube containing the axis of cylindrical symmetry intersects the shell as shown in the lower right corner of Fig. 7.2(b): there are two parallel strips of length L and width d_{sh}.

To obtain the "hoop" stress σ_θ, we follow the same logic as for the sphere. Viewing one half of the cylinder as a free body, the pressure P exerts a force $2rLP$, over the area $2rL$ of the cross section. This force is balanced by the force borne by the shell, which is the product of the mean stress $\langle\sigma_\theta\rangle$ and the area of the two surfaces cut from the shell, $2Ld_{sh}$. Equating these expressions for the forces, the length of the cylinder cancels, leaving

$$\langle\sigma_\theta\rangle = rP/d_{sh} \quad \text{(cylindrical shell, hoop direction)} \tag{7.2a}$$

$$\langle\sigma_z\rangle = rP/2d_{sh} \quad \text{(cylindrical shell, axial direction),} \tag{7.2b}$$

where Eq. (7.2b) is included for completeness. In other words, the stress is twice as great around the hoop direction than along the symmetry axis. That boiled hot dogs and sausages split more readily along their length, than around their girth, may reflect these different stresses (for further reading on stresses in shells, see Flügge, 1973; Fung, 1994, and Landau and Lifshitz, 1986).

The stress σ in Eqs. (7.1) and (7.2) is a three-dimensional quantity with dimensions of energy per unit volume. In Chapter 3, we introduced a two-dimensional tension τ, called a stress resultant, which carries dimensions of energy per unit area and is equal to the product of the mean stress and the shell thickness. Thus, Eq. (7.2) corresponds to the tension

$$\tau_\theta = rP \quad \text{(cylindrical shell, hoop direction)} \tag{7.3a}$$

$$\tau_z = rP/2 \quad \text{(cylindrical shell, axial direction).} \tag{7.3b}$$

A relationship between the components of the tension is available for arbitrary radii of curvature, as we now obtain by analyzing the small

Fig. 7.3. (a) Face view of a
small element of a curved
shell with area $dA = ds_1 \cdot ds_2$
and principal curvatures $1/R_1$
and $1/R_2$. (b) Cross section
through the shell, showing
the tension at two edges
directed into the plane.

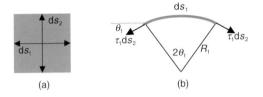

Fig. 7.3. (a) Face view of a
small element of a curved
shell with area $dA = ds_1 \cdot ds_2$
and principal curvatures $1/R_1$
and $1/R_2$. (b) Cross section
through the shell, showing
the tension at two edges
directed into the plane.

section of the curved surface in Fig. 7.3(a). The rectangular shape of the section is taken to coincide with the principal directions 1 and 2 (of the curvature), along which lie arcs of length ds_1 and ds_2, resulting in an area element $dA = ds_1 \cdot ds_2$. The tension components τ_1 and τ_2 may be unequal, as may the radii of curvature R_1 and R_2.

Now consider a section through the surface along the 1-direction, as shown in Fig. 7.3(b), which is a somewhat exaggerated view illustrating R_1 and the angle $2\theta_1$ subtended by arc length ds_1. At each boundary, perpendicular to the plane of the diagram, the 1-component of the tension exerts a force $\tau_1 \cdot ds_2$ in a direction along the tangent to the surface at the edge. The vertical component of the force at each edge is $\tau_1 \cdot ds_2 \sin\theta_1$, which we may approximate by $\tau_1 \cdot ds_2 \cdot \theta_1$ at small angles. Thus, the total vertical force from τ_1 is $2\tau_1 \cdot ds_2 \cdot \theta_1$. The angle θ_1 is related to the arc length ds_1 by the radius of curvature, $2\theta_1 = ds_1/R_1$, permitting the vertical component of the force to be written as $\tau_1 \cdot ds_1 \cdot ds_2/R_1$. This argument can be repeated for a section in the 2-direction, yielding a second contribution to the vertical force of $\tau_2 \cdot ds_1 \cdot ds_2/R_2$. The sum of these forces is balanced by the force from the pressure difference across the surface, $P \cdot ds_1 \cdot ds_2$. Equating the forces and removing the common factor $ds_1 \cdot ds_2$ yields

$$\tau_1/R_1 + \tau_2/R_2 = P, \tag{7.4}$$

one of several results obtained for thin shells in Chapter 2 of Flügge (1973). Eq. (7.4) reduces to Eq. (7.3) upon substituting $R_1 = R_\theta = r$ and $R_2 = R_z = \infty$. For fluid sheets, the tension is isotropic, and Eq. (7.4) becomes the Young–Laplace equation

$$\tau(1/R_1 + 1/R_2) = P \quad \text{(isotropic tension)}, \tag{7.5}$$

a result obtained by Laplace in his study of soap films. Note that bending energy considerations, which we have completely omitted, changes the form of these results (see Flügge, 1973).

Eqs. (7.3) – (7.5) relate the in-plane stretching forces on a thin sheet to a three dimensional stress like the pressure. These relations augment the elastic properties of networks and membranes obtained in Chapters 3 and 5, where we introduced the two-dimensional shear modulus μ, the area compression modulus K_A, and the edge tension λ for membranes with free edges. There, we also expressed the bending energy of a membrane in terms of its local curvature and elastic parameters κ_b and κ_G, the bending rigidity and Gaussian bending rigidity, respectively. In other words, we can now write the energy of a thin shell using a two-dimensional description, without reference to its thickness. By minimizing this energy, the zero-temperature shape of the shell can be found, in principle, as a function of its elastic parameters and other physical characteristics such as its surface to volume ratio. In practice, however, finding energy minima is not always an easy numerical task, and the biological systems treated thus far usually have axial symmetry and shells without shear resistance (i.e. fluid membranes), as demonstrated in Section 7.3. In some applications, it is not overly important to work with the true ground state, and these situations can be treated by numerical simulation; in this vein, Section 7.4 contains examples of simple cells with surface shear resistance and arbitrary shape.

Before making recourse to computer simulations, let us push our analytical formalism a little harder and explore systems restricted to axially symmetric shapes, where the boundary is subject only to bending resistance. Physical examples of such systems include pure lipid bilayer vesicles, which can be manufactured in the laboratory. The simplest model for the energy density $\mathscr{F}$ of a membrane depends quadratically on its local principal curvatures C_1 and C_2 (see Section 6.2) as

$$\mathscr{F} = (\kappa_b/2){\cdot}(C_1 + C_2 - C_o)^2 + \kappa_G C_1 C_2, \tag{7.6}$$

where C_o is a parameter representing the spontaneous curvature that a bilayer may possess arising from compositional inhomogeneities in its two leaflets (a sphere with $C_1 + C_2 - C_o = 0$ has a radius of $2/C_o$). For a spherical shell, the enclosed volume V is related to the surface area A by $V = A^{3/2}/(6\sqrt{\pi})$, and the energy of the shell is uniquely

$$E = 4\pi(2\kappa_b + \kappa_G) \quad \text{(spheres)}, \tag{7.7}$$

as found by integrating Eq. (7.6) over the surface. However, suppose now that the volume is less than $A^{3/2}/(6\sqrt{\pi})$, regarding A as fixed, such that the shell looks like a pancake, a cigar, or something else. How do we determine which shape has the minimum energy?

The energy of a closed surface involves an integral of the energy density $\mathscr{F}$, say the model energy in Eq. (7.6), over the area of the shell. Let's consider an axially symmetric shell, such as that displayed in Fig. 7.4. The shape of the shell can be described by a function $x = f(\ldots)$, where x is the distance from the symmetry axis to the surface, and the argument of f may be just the z-coordinate. In this shell, the area element dA of a ring of radius x and width ds can be easily written as

$$dA = 2\pi x \, ds, \tag{7.8}$$

Fig. 7.4. Shape of a shell with axial or cylindrical symmetry. A position on the surface has normal and tangent vectors **n** and **t**, respectively, and principal curvatures C_m and C_p; C_m is defined by a plane containing the symmetry axis (such as the plane of the diagram) and C_p is defined by the shaded plane perpendicular to the plane of the diagram. The normal vector **n** makes an angle θ with respect to the z-axis; s denotes arc length.

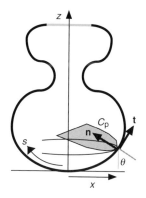

where s is the arc length. The tangent $\mathbf{t}$ and normal $\mathbf{n}$ to the surface can be obtained from derivatives of the position vector $\mathbf{r}(x, z)$ as described in Chapter 2 for polymers; we define θ as the angle between the interior normal and the z-axis. The curvature C_m defined by a plane containing the symmetry axis in Fig. 7.4 can be found from $\partial \mathbf{t}/\partial s = C_m \mathbf{n}$ along the boundary, where the subscript "m" indicates a meridian. In the diagram, $\mathbf{n} = (-\sin\theta, \cos\theta)$ and $\mathbf{t} = (\cos\theta, \sin\theta)$, so that $\partial \mathbf{t}/\partial s = (-\sin\theta, \cos\theta) \, d\theta/ds$. Taking the scalar product $\mathbf{n} \cdot (\partial \mathbf{t}/\partial s)$ yields the curvature

$$C_m = d\theta/ds. \tag{7.9}$$

The second principal curvature C_p is defined by a plane perpendicular to the plane of the diagram and contains the surface normal $\mathbf{n}$. Only if $\mathbf{n}$ were perpendicular to the z-axis would C_p be equal to $1/x$; for a general orientation of $\mathbf{n}$,

$$C_p = \sin\theta /x, \tag{7.10}$$

which is obtained by noting that surface normals on a parallel of latitude intersect at the symmetry axis a distance $x/\sin\theta$ from the surface. Parallels of latitude meet meridians at right angles, and hence the subscript "p" in C_p. Curvatures are positive when the center of curvature is in the same direction as the interior normal.

Combining Eqs. (7.8) – (7.10), the energy $E[S]$ of a particular surface S can be written as

$$E[S] = \pi \kappa_b \int x \, (\sin\theta /x + d\theta/ds - C_o)^2 \, ds +$$
$$2\pi \kappa_G \int (d\theta/ds) \sin\theta \, ds, \tag{7.11}$$

and can be generalized to include shapes with holes (Boal and Rao, 1992b). Eq. (7.11) is a model-dependent expression for the energy of a closed membrane without constraints on the shape other than axial symmetry. As applied to a biological cell, however, constraints such as fixed surface area and volume of the cell must be imposed on the search for energy-minimizing shapes. The search might include conditions such as the smoothness of the shape at the poles, which forces the surface to meet the symmetry axis at right angles (i.e., $\theta = 0$ at the poles). Applications of this approach to cell shape will be reported in Sections 7.3 and 7.4; however, because the solutions are ultimately numerical, we will not derive the intermediate algebraic equations here.

It is not difficult to evaluate Eq. (7.11) for the simplest shapes; as examples, several shells treated in the problem set are displayed in Table 7.1. The table includes the bending energy at $C_o = 0$, as well as the surface area and volume of the cross-hatched regions, both quantities integrated around the rotational axis. These results have familiar limiting situations: for example, the $\rho \rightarrow 0$ limit of shape I is a sphere, whose

Table 7.1. *Bending energy, area and volume of five axisymmetric shells. The surface area of the arc and volume of the shaded region are integrated around the (vertical) symmetry axis. The ratio R/r is defined as* ρ

Shell		Energy	Area	Volume
I		$\pi\kappa_b(8+\pi\rho)+4\pi\kappa_G$ $\rho\gg1$	$2\pi r^2(\pi\rho+2)$	$2\pi r^3(\pi\rho/2+2/3)$
II		$4\pi\kappa_b+2\pi\kappa_G$	$2\pi r^2$	$2\pi r^3/3$
III		$\pi\kappa_b(-8+\pi\rho)-4\pi\kappa_G$ $\rho\gg1$	$2\pi r^2(\pi\rho-2)$	$2\pi r^3(\pi\rho/2-2/3)$
IV		$\pi^2\kappa_b\rho^2/(\rho^2-1)^{1/2}$	$2\pi^2 rR$	$\pi^2 r^2 R$
V		$\pi\kappa_b L/r$	$2\pi rL$	$\pi r^2 L$

energy, volume and area are twice the hemisphere of shape II (the unit-less parameter ρ characterizes the shape and is defined in the table). Similarly, shapes I and III can be joined to give a torus, a section of which is shape IV. We use these results in Section 7.3 to find the bending energies of a variety of simple cell shapes.

7.3 Pure bilayer systems

Mechanically simple cells, which may have a two-dimensional scaffolding attached to their plasma membrane, often adopt smooth symmetrical shapes such as the familiar biconcave disc of the red blood cell, or the spherocylinder of some bacteria. What features of cell shape arise from the properties of the membrane itself, without the presence of a cytoskeleton or cell wall? Controlled experimental studies on cell shape can be performed in the laboratory using bilayer systems called liposomes or artificial vesicles, which are structurally similar to the biological vesicles shown in Fig. 7.5, although liposomes may have diameters two orders of magnitude larger than a vesicle. In cells, vesicles are small cargo-carrying structures that may be absorbed at a membrane or pinch off from it, as described later in this section.

A typical laboratory study of bilayer vesicles records their shape as the surface area is changed while holding the volume roughly constant

Fig. 7.5. Thin section showing a 50-nm diameter vesicle which has come off the outer membrane of the Gram-positive bacterium *Pseudomonas aeruginosa*. The dark areas in the cytoplasm are ribosomes (courtesy of Dr Terry Beveridge, University of Guelph).

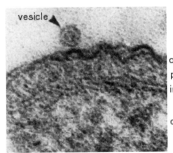

Fig. 7.6. Schematic cross sections of several observed vesicle shapes. For each shape, the axis of rotational symmetry is indicated by the shaded line.

(by adjusting the ambient temperature, for instance). Examples of axially symmetric vesicle shapes that may be observed experimentally are displayed in Fig. 7.6, where the rotational axis is indicated by the shaded line. On the left, a bud is appended to a nearly spherical shell whose enclosed volume is not far below the maximum allowed by the vesicle surface area. The other two sample configurations have volumes well below the maximum value; these shapes are not ellipsoidal, where the curvature would have the same sign everywhere on the surface, but rather display both positive and negative curvature along meridians running from pole to pole. Note that not all of the configurations are "up-down" symmetric. In this section, we examine the origin of these phenomena in terms of the bending properties of fluid membranes.

We work within the spontaneous curvature model for bending, as represented by Eq. (7.11). The model has as its parameters the rigidities κ_b and κ_G, as well as the spontaneous curvature parameter C_o. We return to the physical origin of C_o later; for now, we treat it as a parameter and explore its effect on cell shape. With vanishing C_o, the bending energy of a spherical shell is $E_{sphere} = 4\pi(2\kappa_b + \kappa_G)$. How does this deformation energy change as the spherical shell is distorted, with a resulting loss of volume at fixed surface area? Which generic shape – prolate (like a rod) or oblate (like a disk) – has the lower energy for a given volume and surface area? To address this question, we use two configurations, the spherocylinder and the pancake, to represent prolate and oblate shapes. Referring to Fig. 7.7, these shapes are described by length parameters r, R, and L, and have energies

$$E_{spherocylinder} = 8\pi\kappa_b + \pi\kappa_b(L/r) + 4\pi\kappa_G \tag{7.12a}$$

$$E_{pancake} = \pi\kappa_b(8 + \pi R/r) + 4\pi\kappa_G \quad (R/r \gg 1), \tag{7.12b}$$

Fig. 7.7. The spherocylinder (a) and pancake (b) are axially symmetric shapes whose bending energy is easy to evaluate in the spontaneous curvature model.

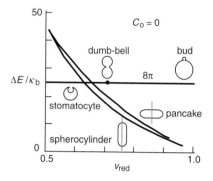

Fig. 7.8. Difference in bending energy ΔE relative to a sphere for three axisymmetric shapes – two touching spheres, a spherocylinder and a pancake, all with $C_o = 0$. The energies are displayed as a function of reduced volume v_{red} and normalized by κ_b. Shaded lines indicate axes of rotational symmetry.

when $C_o = 0$, as can be constructed from Table 7.1. Eq. (7.12b) is an approximation valid for $R/r \gg 1$; the exact result is more cumbersome.

Displayed in Fig. 7.8 are the energies of the two representative shapes compared with E_{sphere}, as in $\Delta E_{pancake} = E_{pancake} - E_{sphere}$, where the exact result is used for the pancake energy, not Eq. (7.12b). The energy difference has been divided by κ_b to produce a unitless observable. To quantify the volume, we introduce the unitless combination v_{red} given by

$$v_{red} = 6\sqrt{\pi}\, V/A^{3/2}, \tag{7.13}$$

which equals unity for a sphere. When $C_o = 0$, Eqs. (7.12) do not depend on the absolute size of the shell, only on the ratios L/r and R/r; hence, ΔE is unique for each v_{red} when $C_o = 0$. The figure shows that the spherocylinder is energetically favored over the pancake in the volume range $0.6 < v_{red} < 1$, with an energy difference as high as $3\kappa_b$ at $v_{red} \sim 0.7$. However, the energies converge and then reverse as v_{red} passes below 0.5, with the pancake becoming the favored shape for $v_{red} < 0.5$.

Also displayed on the figure is the energy of two touching spheres joined by a narrow neck with $E_{doublet} \sim 16\pi\kappa_b + 4\pi\kappa_G$, an approximation valid at $C_o = 0$ for two shells joined externally as a bud or internally as a pocket, illustrated by the two shapes in Fig. 7.9(a). Fourcade *et al.* (1994) have shown that the bending energy of the neck region may not be important because the principal curvatures have opposite sign (resulting in a small mean curvature). In fact, shape III of Table 7.1 could be regarded as a section of a neck whose bending energy can be made small by a suitable choice of curvatures. With $\Delta E_{doublet}/\kappa_b \sim 8\pi$,

Fig. 7.9. Axially symmetric shapes whose bending energies are easy to calculate. (a) Doublet and extreme stomatocyte. (b) Stomatocyte with simple curvatures.

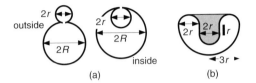

(a)   (b)

the energy of these configurations is independent of v_{red}; the reduced volume of the stomatocyte shape can range from 0 to 1, but the doublets obey $1/\sqrt{2} \leq v_{\mathrm{red}} < 1$. Higher multiplet states with n externally connected spheres can attain lower values of v_{red} (down to $1/\sqrt{n}$) at the cost of adding another $8\pi\kappa_b$ to the bending energy for each new sphere, at vanishing C_o. Shapes with small buds are clearly not favored in Fig. 7.8, and hence processes like the formation of a vesicle in Fig. 7.5 must involve more than just membranes with $C_o = 0$.

Figure 7.8 predicts that, of the shapes considered, the prolate spherocylinder is favored for $v_{\mathrm{red}} > 0.63$ and the extreme stomatocyte (internal cavity) for $v_{\mathrm{red}} < 0.63$. However, these shapes may not be true energy minima, as they have been selected only for numerical convenience. Softening the curves may raise or lower the energy of the configurations, and affect which shape has the lowest energy. The gentle stomatocyte in Fig. 7.9(b), for instance, has a bending energy of $\pi\kappa_b(9 + 4\pi/\sqrt{3}) + 4\pi\kappa_G$ at $C_o = 0$, as established in the problem set. The energy difference between this shape and E_{sphere} is $\pi\kappa_b(1 + 4\pi/\sqrt{3}) = 8.255\ \pi\kappa_b$, barely more than the doublet energy, and has a reduced volume $v_{\mathrm{red}} = 0.67$, close to the transition region in Fig. 7.8. Numerical searches have been performed within the spontaneous curvature model, and others, for the true minimal energy states, assuming axial symmetry and continuity of the curvature at the poles of the configuration. A selection of shapes from one such search at $C_o = 0$ are displayed in Fig. 7.10 as a function of v_{red} (Seifert *et al.*, 1991; specific configurations also have been considered by Canham, 1970, and Deuling and Helfrich, 1976). Although similar to our approximate boundary between stomatocytes and prolates at $v_{\mathrm{red}} = 0.63$, the numerical results yield a narrow range of reduced volume around $0.59 \leq v_{\mathrm{red}} \leq 0.65$ where the oblate shapes have the lowest energy. Further, these shapes are not ellipsoidal but rather biconcave, like the red blood cell.

Fig. 7.10. Energy-minimizing shapes in the spontaneous curvature model with $C_o = 0$, shown for selected values of the reduced volume v_{red}. Note how the shape changes dramatically at $v_{\mathrm{red}} = 0.59$ and 0.65. The axis of symmetry for all shapes is vertical (reprinted with permission from Seifert *et al.*, 1991; © 1991 by the American Physical Society).

stomatocyte   oblate   prolate

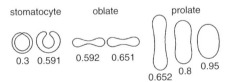

0.3   0.591   0.592   0.651   0.95
                                      0.652   0.8

What effect does a non-zero value for C_o have on the configuration energy? First, it introduces a length scale C_o^{-1}. For example, the bending energy of a spherical shell with radius r is $8\pi\kappa_b(1 - rC_o/2)^2 + 4\pi\kappa_G$ at $C_o \neq 0$. The first term vanishes at $r = 2/C_o$, with the result that the spherical shell with the lowest deformation energy has a particular size at $C_o \neq 0$. This conclusion is true for arbitrary shapes, meaning that the bending energy is a function of cell shape *and* size at $C_o \neq 0$. In addition, the sign of C_o influences the favored shape, as we can see by considering the two axisymmetric shells in Fig. 7.9(a). The doublet shape on the left has curvatures $1/r$ in the small bud and $1/R$ in the main body, whereas the extreme stomatocyte to its right has $-1/r$ in the cavity and $1/R$ on the exterior surface. As a result, the bending energies are simply

$$E_{outside} = 8\pi\kappa_b[(1 - rC_o/2)^2 + (1 - RC_o/2)^2] + 4\pi\kappa_G \qquad (7.14a)$$

$$E_{inside} = 8\pi\kappa_b[(1 + rC_o/2)^2 + (1 - RC_o/2)^2] + 4\pi\kappa_G, \qquad (7.14b)$$

where the labels refer to the position of the smaller shell and where the energy of the neck has been neglected. Let's take RC_o and rC_o to be the same for both configurations, meaning that the shells have the same areas but different enclosed volumes. The difference in their energies is then

$$E_{outside} - E_{inside} = -16\pi\kappa_b(r/R)RC_o, \qquad (7.15)$$

indicating that the doublet configuration is favored ($E_{outside} < E_{inside}$) if $C_o > 0$ and the stomatocyte shape is favored if $C_o < 0$. In other words, negative values of C_o favor shapes with regions of negative curvature.

What determines the magnitude and sign of C_o? We recall from Section 5.2 how the molecular shape parameter was related to the stability of structures like micelles and bilayers: lipids with head groups large in cross section compared to their hydrocarbon regions tend to form micelles which match the natural curvature of the lipids' molecular shape. A lipid monolayer, then, may have a spontaneous curvature arising from its molecular composition; measurements of this curvature can be found in Marsh (1996) or Chen and Rand (1997), for example. Similarly, spontaneous curvature can arise in lipid bilayers, either because the composition of each leaflet or its environment is inequivalent (see Döbereiner *et al.*, 1999, and references therein), or because of molecular packing arrangements within leaflets (Safran *et al.*, 1990). The "phase diagram" for vesicle shapes at non-zero C_o has been partially explored (Seifert *et al.*, 1991; Miao *et al.*, 1991) and is displayed in Fig. 7.11. The shapes along the $C_o = 0$ line correspond to Fig. 7.10, with the narrow domain of oblates visible at $v_{red} \sim 0.6$. Having regions of negative curvature, stomatocytes are more favored at negative C_o; in contrast,

Fig. 7.11. Partial phase diagram for shapes with spherical topology as a function of reduced volume v_{red} and spontaneous curvature C_o, written as the unitless combination $C_o(A/4\pi)^{1/2}$, where A is the surface area (redrawn with permission from Seifert *et al.*, 1991; see original paper for exact position of phase boundaries; © 1991 by the American Physical Society).

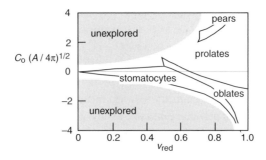

prolate ellipsoids and pears (doublets with smooth necks) are favored at large positive C_o. If the composition of a leaflet is multicomponent, there is a possibility that C_o is not uniform over the shell; one can imagine that molecules with conical shapes might migrate to regions of high curvature, such as the folds in Fig. 1.11(b). With C_o being a function of position on the membrane, the energy minimization problem becomes more involved (Seifert, 1993; Kumar *et al.*, 1999).

The spontaneous curvature model recognizes that the inner and outer leaflets of a bilayer may be compositionally inequivalent, or may be embedded in differing environments. However, it does not recognize that the two leaflets may be mechanically decoupled, in the sense that they can slide past one another to relieve stress that would arise if the membrane were a single continuous sheet like a metal plate. Specifically, consider the inner and outer leaflets of the small vesicle in Fig. 7.5. The lipid head groups in the outer leaflet form a spherical shell whose diameter is roughly double that of the inner shell of lipid headgroups. If the leaflets of this vesicle were mechanically coupled, such that they contained the same number of lipids, then the outer leaflet would have only a quarter the number of lipids per unit area as the inner leaflet, clearly a high-stress situation. The fact that the leaflets are not coupled reduces this stress, as extra material from the outer leaflet of the parent cell can be pulled onto the daughter vesicle largely independent of the make-up of the inner leaflet.

Fig. 7.12. Two arcs with a common center of curvature. The ratio of the outer to inner arc lengths, indicated by the bold lines, is equal to $(1 + Cd_{bl}/2)/(1 - Cd_{bl}/2)$, where C is the curvature of the mid-line between the arcs.

The difference in area of the leaflets ΔA can be written in a two-dimensional representation, so that it is easily incorporated into the spontaneous curvature model. Consider first the two arcs of a circle displayed in Fig. 7.12 (in bold) separated by a distance d_{bl} and having a common center of curvature; the curvature of the mid-line between the arcs is defined as C. By geometry, the ratio of the outer arc length to the inner one is just $(1 + Cd_{bl}/2)/(1 - Cd_{bl}/2) \approx 1 + Cd_{bl}$. The same result holds in an orthogonal direction, so that the ratio of the corresponding outer and inner areas is just $A_{outer}/A_{inner} = (1 + C_1 d_{bl})\cdot(1 + C_2 d_{bl}) \approx 1 + C_1 d_{bl} + C_2 d_{bl}$, where C_1 and C_2 are the principal curvatures. Thus, the difference between the areas of the outer and inner boundaries of a membrane segment (of area dA) is approximately $(C_1 d_{bl} + C_2 d_{bl}) dA$, and the total area difference, integrated over the surface, is

$$\Delta A \approx d_{bl} \int (C_1 + C_2) dA. \tag{7.16}$$

As expected, ΔA vanishes when $d_{bl} = 0$ or the integrated mean curvature vanishes. To be clear, even though it is obtained by summing over local differences in leaflet area, ΔA is a global quantity characteristic of the surface as a whole.

Now, ΔA for the vesicle shown in Fig. 7.5 is about half of the mean area of the inner and outer leaflets; in other words, ΔA need not vanish for an unstressed system. Thus, the energy associated with the leaflet area difference is proportional to the (square of the) *deviation* of ΔA from its unstressed value ΔA_o. The bending energy of the spontaneous curvature and area difference contributions can be parametrized as

$$E = (\kappa_b/2) \int (C_1 + C_2 - C_o)^2 \, dA + \kappa_G \int C_1 C_2 \, dA$$
$$+ (\kappa_{nl}/2) \cdot (\pi / A d_{bl}^2) \cdot (\Delta A - \Delta A_o)^2, \tag{7.17}$$

which is referred to as the ADE model (for *area difference elasticity*). The constant κ_{nl} is a non-local bending resistance, carrying units of energy, and can be related to the area compression modulus of the leaflets (Svetina *et al.*, 1985; Miao *et al.*, 1994). Two limiting cases of Eq. (7.17) include the spontaneous curvature model at $\kappa_{nl} = 0$ and the bilayer couple (or ΔA) model at $\kappa_{nl} / k_B T \to \infty$, where ΔA is driven to the fixed value ΔA_o. This latter approach was developed more than a decade ago and underlies the calculations appearing in Fig. 7.14 (Sheetz and Singer, 1974; Svetina and Zeks, 1985; see also Evans, 1974, and Helfrich, 1974a, b). One measurement of κ_{nl} (Waugh *et al.*, 1992) and estimates of κ_{nl} from the underlying bilayer deformation (Miao *et al.*, 1994) argue that κ_{nl}/κ_b is of order unity.

It can be shown that the set of stationary shapes in the ADE model (pears, ellipsoids, dumb-bells etc.) is the same as the spontaneous curvature approach (see Miao *et al.*, 1974); however, the energy of a given configuration will vary according to the values of C_o, κ_{nl}/κ_b etc. Thus, the phase diagram will be different from one model to the next. A section of the phase diagram for $\kappa_{nl}/\kappa_b = 4$ at $C_o = 0$ is displayed in Fig. 7.13, showing several shapes familiar from Fig. 7.10. The fixed area difference ΔA_o is plotted as the unitless quantity $\Delta A_o/[2d_{bl}(4\pi A)^{1/2}]$, which is equal to unity for a sphere. In bilayer language, adding excess lipid to the outer leaflet increases ΔA_o. From the configurations in the diagram, one can see how the area of the outer leaflet increases with respect to the inner leaflet as the shape changes from prolate ellipsoids to pears to multiplets at fixed volume. The nature of the transition from

Fig. 7.13. Partial phase diagram in the ADE model for the ratio of bending parameters $\kappa_{nl}/\kappa_b=4$, as a function of the unitless area difference and reduced volume v_{red}. At fixed v_{red}, the minimal energy shape goes from prolate to pear to multiplet as the area difference increases. The dashed lines are trajectories that a vesicle would follow as it is heated, causing v_{red} to rise (redrawn with permission from Miao *et al.*, 1994; © 1994 by the American Physical Society).

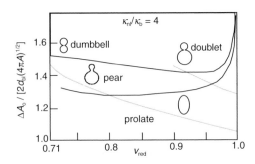

one generic shape to another (e.g. ellipsoids to pears) also depends upon the model: the transition may be smooth (continuous) in some cases while abrupt (discontinuous) in others.

The spontaneous curvature model and its variants allow the prediction of energy-minimizing shapes for a given parameter set, as demonstrated in Figs. 7.8, 7.10, 7.11 and 7.13. In addition, one can analyze the shapes of specific cells, vesicles or liposomes and determine the relevant parameters describing the shape within a given model. Of course, not all cells of a given type have exactly the same shape: for instance, even the blood cells drawn from one person vary in size and shape, potentially requiring different values of C_o or ΔA for their description (see Bull, 1977, for examples). Does this mean that our models are nothing more than parametrizations without predictive power? Certainly not. Once the characteristics such as C_o and ΔA have been determined for a specific cell, other attributes such as the cell's reduced volume can be varied (by heating, for example) and the resulting change in shape can be both predicted and compared against experiment. As an illustration, Figure 7.13 displays trajectories that specific cells should follow in configuration space under the ADE model (Miao *et al.*, 1994). Fig. 7.14 contains a specific sequence from the ΔA model, plotted over the very narrow temperature range 43.8–44.1 °C; the predicted shapes reproduce the observed ones very well (Berndl *et al.*, 1990; see also Käs and Sackmann, 1991). The membrane area of a vesicle increases faster with temperature than does its enclosed volume, so that the reduced volume decreases with rising temperature. By following the trajectories and determining the stability range of specific shapes and the nature of the changes between shapes, differences between models for the bending energy can be exposed; for instance, the spontaneous curvature model does not capture all of the physics of the shape transitions (for example, Döbereiner *et al.*, 1997).

Fig. 7.14. Predicted shapes as a function of temperature within a curvature-based energy description of vesicle shape (reprinted with permission from Berndl *et al.*, 1990; © 1990 by EDP Sciences). The cross sections have axial symmetry and the reduced volume v_{red} increases from left to right.

44.1 °C   44.0 °C   43.9 °C   43.8 °C

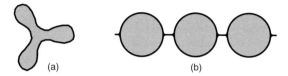

(a)   (b)

Fig. 7.15. (a) Stable starfish shape for $v_{red}=0.5$ (redrawn from Wintz et al., 1996). (b) Representation of pearls (spheres connected by narrow capillaries) appearing in tubular membranes under tension (after Bar-Ziv and Moses, 1994).

At ever smaller values of reduced volume v_{red}, the menu of available shapes grows larger, and new phenomena appear. One of these is the loss of axial symmetry, as exemplified by the starfish shape in Fig. 7.15(a) (Wintz et al., 1996; see also Bar-Ziv et al., 1998). The configuration is a locally stable shape in a curvature-based model for the membrane energy, and it mirrors the observed shape very well. Within the spontaneous curvature model at vanishing C_o, this shape has an energy difference with respect to the sphere of $1.90 \cdot 8\pi\kappa_b$ and a reduced volume of $v_{red}=0.5$, so that it lies close to the pancake and spherocylinder energies in Fig. 7.8 at the same v_{red}.

Vesicles with non-spherical topology, for instance doughnut shapes with one or more holes, also have been discovered (Mutz and Bensimon, 1991; Michalet et al., 1994). As determined in the problem set, an axially symmetric doughnut with a single hole (a torus) may have an energy as low as $4\pi^2\kappa_b$ in the spontaneous curvature model with $C_o=0$, there being no Gaussian rigidity contribution to this topology. This is for a torus with a specific shape, called a Clifford torus, having a reduced volume $v_{red}=0.71$; its bending energy is $(\pi/2)E_{sphere}=1.57E_{sphere}$ when $\kappa_G=0$. Referring to Fig. 7.8, the energy of the Clifford torus is comparable to, or lower than, the energies of the spherical topologies in the region $v_{red}\sim 0.7$, demonstrating that shapes with more exotic topologies than spheres are energetically accessible at lower values of v_{red}. Within the spontaneous curvature model, energy-minimizing shapes of toroidal vesicles have been investigated by Seifert (1991), including non-axisymmetric shapes; even more exotic topologies have been examined by Jülicher (1996).

Tensile stress also affects membrane behavior. Applied to tubular configurations like straws, where the membrane forms a fluid boundary, tension may induce "pearling", in which the fluid contents of a tube becomes segregated into regularly spaced spheres linked by narrow capillaries, as shown in Fig. 7.15(b) (Bar-Ziv and Moses, 1994). Although instabilities against undulations for tubes are expected from energy-minimizing shapes within the spontaneous curvature model (Bar-Ziv and Moses, 1994), the dynamic flow of material also contributes to the formation of pearls (Nelson et al., 1995; Goldstein et al., 1996).

To summarize, we have adopted the spontaneous curvature model as a simple representation for the bending energy of a fluid membrane and shown how energetics affects the equilibrium shape of vesicle-like systems. The model allows us to understand analytically many qualitative features of artificial vesicles manipulated under controlled conditions. However, the model does not capture all the features of shape changes observed experimentally, necessitating the inclusion of effects arising from, for example, the deviation from equilibrium of the mean area per lipid in each leaflet. In the next section, we investigate simple composite systems consisting of fluid sheets and shear-resistant networks such as the cytoskeleton or cell wall.

7.4 Vesicles and red blood cells

The membrane shapes introduced in Section 7.3 are among those adopted by cells and their compartments: red blood cell shapes include biconcave discs while some bacteria resemble spherocylinders. However, biological systems have more structural components than just a membrane, forcing us to ask which aspects of cell shape are determined by membrane bending energy alone, and which arise from other mechanical elements. In this section, we discuss the architecture of two simple systems – small structureless vesicles and mammalian red blood cells with a membrane-associated cytoskeleton.

Vesicles

The biochemical factories of eucaryotic cells tend to be concentrated in localized industrial districts such as the extensive protein production facilities found on the endoplasmic reticulum. Transporting biochemical products from their point of manufacture to their usage sites involves several familiar steps:

(i) packaging the products to prevent loss during transport
(ii) labelling the package so it can be recognized at its destination
(iii) shipping the package along an efficient transportation route.

Here, we describe step (i), the packaging process itself. Item (ii) is beyond the scope of this book; however, the third topic is covered in Chapter 9 under molecular motors.

   The transport containers of the cell are small membrane-bound vesicles, typically 100 nm in diameter. When produced, a vesicle pinches off from an existing membrane by the process of endocytosis, capturing material from the fluid environment on the opposite side of the parent membrane. Upon arrival at its destination, the vesicle may fuse with another membrane *via* exocytosis, releasing its contents to the medium on the opposite side; budding and fusion of vesicles are illustrated in Fig.

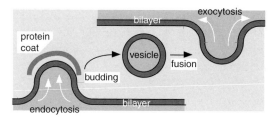

Fig. 7.16. Endocytosis and exocytosis involve the creation of a vesicle and its subsequent fusion after transport within a cell. Often, the budding process is aided by the attachment of coating proteins (e.g. clathrin or coatomer) to the parent membrane; clathrin, but not coatomer, is released once the vesicle pinches off.

7.16. Both the vesicle's fluid interior and its membrane provide locations for carrying cargo to a new location. Looked upon solely as a bilayer, the bending energy of a spherical vesicle in the ADE model of Eq. (7.17) is

$$E = 8\pi\kappa_b + 4\pi\kappa_G + (\kappa_{nl}/2)\cdot(\pi/Ad_{bl}^2)\cdot(\Delta A - \Delta A_o)^2, \qquad (7.18)$$

when the spontaneous curvature C_o vanishes. Depending on the flow of the inner and outer leaflets of the membrane during the budding process, the bending energy associated with individual leaflets may be augmented by the non-local contribution of the area difference $\Delta A - \Delta A_o$.

In the laboratory, the formation of buds has been studied for pure bilayer vesicles, where the bending energy is unquestionably important. When the reduced volume v_{red} is less than unity, budding appears in the phase diagram of Fig. 7.11 at sufficiently positive spontaneous curvature, and in Fig. 7.13 when the excess area in the outer leaflet is large enough (Miao *et al.*, 1991, 1994; Seifert *et al.*, 1991; Döbereiner *et al.*, 1997). In cells, the formation of vesicles is mediated by proteins such as clathrin, coatomer and others (see Chapter 13 of Alberts *et al.*, 1994). A three-armed protein with the appearance shown in Fig. 7.17 (Ungewickell and Branton, 1981), clathrin self-assembles to form regular two-dimensional arrays of hexagons about 25 nm across, with a small fraction of penta-gons. There is one clathrin molecule per vertex in the hexagonal array (Crowther *et al.*, 1976); at six segments per molecule and 3/2 lattice "bonds" per lattice vertex, there must be four segments lying along each bond. Indirectly linked to the membrane by the protein adaptin, the curved clathrin coat deforms the parent membrane to produce buds of relatively uniform curvature and size (Heuser, 1980). The coat does more than induce curvature, however; adaptin is not attached to membrane lipids but rather to a transmembrane receptor protein capable of holding a cargo molecule on the opposite side of the bilayer. Thus, the coat both induces budding and also helps organize the contents of the vesicle. Once the vesicle is free of the parent membrane, the clathrin coat is released by hydrolysis of ATP. In contrast, the formation of coatomer coats is driven by ATP hydrolysis, and the coat is not shed until the vesicle arrives at its destination. The role of lattice defects in generating curvature of the clathrin meshwork is described in Mashi and Bruinsma (1998).

Fig. 7.17. The protein clathrin has a three-legged shape called a triskelion; six segments of these arms lie in a plane and are available to bond with neighboring molecules, while the three terminal segments lie below the plane of the drawing.

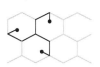

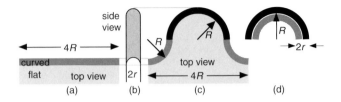

The curvature of the clathrin coat provides a natural length scale for the size of a vesicle budding from a planar parent membrane. Budding from a curved membrane, such as the perimeter of a pancake in the Golgi stack (see Figs. 1.8 and 1.9) may also involve a preferred vesicle size, even if C_o vanishes. Although the bending energy of a vesicle is independent of its radius in the spontaneous curvature model with $C_o=0$, the energy of the path taken by the bud depends upon its curvature. In Fig. 7.18, we show a particular path for a bud at the edge of a pancake. The surface consists of two planar membrane regions (the top one is indicated by light shading in part (a)) connected by half a cylinder of radius r and length $4R$ (indicated by the darker shading); a cross sectional view of this curved edge can be seen in the lower half of part (b). We let the bud push out in the same plane as the pancake, so its top and bottom sections remain flat, as seen by the top view in (c) and side view in (b); its edge is characterized by curvatures $C_m=r^{-1}$ along a meridian and $C_p=R^{-1}$ in the plane, as indicated. The curved regions of the bud can be assembled into half a torus, with radii r and R, as seen in (d).

With their simple geometry, the bending energy of configurations (a) and (c) in Fig. 7.18 are easily calculated in the spontaneous curvature model with $C_o=0$: as half a cylinder, initial configuration (a) has bending energy $2\pi\kappa_b\rho$, where $\rho=R/r$, while configuration (c) has half the energy of a torus, or $\pi^2\kappa_b\rho^2/(\rho^2-1)^{1/2}$ (see problem set). The smallest difference in energy between these two configurations is about $10\kappa_b$ for bud radii near $R\sim2r$. Although we have not followed the pathway to completion, nor have we considered arbitrary shapes, the bending energy suggests that buds and vesicles with radii on the order of the thickness of the pancake will be favored. In a cell, vesicles in this size range are observed to break off from edges of the Golgi stack; even here, however, coating proteins play a role in vesiculation, so bending energy may not be the primary factor in determining the size of these vesicles.

Red cell shapes

Their biological importance, their easy availability and their mechanical simplicity has made the mature human erythrocyte an attractive cell to study both experimentally and theoretically. In our circulatory system, a

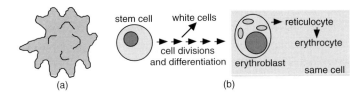

(a)   (b)

Fig. 7.19. (a) Among the shapes adopted by human erythrocytes at lower reduced volumes are lumpy echinocytes. (b) From common stem cells, blood cells differentiate by successive division; in the last step, an erythroblast loses its nucleus to become a reticulocyte, changing to an erythrocyte with the loss of other organelles.

typical red cell has a surface area of $\sim$135 μm^2 and a volume of $\sim$95 μm^3 with 10–20% variation (see Steck, 1989), corresponding to a reduced volume $v_{red} = 6\sqrt{\pi}\, V/A^{3/2} = 0.64$; note that red cells in other animals may be up to five times smaller or ten times larger in linear dimension than in humans. Referring to Figs. 7.8 and 7.10, this region of v_{red} is something of a cross roads in the phase diagram, with a variety of shapes having similar bilayer bending energy. It should be no surprise, then, that altering the environment of a blood cell can drive the cell away from its customary discocytic or biconcave shape. Catalogs of red cell shape exist (Bessis, 1973; see also Steck, 1989) and include the stomatocytes of Fig. 7.6(a) in addition to the more exotic echinocytes of Fig. 7.19(a), whose bumps may be gentle or rough, depending on conditions.

What factors contribute to this assembly of diverse shapes? First, the bending energy of the plasma membrane must be taken into account, recognizing that the molecular composition of its two leaflets is not symmetric – the inner leaflet tends to be richer in anionic phospholipids such as phosphatidylethanolamine (PE) and phosphatidylserine (PS), while the outer leaflet contains principally uncharged lipids such as phosphatidylcholine (PC), among many other molecular components. Thus, it would not be surprising if the red cell membrane had a non-zero spontaneous curvature C_o. Canham (1970) analysed a collection of more than 30 human red cells (flaccid, swollen and stomatocytes) within the spontaneous curvature model with $C_o = 0$, confirming that these shapes were close to those which minimized the model bending energy at fixed volume. Of course, the mean size of a red cell is not predicted by the bending energy, but presumably reflects the ease of transporting the cell through the circulatory system, and the efficiency of its operation in a capillary. The cell volume is maintained osmotically through the active regulation of sodium and potassium ion concentrations across the plasma membrane (see Beck, 1991).

The membrane-associated cytoskeleton also provides deformation resistance in a red cell, although the structure of the cytoskeleton may evolve during the maturation of the cell. Blood cells have a sufficiently short lifetime that they must be continuously and copiously produced in the body. Descendants of self-renewing stem cells, blood cells differentiate during successive cell divisions into families of white and red cells, as depicted in Fig. 7.19(b). Even the last cell in the sequence that yields an erythrocyte undergoes dramatic changes in its lifetime: born as a nucleated erythroblast, the cell expels its nucleus to produce an irregularly shaped reticulocyte. Shortly after abandoning the bone marrow for the circulatory system, the reticulocyte loses its ribosomes and remaining organelles to become a mature erythrocyte with its familiar discocytic shape (see Chapter 22 of Alberts *et al.*, 1994). The shape of the cytoskeleton is similar to the discocyte, although independent studies in which the plasma membrane is washed away by a detergent have found that the cytoskeleton may be somewhat smaller (Lange *et al.*, 1982; for a review, see Steck, 1989) or larger and more spherical, than the parent cell (Svoboda *et al.*, 1992). The contribution of the cytoskeleton to the erythrocyte rest shapes remains a topic of active study.

Aspirated red cells

Irrespective of its contribution to the rest shape of the red cell or to elasticity under weak deformations, the cytoskeleton dominates the cell's resistance to strong deformation, as we now demonstrate by considering a specific situation commonly used in mechanical studies of the cell. Micropipette aspiration involves the application of a suction pressure to a cell via a micropipette 1 μm or so in diameter; experimental images of aspirated red cells can be seen in Figs. 1.14 and 3.21(a), and of an auditory outer hair cell in Fig. 3.22. The deformation in the figures principally involves bending the membrane, and stretching and shearing the network, while simultaneously preserving the overall areas of the cytoskeleton and membrane (the latter is relatively incompressible). To estimate the relative energies of bending and stretching, let's examine the tip of the aspirated segment shown in Fig. 7.20(a), where the network deformation is a dilation. In experiments, the stretched area at the tip may be twice as large as the unstressed area (Discher *et al.*, 1994). From Eq. (3.16), the energy density of a two-dimensional isotropic material under dilation is $(K_A/2) \cdot (u_{xx} + u_{yy})^2$, where K_A is the area compression modulus and u_{ij} is the strain tensor; the sum $u_{xx} + u_{yy}$ equals the relative

Fig. 7.20. (a) Simulation of the aspiration of a red cell, showing the dilution of the cytoskeleton at the tip of the aspirated segment. The micropipette is represented by the grey bars at the sides. (b) Face view of a schematic network being drawn up a pipette of radius R_p perpendicular to the plane. When aspirated, the shaded region of the radius r_o is reduced to an annulus of radius r, plus the cylinder and hemisphere inside the pipette.

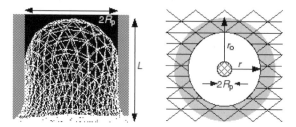

change in area $\Delta A / A$. Choosing $u_{xx} + u_{yy} = 1/2$ and $K_A = 10^{-5}$ J/m² (see Section 3.5) as representative of moderate deformations, we expect the energy density of the stretched network to be $\sim 10^{-6}$ J/m². Now, the energy density of the bending mode of the spherical cap in Fig. 7.20(a) is $(\kappa_b/2)(2/R_p)^2$, where R_p is the pipette radius (note that we have ignored the Gaussian and area-difference contributions). With $R_p = 1$ μm and $\kappa_b = 20 k_B T$ as a typical experimental situation, the bending energy density is 1.6×10^{-7} J/m². The order of magnitude difference between stretching and bending depends upon our choice of geometry, etc., but is not atypical of the configuration as a whole; in general, we find that the bending energy of the membrane is not as important as the stretch and shear of the cytoskeleton at large deformations.

The pure dilation of the network at the tip of the aspirated segment is a special situation; in general, the network is subject to shear whose magnitude can be estimated from Fig. 7.20(b). When material is drawn up into the pipette, the remaining network in the plane outside is strained as it moves closer to the pipette entrance. In Fig. 7.20(b), for example, the material at the boundary of the shaded region is moved from a distance r_o to a distance r of the pipette center. This results in a strain because the distance between network vertices along the arc at r falls to r/r_o of its original, unstrained value. The strain along the arc is then $\Lambda - 1$, where $\Lambda = r_o/r$. The area of the network within r_o before aspiration is πr_o^2, but this section covers $\pi[r^2 + 2R_p L - R_p^2]$ afterwards (a sum of $\pi[r^2 - R_p^2]$ from the plane, $2\pi R_p[L - R_p]$ from the cylinder in the pipette and $2\pi R_p^2$ from the hemispherical cap). Taking the overall area to be unchanged by the deformation, then $\pi r_o^2 = \pi[r^2 + 2R_p L - R_p^2]$ and

$$\Lambda^2 = 1 + (R_p/r)^2 (2L/R_p - 1). \tag{7.19}$$

As expected, this part of the strain vanishes (i.e., $\Lambda \to 1$) as $r \to \infty$, and has its maximum value $\Lambda^2 = 2L/R_p$ at the pipette entrance $r = R_p$. Demanding that area be conserved locally, the in-plane deformation is a shear, with network elements moving further apart in a radial direction, but closer together around a hoop.

By imposing a shear on the network, the aspiration deformation can be used to extract its shear modulus μ. From Laplace's law, Eq. (7.5), the force per unit length pulling the cylindrical section of the network up the pipette is $\Delta p R_p/2$, as visualized for a hemispherical cap in Fig. 7.2, where Δp is the pressure difference applied by the pipette. Balancing forces leads to the expression (Waugh and Evans, 1979; earlier work includes Evans 1973a,)

$$\Delta p = (2\mu/R_p) \int_{R_p}^{\infty} (\Lambda_p^2 - 1/\Lambda^2) r^{-1} \, dr. \tag{7.20}$$

This integral can be evaluated exactly to yield

$$\Delta p = (\mu / R_p)[(2L/R_p - 1) + \ln(2L / R_p)]. \tag{7.21}$$

Several of the elastic moduli quoted in Chapter 3 were obtained by analyzing the aspiration length as a function of pressure using Eq. (7.21). Although this particular deformation mode is of more interest in the laboratory than in the circulatory system, it illustrates the importance of the cytoskeleton to the cell's deformation resistance.

7.5 Simple bacteria

With no cytoskeleton to distort their membranes in peculiar ways, bacteria tend to choose from among a small set of basic cell shapes, although, as always in biology, there are many exceptional cases. Common bacterial shapes are displayed in Fig. 7.21. Cells in one category of shapes are approximately spherical; referred to as *cocci* (plural of coccus) they typically have diameters of a few microns. Rods or *bacilli* (plural of bacillus) make up a second category; these are often 0.5–1 μm in diameter and several times that in length, although they may range up to 500 μm long for the cigar-shaped *Epulopiscium fishelsoni*. A third category includes corkscrew-shaped bacteria such as rigid *spirilla* and flexible *spirochetes*, whose tubular cross section may have a diameter of just 0.1 μm and an overall length of a few microns (although, again, some may approach hundreds of microns). In addition to these main classifications, there are more exotic shapes such as flat plates or stalks with bulbous ends.

What factors influence bacterial shape? Most bacteria support an elevated osmotic pressure which may be as high as a few tens of atmospheres, easily sufficient to inflate the cell into a sphere if its envelope were isotropic. This pressure difference P across the envelope generates a two-dimensional tension within it which, from Eq. (7.3), is equal to the product RP, give or take a factor of two, where R is the radius of the sphere or cylindrical cross section. For instance, with $P = 10^6$ J/m^3 (10 atm) and $R = 10^{-6}$ m, the tension is 1 J/m^2, well above the bilayer rupture tension of $\sim 10^{-2}$ J/m^2 on laboratory time-scales (see Sec. 5.3). As a consequence, the bilayer of micron-sized bacteria under pressure must be mechanically reinforced by a cell wall. Notable exceptions are tiny mycoplasmas just 0.3 μm in diameter, which could operate at half an

Fig. 7.21. Bacterial shapes include spheres (or cocci), rods (or bacilli) and spirals (spirilla and spirochetes), among others. Typical dimensions for each shape are quoted in the text; the configurations are not drawn to the same scale.

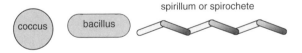

spirillum or spirochete

coccus   bacillus

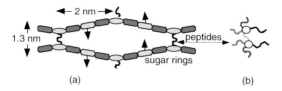

(a)   (b)

Fig. 7.22. (a) Face view of a section of a peptidoglycan network: polymeric ropes of sugar rings run in one direction and peptide strings form transverse links. (b) As seen along the glycan axis, peptides can link glycan ropes in several directions.

atmosphere without a cell wall by virtue of their small radius and osmotic pressure (and hence small RP). As can be seen from the image in Fig. 1.6, the shape of a mycoplasma is more irregular than the pressurized *E. coli* in Fig. 1.1(c).

The tension τ borne by the cell boundary is isotropic in spherical cocci, but depends on direction for rod-shaped bacilli: from Eq. (7.3), τ at the end caps and along the axial direction is $RP/2$, but τ_θ around the girth of the cell (hoop direction) is RP, twice as large. Not surprisingly, the cell accommodates this anisotropic stress with an anisotropic cell wall. The molecular composition of the bacterial cell wall (plant walls are different!) are chains of sugar rings running in parallel, linked transversely by floppy peptide strings, to form peptidoglycan. A commonly accepted model for the connectivity of the network, illustrated in Fig. 7.22, has the amino-acid linkages at every second sugar arranged in a helix, such that every fourth linkage points in the same direction.

The glycan ropes have one cross-link available on each of the 1.0 nm long disaccharide units, although not all links are attached to neighboring glycan strands. Looking along the axis of a glycan rope, the peptides radiate like a helix, as in Fig. 7.22(b), such that there are four disaccharides, or 4 nm, between peptides linking the same pair of glycans. The peptide strings themselves have a contour length of 4.3 nm (Braun *et al.*, 1973) but an end-to-end displacement of just 1.3 nm (Burge *et al.*, 1977), a reduction that one would expect for floppy chains. As described in Section 3.1 and shown in Fig. 1.7, the cell wall of a Gram-negative bacterium is relatively thin; a thickness of 3–8 nm would correspond to 2–6 glycan strands, with an interchain separation of 1.3 nm. In contrast, Gram-positive cell walls are much thicker at 25 nm, corresponding to 20 or so glycans. It's no surprise, then, that Gram-positive bacteria can support more than 20 atmospheres pressure, compared with 3–5 atmospheres for Gram-negative bacteria (as low as 0.85–1.5×10^5 J/m^3 in some cases, see Arnoldi *et al.*, 2000). Clearly, there must be unconnected peptides in gram-negative walls owing to their small thickness, and this is observed for gram-positive walls as well (Zipperle *et al.*, 1984).

Measurements of the elastic moduli of the bacterial cell wall have concentrated on obtaining direction-dependent Young's moduli Y. In one approach, a sample of highly aligned walls from the Gram-positive bacterium *Bacillus subtilis* was found to have $Y = 1.3 \pm 0.3 \times 10^{10}$ J/m^3

(dry) and 3×10^7 J/m³ (hydrated) from an analysis of stress/strain curves (Thwaites and Surana, 1991). The wall material in these experiments was oriented with the peptide strings largely parallel to the axis of applied tension, so $Y \sim 3 \times 10^7$ J/m³ reflects the "soft" direction of peptidoglycan. Another approach using an atomic force microscope to deform peptidoglycan from the gram-negative bacterium *E. coli* gave a similar value for the average Young's modulus of 2.5×10^7 J/m³ for hydrated material, which varied by a factor of two according to direction (Yao *et al.*, 1999). Dried samples of the *E. coli* wall were more rigid at 3–4×10^8 J/m³, although not as stiff as *B. subtilis*.

We would expect the cell walls of bacilli and other non-spherical bacteria to be anisotropic, given the stress relationship of Eq. (7.3), where the tension along the axis of symmetry is half that of the hoop. Young's moduli for *E. coli* walls are softer along the cylindrical axis, and it has been found in other studies of *E. coli* that the peptide strands tend to be oriented along this axis (Verwer *et al.*, 1978), while the glycans lie in the hoop direction. This design strategy has been employed elsewhere in the bacterial world in the construction of tubular sheaths within which bacteria elongate and divide. There, microscopic rings of material from the sheath have been isolated and imaged for the bacterium *Methanospirillum hungatei* (Southam *et al.*, 1993), although one should note that this sheath material is much stiffer than peptidoglycan (Xu *et al.*, 1996).

Let's now try to understand the elastic behavior of peptidoglycan at a microscopic level, beginning with the properties of the individual chains. The peptide appears to have the classic properties of an entropic spring, given that its end-to-end length in the network ($r_{ee} = 1.3$ nm) is much less than its contour length ($L_c = 4.2$ nm). We analyze the peptide strand using two relationships valid for individual strings, recognizing that a filament may have different behavior in isolation than in a network. The persistence length ξ_p of the string can be obtained from $\langle r_{ee}^2 \rangle = 2\xi_p L_c$, yielding $\xi_{p,pep} = 0.2$ nm for peptides; this value is somewhat lower than alkanes, perhaps indicating the presence of other attractive interactions within the chain. The effective spring constant for an ideal chain in three dimensions is $k_{sp} = 3k_B T / \langle r_{ee}^2 \rangle$, which gives $k_{sp,pep} = 7.1 \times 10^{-3}$ J/m² for peptides. The glycan chains are much stiffer: $\xi_p \sim 10$ nm for a typical unbranched polysaccharide (Stokke and Brant, 1990; Cros *et al.*, 1996).

Suppose we assume the filaments in peptidoglycan are sufficiently flexible that the material behaves like the networks of floppy chains studied in Chapters 3 and 4. There, we showed that the shear modulus μ was close to $\rho k_B T$ for both two- and three-dimensional polymer networks, where ρ is the (2D or 3D) density of chains. To apply this benchmark to peptidoglycan, we evaluate the chain density using the peptide connectivity in the three-dimensional model of Fig. 7.23, where the network "bonds" are drawn as heavy lines and the glycan–peptide junctions are shown as disks. Four glycan chains, with disaccharide length b = 1 nm, are displayed in the figure, linked by five peptide strings of length a = 1.3 nm. All elements are drawn as straight lines for convenience. The rectangular box of volume $a \times a \times 4b$ in the diagram effectively contains four vertices; the eight vertices at the corners are each shared with eight adjoining boxes, while the twelve vertices along the edges are shared with four adjoining boxes, giving a net total of $8/8 + 12/4 = 4$ vertices. Thus, the density of vertices is $1/a^2b$. A vertex joins three bonds – two glycans and one peptide – each of which is shared by another vertex; consequently, there are 3/2 bonds per vertex, and the bond density is $\rho = 3/2a^2b$. Using the bond lengths from the figure, the benchmark modulus is $\mu = \rho k_B T = 3.6 \times 10^6$ J/m³. The relationship between Y and μ for an isotropic system can be obtained from Eqs. (D.32)–(D.35), where we show that for many materials, $Y = K_V = (8/3)\mu$. Thus, we expect Y to be about 1×10^7 J/m³ in this representation, which is surprisingly close to $Y = 2$–3×10^7 J/m³ observed experimentally. The accuracy of this model can be improved by recognizing that the glycan strands do not extend unbroken around the equator of the bacterial cylinder; instead, the strands are rather short with a mean length of 5–10 disaccharide units (or 5–10 nm; Höltje, 1998). Fragmentation of the glycan reduces the anisotropy of the Young's modulus, and helps explain why a floppy network provides a good approximation for peptidoglycan.

Just a few nanometers thick, Gram-negative cell walls appear somewhat like Fig. 7.23 in cross section: many glycan strands are laced together in an extended mat that is only a few strands in depth. Thus, there are many disconnected peptide links and the two-dimensional representation of network elasticity is easily determined. For example, the planar rectangular network in Fig. 7.24(b) has two inequivalent springs of force constant k_x and k_y, with rest lengths L_x and L_y. As established in Problem 3.16, the area compression modulus of this network is

$$K_A^{-1} = (L_y/L_x)k_x^{-1} + (L_x/L_y)k_y^{-1}. \tag{7.22}$$

which simplifies when $L_x \sim L_y$ and one spring is much stiffer than the other. As a model for a single layer of peptidoglycan, Eq. (7.22) predicts

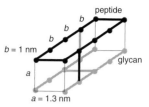

Fig. 7.23. Section of an idealized peptidoglycan network; heavy lines are peptide or glycan chains and disks designate the junctions between chains. The segment lengths are $b = 1$ nm for dissacharides and $a = 1.3$ nm for peptides.

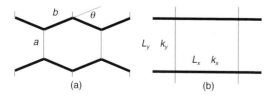

(a)   (b)

Fig. 7.24. Examples of planar networks capturing some features of Gram-negative cell walls.

$K_A = 1.3k_{sp,pep} \sim 0.9 \times 10^{-2}$ J/m^2. Invoking the relationship $K_V \sim K_A/d$ for a thin plate of thickness d, this value of K_A returns us to a prediction of K_V in the few times 10^7 J/m^3 range when d is a few nanometers. This prediction is close to the benchmark $K_A \sim 2\mu \sim 2\rho k_B T$ for two-dimensional networks, which yields $K_A = 0.5 \times 10^{-2}$ J/m^2 using an in-plane area of 2.5 nm^2 per disaccharide unit (Wientjes *et al.*, 1991), at 1–1.5 effective chains per unit. Yet another approach may include the bending resistance of the glycan strands. For example, the planar network shown in Fig. 7.24(a) is treated in Problem 3.15 for a filament bending energy of $k_{bend}\theta^2/2$ at each vertex, leading to

$$K_A = 4ak_{bend}/b^3, \tag{7.23}$$

where the peptide cross links are taken to be rigid. Variations of this model with soft peptides and flexible glycans are easily evaluated.

Our microscopic view of the origin of cell elasticity is only part of the picture: to accommodate cell growth, the wall must be able to continuously incorporate new strands of peptidoglycan (see Goodell, 1985). This is a rather delicate operation for Gram-negative cell walls, where the lateral stress may be borne by just a single layer of peptidoglycan. A number of possible mechanisms for wall growth have been advanced (see reviews by Höltje, 1993; Koch, 1993; and Thwaites, 1993), which generally recognize the importance of forming new network bonds before breaking old ones (see Koch, 1990, for an overview of wall growth issues). New strands may be added with random orientation to spherical cocci, as the network is presumably isotropic. However, strands must be added with a preferred direction to rod-shaped bacteria, both to accommodate the anisotropic tension and to direct growth along the cylindrical direction of the cell. That is, rod-shaped bacteria do not expand like a balloon, but rather elongate and divide like a string of sausages. The mechanism of wall growth and failure remains an area of active study.

7.6 Summary

The mechanical elements of a simple cell or vesicle are just a fluid membrane and perhaps a cell wall or cytoskeleton forming a compound envelope, whose elastic moduli, including bending, compression and shear resistance, strongly influence cell shape. A pressure difference P across the boundary gives rise to a lateral tension whose magnitude and direction-dependence is determined by the local curvature of the surface and which obeys the Young–Laplace equation $\tau_1/R_1 + \tau_2/R_2 = P$, where R_1 and R_2 are the local radii of curvature along directions 1 and 2. For example, the tension around the ring direction of a cylinder of radius R is RP, while the corresponding tension along the axis is $RP/2$. The bending energy of many shapes with constant, but not isotropic, local curvature can be evaluated analytically for simple models of the energy density. The most thoroughly investigated approach is the spontaneous curvature model introduced in Section 6.2, with a free energy density given by $\mathcal{F} = (\kappa_b/2) \cdot (C_1 + C_2 - C_o)^2 + \kappa_G C_1 C_2$, where κ_b and κ_G are the bending rigidity and Gaussian rigidity, respectively. Unrestricted in sign, the spontaneous curvature parameter C_o may arise from the chemical inequivalence of the bilayer leaflets or their environment. Integrated values of $\mathcal{F}$ for a selection of surfaces are summarized in Table 2.1; these expressions can be used to construct the bending energies of a variety of surfaces with axial symmetry.

Our understanding of membrane energetics is applied to three situations: pure bilayer vesicles, the red blood cell, and rod-shaped bacteria. Synthetic vesicles of micron dimensions have been well-studied experimentally, although their membranes do not possess the protein components of 100 nm vesicles in cells. The phase diagrams of spherical or toroidal vesicles have been evaluated within both the spontaneous curvature and area-difference elasticity models. Although the true energy-minimizing shapes usually must be found numerically, analytical approximations are available which allow us to determine the dependence of vesicle shape on its reduced volume $v_{red} = 6\sqrt{\pi} \, V/A^{3/2}$, where A and V are the vesicle's surface area and volume, respectively. In the spontaneous curvature model with $C_o = 0$, the energetically favored state with axial symmetry is prolate (cigar-shaped) as v_{red} initially decreases from unity, changing to a biconcave disk near $v_{red} = 0.65$, then becoming a stomatocyte at v_{red} less than 0.59. Interestingly, shapes such as dumb-bells and doughnuts, which have been observed experimentally, also become energetically competitive around $v_{red} \sim 0.65$. We note that the red cell operates close to this volume, permitting shape change with little cost in energy. The phase diagram of vesicle shapes has been investigated experimentally by following trajectories in v_{red} induced by changing the temperature of synthetic vesicles. The shapes of vesicles as they pinch off from a membrane can be studied within the spontaneous curvature model, and we show how the vesicle radius is of the order of the inverse curvature of the parent membrane, in the absence of other factors. Of course, other factors abound, including the presence of coating proteins such as clathrin (some of whose properties are examined in the problem set).

The deformation resistance of a red cell is strongly influenced by the presence of its cytoskeleton, many aspects of which are covered in Chapter 3. Here, we describe the deformation of a compound membrane induced by micropipette aspiration, which can be used to determine the cytoskeleton shear modulus. As a last example, we examined the shapes and mechanical properties of the bacterial cell wall. For micron-sized cells, the plasma membrane is able to sustain a pressure difference of just a tenth of an atmosphere, and must be strengthened by a cell wall to withstand internal pressures of 3–5 atmosphere in a Gram-negative bacterium (thin wall) or more than 20 atmospheres in a Gram-positive bacterium (thick wall). The wall material is peptidoglycan, an anisotropic fabric consisting of short glycan ropes (5–10 dissacharide units) laid out in parallel and cross-linked at 1 nm intervals by strings of amino acids. These cross-linking peptides have a contour length of 4.2 nm, but a much shorter end-to-end displacement of 1.3 nm. The convoluted peptide strings, and relatively short glycan chains suggests that entropy may strongly influence the elastic properties of the wall. In fact, the prediction from isotropic polymer networks is that the Young's modulus $Y \sim 3\rho k_B T$, where ρ is the density of polymer chains, is within a factor of two or three of the experimentally measured value of $Y \sim 2$–3×10^7 J/m^3. Of course, this simple model is not the whole story, in that the fabric in a rod-shaped bacterial cell wall must be at least partially aligned to accommodate its anisotropic stress, and to encourage cell growth along the cylindrical axis.

Problems

Biological applications

7.1 Suppose that we have a synthetic solid material with volume compression modulus $K_V = 3 \times 10^9$ J/m^3, out of which we wish to make membranes to enclose

(i) a spherical cell of radius 10 μm
(ii) a spherical weather balloon of radius 10 m.

Both membranes must support a pressure difference of one atmosphere (10^5 J/m^3). Assuming that the membrane fails if the lateral tension exceeds 5% of its area compression modulus K_A, what minimum membrane thicknesses are required for the cell and balloon? (*An approximate relationship between K_A and K_V for solids is given at the end of Section 5.3.*)

7.2 The inner and outer membranes of the nuclear envelope are joined by circular, protein-lined pores with the approximate dimensions indicated in the cross section:

The pore has axial symmetry with the bilayer joined smoothly around the ring, as shown by the dashed lines (see also Fig. 3.2). For a nucleus with 3500 pores, typical of a mammalian cell, what is the total bending energy associated with the pores (in k_BT; use Table 7.1 and take $\kappa_b = 20k_BT$, $\kappa_G = 0$). If this energy were provided by the hydrolysis of ATP to ADP, how many ATP molecules would be required?

7.3 Shown below is a cross section through a simplified model for the three-dimensional stacks in the Golgi apparatus (see also Fig. 1.8). The length scale is set by the radius of curvature r in the plane of the drawing (i.e. along meridians). Each long horizontal region is a pancake perpendicular to the plane linked by axially

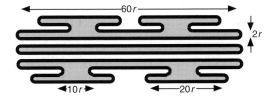

symmetric necks. There are four pancakes at the top and bottom of diameter $20r$ and three pancakes in the middle with diameter $60r$. The four necks have diameter $10r$. Using results from Table 7.1, determine the bending energy of this structure in the spontaneous curvature model with $C_o = 0$. Assuming $\kappa_b = 20k_BT$ and $\kappa_G = 0$, express this energy in k_BT at $T = 300$ K.

7.4 Some cells can adhere to a substrate by spreading, as shown in the idealized shape:

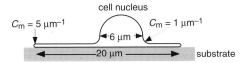

Assume that the thin region adhering to the substrate is a circular pancake of diameter 20 μm. Note that for many cells, the curvature at the edge may be a factor of two higher than indicated.

(i) Using the dimensions and curvatures along the meridians (C_m) in the diagram, what is the membrane bending energy in terms of κ_b and κ_G (use Table 7.1).

(ii) What is the energy difference $\Delta E / k_B T$ between this shape and a sphere, assuming $\kappa_b = 20 k_B T$?

7.5 The clathrin network in a cell has a honeycomb structure with dimensions as indicated:

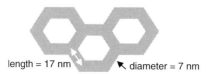

length = 17 nm   diameter = 7 nm

Take the network elements to be cylinders of diameter 7 nm with a Young's modulus of $Y = 10^9$ J/m³.

(i) Find the effective spring constant of the network elements using the result from Problem 3.17.
(ii) What is the area compression modulus K_A of this network according to Problem 7.16? Compare your result with $K_A = 10^{-5}$ J/m² observed for the human red cell cytoskeleton.

7.6 Suppose that the area expansion of a thin layer of peptidoglycan is described by the honeycomb network in Problem 7.16, with $k_{sp} = 10^{-2}$ J/m² and $s_o = 1$ nm. At what tension τ_{exp} does the network expand without limit? If a spherical bacterium of radius 2 μm is enclosed by this network, what is its interior pressure relative to its environment at τ_{exp} (quote your answer in J/m³ and in atmospheres)?

7.7 The area compression modulus K_A of a plate is related to its volume compression modulus K_V by the approximate relationship $K_A = d K_V$, where d is the plate thickness. Assuming that $K_V = 2 \times 10^7$ J/m³ for peptidoglycan, find the change in relative area of a cell wall encasing a sphere of radius 2 μm sustaining 25 atm internal pressure. Plot the area strain as a function of wall thickness for $5 \leq d \leq 25$ nm. Areal stress and strain are described in Section 5.3; see also Eqs. (D.28) and (D.29)).

7.8 Plot the bending energy (in units of κ_b) of an axially symmetric torus in the spontaneous curvature model as a function of the reduced volume v_{red}. Use results from Problem 7.10 for the area and volume, and obtain the bending energy from Problem 7.13(i). Over what range of v_{red} are doublet shapes with $E = 16 \pi \kappa_b$ favored over tori (see Fig. 7.8)?

Formal development and extensions

7.9 The pancake shape shown in the cross section (where the dashed line is the axis of rotational symmetry),

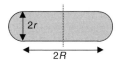

is described by two geometrical parameters r and R. Prove that the surface area A and enclosed volume V of this shape are given by

$$A = 2\pi r^2 \left[(R/r)^2 + \pi R/r + 2 \right]$$
$$V = 2\pi r^3 \left[(R/r)^2 + (\pi/2)R/r + 2/3 \right].$$

7.10 The torus displayed in the cross section (where the dashed line is the axis of rotational symmetry),

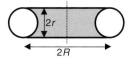

is described by two geometrical parameters r and R.

(i) Prove that the surface area A and enclosed volume V of this shape are given by

$$A = 4\pi^2 r R$$
$$V = 2\pi^2 r^2 R.$$

(ii) Express the reduced volume v_{red} in terms of r/R, and find its allowed range.

7.11 As a simple application of Eqs. (7.9) to (7.11), calculate the bending energy of a spherocylinder of radius R and total length $L + 2r$.

(i) For the lower hemisphere, obtain $\theta(s)$ and $x(s)$ with the coordinate origin at the south pole. Prove that the principal curvatures are $C_m = C_p = 1/r$ by evaluating Eqs. (7.9)–(7.10). Integrate Eq. (7.11) to obtain the energy of the south pole.

(ii) Repeat the steps in part (i) for the cylindrical section in the middle of the shell.

(iii) Add the sections together to obtain

$$E = 2\pi\kappa_b (2 - rC_o)^2 + \pi\kappa_b (L/r) \cdot (1 - rC_o)^2 + 4\pi\kappa_G.$$

Note that the Gaussian term is the same for a sphere or a pancake (Problem 7.12).

7.12 For the pancake shape shown in Problem 7.9, establish that the bending energy is given by

$$E_{pancake} = 8\pi\kappa_b + 4\pi\kappa_b\rho^2 \, (\rho^2-1)^{-1/2} \, \{ \, arctan[(\rho+1)(\rho^2-1)^{-1/2}]$$
$$- \, arctan[(\rho^2-1)^{-1/2}] \, \} + 4\pi\kappa_G \, (\rho>1),$$

in the spontaneous curvature model when $C_o = 0$ (the relevant integral can be found in Sec. 2.552 of Gradshteyn and Ryzhik, 1980). Here, we have put $\rho = R/r$. Show that this energy becomes

$$E_{pancake} = \pi\kappa_b(8 + \pi\rho) + 4\pi\kappa_G$$

in the limit where ρ is large.

7.13 Apply the spontaneous curvature model to the torus in Problem 7.10 without resorting to Table 7.1.

(i) Prove that the bending energy is given by

$$E_{torus} = 2\pi^2\kappa_b \, (R/r)^2/[(R/r)^2-1]^{1/2} \, (R>r)$$

when $C_o = 0$. Confirm that the Gaussian contribution to the bending energy vanishes.

(ii) Show that the minimum value of E_{torus} occurs at $R/r = \sqrt{2}$, a special geometry called the Clifford torus. Calculate E_{torus}/E_{sphere} (at $\kappa_G = 0$) for this shape and determine its reduced volume v_{red} using results from Problem 7.10.

7.14 Consider the specific stomatocyte shape which is assembled from sections of spheres, cylinders and tori.
To simplify the algebra, the curvatures along the meridians are chosen to be 0, r^{-1}, or $(3r)^{-1}$.

(i) Prove that the area A and volume V are given by

$$A = 2\pi(11 + 2\pi)r^2$$
$$V = (49\pi/3 + 2\pi^2)r^3.$$

(ii) What is the reduced volume v_{red} of this configuration?

(iii) Using results from Table 7.1, show that the bending energy of this shape in the spontaneous curvature model is

$$E_{\text{special}} = \pi \kappa_b (9 + 4\pi/\sqrt{3}) + 4\pi \kappa_G$$

when $C_o = 0$. What is the ratio of this energy to that of a sphere if $\kappa_G = 0$?

7.15 Consider the uniformly curved surface joining two flat regions, as might appear in the neck of the doublet of spheres in Fig. 7.8.

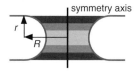

symmetry axis

(i) Prove that the area of the curved region is $2\pi r^2 [(\pi R/r) - 2]$.
(ii) Defining $\rho = R/r$, establish that the bending energy of the curved region is

$$E_{\text{pore}} = -8\pi\kappa_b + 4\pi\kappa_b\rho^2 (\rho^2 - 1)^{-1/2} \{\arctan[(\rho - 1)\cdot(\rho^2 - 1)^{-1/2}] + \arctan[(\rho^2 - 1)^{-1/2}]\} - 4\pi\kappa_G$$

when $\rho > 1$. Show that this energy becomes

$$E_{\text{pore}} = \pi\kappa_b(-8 + \pi\rho) - 4\pi\kappa_G$$

in the limit where ρ is large.

7.16 The clathrin network of a cell has the appearance of a two-dimensional honeycomb with three-fold connectivity, as in the diagram below.

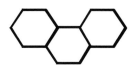

Suppose that each network element is a spring with force constant k_{sp} and rest length s_o. If the network is placed under a tension τ, show that

(i) the enthalpy per network vertex is $H_v = (3/4)k_{\text{sp}}(s - s_o)^2 - (3\sqrt{3}/4)\tau s^2$.
(ii) the spring length under tension is $s_\tau = s_o/(1 - \sqrt{3}\,\tau/k_{\text{sp}})$.
(iii) the area compression modulus K_A is given by $K_A = (k_{\text{sp}}/2\sqrt{3}) \cdot (1 - \sqrt{3}\,\tau/k_{\text{sp}})$.

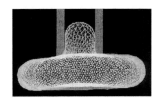

Chapter 8

Intermembrane forces

The variety of single cells capable of living independently is truly impressive, from small featureless mycoplasmas just half a micron across, to elegant protists a hundred times that size, perhaps outfitted with tentacles and even a mouth. Organisms such as ourselves, however, are multicellular, emphatically so at 10^{14} cells per human. How do our cells interact, and adhere when appropriate? In this chapter, we investigate intermembrane forces at large and small separations, beginning with a survey of the different features of membrane interactions in Section 8.1. The Poisson–Boltzmann equation, which we derive in Section 8.2, provides our primary framework for treating membrane interactions at non-zero temperature. This approach is applied to the organization of ions near a single charged plate and to the electrostatic pressure between charged walls in Sections 8.2 and 8.3, respectively. Two other contributors to membrane interactions also are treated in some mathematical detail here: the van der Waals attraction between rigid sheets (Section 8.3) and the steric repulsion experienced by undulating membranes (Section 8.4). The adhesion of a membrane to a substrate, and its effect on cell shape, is described in Section 8.5. Our presentation of membrane forces is necessarily limited; the reader is referred to Israelachvili (1991) or Safran (1994) for more extensive treatments.

8.1 Interactions between membranes

Let's briefly review the molecular composition of a conventional plasma membrane. Fundamental to a biomembrane is the lipid bilayer, a pair of two-dimensional fluid leaflets described at length in Chapter 5. The lipid composition of the leaflets is inequivalent: in the human red blood cell, one of the best-characterized examples, the phospholipids in the outer leaflet are predominantly cholines, while lipids containing ethanolamine or serine dominate the inner leaflet facing the cytosol (see Appendix B for head-group structures). Cholesterol also may be abundant in the

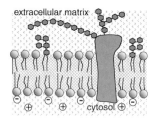

Fig. 8.1. Schematic representation of a plasma membrane. The large blob is an embedded protein; attached to the protein and to two lipids are linear and branched polysaccharides, forming part of the extracellular matrix. Charged lipids, such as phosphatidylserine, and their counterions are indicated by the circles in the cytosol.

plasma membrane. The head-group of phosphatidylserine, but not the others, is negatively charged, providing a surface charge density ranging up to 0.3 C/m^2 if every lipid carried a single charge.

Sometimes embedded in a bilayer, sometimes lying adjacent to it and attached by other means, is a variety of proteins, as displayed in Fig. 8.1. Although proteins make up 50% of a membrane's mass, they are outnumbered by lipids fifty-to-one. Lipids and membrane-bound proteins both are important functional elements of the cell, providing leakage resistance and controlling access, among their other tasks. Not surprisingly, the cell protects these valuable components with the glycocalyx, a sugary coat made from linear and branched polysaccharides attached to lipids (glycolipids) or proteins (glycoproteins and proteoglycans); this coating may extend several nanometers away from the head-group plane of the bilayers. Lastly, sandwiching the membrane and permeating the glycocalyx are aqueous solutions of mobile counterions (for the charged lipids), ions from dissolved salts, and sundry other chemical building blocks.

What is the physical origin of the interaction between membranes? A sample of the forces to be described in this chapter is illustrated in Fig. 8.2. At separations of the order 5 nm or more, contributions are made by the van der Waals forces of molecular polarizability, and by electrostatic forces from charged surfaces. At shorter distances, steric conflict between the extracellular matrix of each membrane adds to the intermembrane force, which is further augmented by the effort needed to expel the solution adjacent to the bilayers. In addition, the membrane itself experiences gentle thermal undulations, which provide entropic resistance against membrane adhesion even at distances of several nanometers. What is the strength of each contribution, and how does it vary with distance? We now provide a synopsis of several forces, a selection of which are described more thoroughly in Sections 8.2–8.4.

van der Waals forces

Even if it is electrically neutral overall, the charge separations within a molecule, say between its electrons and nucleus, provide the basis for an intermolecular force. Collectively called van der Waals forces, the interaction energy between permanent and instantaneous electric dipole

Fig. 8.2. Contributors to the force between membranes include: (a) van der Waals attraction between electric dipoles, (b) electrostatic repulsion between charged bilayers, and (c) entropic repulsion between undulating membranes.

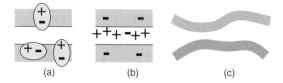

(a)   (b)   (c)

moments decreases with distance as $1/r^6$, much faster than the $1/r$ behavior of Coulomb's potential for point charges. For two rigid slabs, an attractive pressure proportional to D_s^{-3} arises from the van der Waals interaction when the thickness of the slabs is appreciably larger than the distance D_s between them. If the contributions from each molecular pair add independently, this pressure is $(\pi^2 C_{vdw}\rho^2)/6\pi D_s^3$ where C_{vdw} is a constant and ρ is the number of molecules per unit volume in the material. The combination $\pi^2 C_{vdw}\rho^2$ is called the Hamaker constant, with a value of about 2–5×10^{-20} J ~5–$10\,k_B T$ for many types of material separated by a vacuum. At large distances compared to the thickness of two interacting sheets or films, the van der Waals pressure decreases more rapidly as D_s^{-5}.

Electrostatic forces

According to Coulomb's law, two identical point charges of magnitude q experience a repulsive potential energy of $q^2/4\pi\varepsilon r$, where ε is the permittivity of the medium surrounding the charges. The permittivity of vacuum (ε_o) is 8.85×10^{-12} C^2/N·m^2, while the permittivities of water and various hydrocarbons are $80\,\varepsilon_o$ and 2–$3\,\varepsilon_o$, respectively. The interaction between charged membranes at ambient temperatures is considerably more complex than point charges separated by vacuum. First, charged counterions must be present in solution to compensate for the membrane's charge. Being mobile, the counterions may stay close to the membrane at low temperature because of coulombic forces, but may range far and wide at high temperature, enticed by entropy to explore their surroundings. The Poisson–Boltzmann equation provides a framework for describing such charged systems at non-zero temperature. Applied to charged plates separated by a distance D_s, the solution to this equation yields a repulsive pressure of $\pi k_B T/2\ell_B D_s^2$, the Bjerrum length ℓ_B reflecting the properties of the medium through $\ell_B \equiv q^2/4\pi\varepsilon k_B T$ for counterions of charge q. However, this calculated pressure will be strongly suppressed (or *screened*) by the presence of dissolved salts over a distance scale given by the Debye length $\ell_D \equiv (8\pi\ell_B\rho_s)^{-1/2}$, where ρ_s is the molecular number density of the salt (this expression is valid for one species of monovalent salt). Depending on the magnitude of the Debye length, the combined electrostatic and van der Waals force may be attractive at long *and* short distances because of the van der Waals contribution, with electrostatic repulsion emerging only at intermediate separations.

Steric interactions

The structure and movement of the plasma membrane and its polymer coat, the extracellular matrix, provide an often repulsive interaction between cells. Seen in Fig. 8.1, the exterior of many cells is covered by a layer of polymers that can be described as twigs, mushrooms or brushes, depending upon their stiffness and mean separation. The construction of this polymer labyrinth is sufficiently complex that a transparent analytical description of its elastic properties is difficult to achieve; however, the entropic repulsion of the polymers can be felt over distances of many nanometers, obviously dependent on the thickness of the coat. Equally important, the plasma membrane itself experiences thermal undulations that generate a repulsive pressure. This pressure is entropic, and has the form $P = 2c_{fl} (k_B T)^2 / \kappa_b D_s^3$, where κ_b is the bending rigidity (10–$25\,k_B T$ for common bilayers) and c_{fl} is a numerical constant equal to about 0.1. At large values of D_s this pressure may exceed both the van der Waals attraction of thin films and the electrostatic repulsion of screened charges.

Solvation forces

The above forces are most important at distances of several nanometers or more, with the dominant force depending strongly on conditions. However, at distance scales of less than a few nanometers, the solvent becomes important because:

- solvent must be removed from the space between membranes for them to touch
- solvent molecules near a flat surface may be organized into an irregular structure.

Owing to molecular organization, the solvation force between rigid plates may be oscillatory, with a wavelength comparable to the molecular diameter. Where present, such oscillations may be superimposed on an overall trend that is repulsive or attractive, depending upon the nature of the solvent and boundary surfaces. In the restricted case where the interaction between plate and solvent is ignored, the solvation pressure is attractive as $D_s \rightarrow 0$, equal to $k_B T$ times the concentration of the solvent at the plate as $D_s \rightarrow \infty$ (see Sec. 13.3 of Israelachvili, 1991); for an aqueous medium, this calculated pressure is about a thousand atmospheres. However, if there is an attractive interaction between plate and fluid, the liquid density at the plate is higher than in the bulk, and the force is generally repulsive. Owing to the aqueous medium, the solvation force between lipid bilayers may be referred to as a hydration force (see Rand and Parsegian, 1989, for a review). Given its flexible headgroup structure, the bilayer is rough on length scales of tenths of nanometers, potentially washing out the ordered solvent layer that may be present near atomically smooth surfaces. As a consequence, the oscillating behavior of the force is suppressed and it tends to be monotonically repulsive.

Fig. 8.3. Adhesion energies are large enough to influence the shape of a cell attached to a substrate or of two cells in close contact.

Adhesion

At very small separations, membranes may adhere through direct molecular contact, with binding energies that are significant, but not overwhelming. For example, the protein avidin and the vitamin biotin display one of the strongest association energies at $35\,k_B T$ per bond, which is still less than a typical covalent bond energy of $140\,k_B T$ (e.g., $C-C$ or $C-O$). The difference between the surface energy density of two membranes in contact, or separated by an intervening medium like water, is termed the adhesion energy density. Taken together with the other interactions described above, the adhesion energy density between membranes is about 10^{-5} J/m²; for instance, the adhesion energy of two membranes with a contact area of 10 µm² is 10^{-16} J. What effect does adhesion have on cell shape? We recall from Section 5.4 that the bending energy of a spherical shell is $8\pi\kappa_b + 4\pi\kappa_G$; for many bilayers, the bending rigidity κ_b is around $10\,k_B T$, so that the bending energy is in the range 10^{-18} J, ignoring the contribution of the Gaussian rigidity (which is shape-invariant for a given topology). Thus, the adhesion energy is sufficient to overcome the change in bending energy arising from shape changes such as those shown in Fig. 8.3.

We return to the shapes of adhering cells in Section 8.5; let us now formally determine the van der Waals, electrostatic and undulation forces between membranes in Sections 8.2–8.4.

8.2 Charged plate in an electrolyte

The phosphate group of a phospholipid carries a negative charge which may be balanced by the positive charge borne by many lipid headgroups. An exception is the serine group, which is electrically neutral, leaving the phospholipid negative overall (see Appendix B). Thus, a bilayer made from the common phospholipids may have a negative surface charge density. For an area per head-group of 0.5 nm² in the bilayer plane, this charge density could be as high as 0.3 C/m² if each lipid carried a single electron charge (1.6×10^{-19} C). Overall, the membrane and its environment is electrically neutral, so that positive counterions must be present in the media adjacent to the membrane. Although lipids may diffuse laterally within the bilayer, their charges can be regarded as fixed in the sense that they are confined to a plane.

In contrast, the counterions are mobile and may hug the bilayer or diffuse away from it, depending on conditions.

Our task in this chapter is to develop a formalism for calculating the distribution of mobile ions surrounding a charged object with fixed geometry. The formalism must recognize the effects of entropy, which encourages mobile ionic species to explore the physical space available to them, a tendency that increases with temperature. We adopt a mean field approach which ignores local fluctuations in charge, leading to the Poisson–Boltzmann equation. As a first application, we obtain the ion distribution generated on one side of a rigid charged plate, which we interpret as a model system for how mobile ions arrange themselves near a charged bilayer. In Section 8.3, we tackle the more general problem of the forces between two charged plates when there are counterions and perhaps other species present.

Poisson–Boltzmann equation

A widely applicable approach to calculating charge distributions at non-zero temperature is based on a mean field approximation which ignores local variations in the charge density. We start with Gauss's law from introductory physics, which, in its integral form, reads

$$\int \mathbf{E} \cdot d\mathbf{A} = (1/\varepsilon) \int \rho_{ch} \, dV, \tag{8.1}$$

where ε is the permittivity of the medium (in vacuum, $\varepsilon_0 = 8.85 \times 10^{-12} \, \mathrm{C^2/Nm^2}$). The integrals are performed over a closed surface with area $\int dA$ and enclosed volume $\int dV$; as in Chapter 5, $d\mathbf{A}$ is a vector representing an infinitesimal area element where the orientation of $d\mathbf{A}$ is locally normal to the surface. On the right-hand side, the integral over the charge density ρ_{ch} yields the net charge enclosed by the surface. For example, if the mathematical surface is a spherical shell of radius R surrounding a point charge Q, Eq. (8.1) gives the familiar $E = Q/4\pi\varepsilon R^2$. The divergence theorem allows the left-hand side to be written as $\int \nabla \cdot \mathbf{E} \, dV$, where $\nabla \cdot \mathbf{E} = \partial_x E_x + \partial_y E_y + \partial_z E_z$. With this substitution, both integrals of Eq. (8.1) are of the form $\int \ldots dV$, and the integrands must be equal,

$$\nabla \cdot \mathbf{E} = \rho_{ch}/\varepsilon, \tag{8.2}$$

because the equation is valid for any arbitrary closed surface. This is the differential form of Gauss's law.

Now, just as force $\mathbf{F}$ and potential energy $V(\mathbf{r})$ are related by $\mathbf{F} = -\nabla V$ for a conservative force, so too electric field $\mathbf{E}$ and potential ψ are related by $\mathbf{E} = -\nabla \psi$, since $\mathbf{E}$ and ψ are just $\mathbf{F}$ and V divided by charge. Replacing

E by the negative gradient of ψ permits Eq. (8.2) to be recast in a form called Poisson's equation

$$\nabla^2\psi = -\rho_{ch}/\varepsilon, \tag{8.3}$$

which relates the charge density to the electric potential (for applications, see Chapter 4 of Lorrain *et al.*, 1988). As we stated above, the charge distribution reflects a competition between energy and entropy, with the entropic contribution becoming more important with rising temperature. For instance, positive ions cluster near a negatively charged bilayer at low temperature, but roam further afield at high temperature. In our treatment of the harmonic oscillator in Appendix C, we show that the relative probability of the oscillator having a displacement x is proportional to the Boltzmann factor $\exp(-V(x)/k_BT)$, where $V(x)$ is the potential energy. The same reasoning applies here: the density of counterions at a given position is proportional to the Boltzmann factor with a potential energy evaluated at that position. We assume for now that only one species of mobile ion is present, each ion with charge q experiencing a potential energy $V(\mathbf{r}) = q\psi(\mathbf{r})$. Defining ρ_o to be the ion density at the reference point $\psi = 0$, the Boltzmann expression for the density profile is

$$\rho(\mathbf{r}) = \rho_o \exp(-q\psi(\mathbf{r})/k_BT). \tag{8.4}$$

Here, the electrostatic potential ψ represents an average over local fluctuations in the ion's environment. This form for $\rho(\mathbf{r})$ can be substituted into the Poisson expression of Eq. (8.3) using $\rho_{ch}(\mathbf{r}) = q\rho(\mathbf{r})$ to give a differential equation in ψ

$$\nabla^2\psi = -(q\rho_o/\varepsilon)\exp(-q\psi(\mathbf{r})/k_BT), \tag{8.5}$$

known as the Poisson–Boltzmann equation. Eq. (8.5) can easily be generalized to include more than one species of charge (e.g., Chapter 12 of Israelachvili, 1991).

Charged plate with one counterion species

In this section and Section 8.3, the Poisson–Boltzmann equation is applied to several situations involving very large, but thin, charged plates surrounded by a medium of permittivity ε. Let's briefly review the electrostatics of a single plate with charge per unit area σ_s in the absence of counterions, using Eqs. (8.1) or (8.2). Because the plate is large, the electric field is uniform across its face, except near the plate boundary. The contributions to **E** from each element of charge on the plate can be resolved into components parallel and perpendicular to the plane of the

Fig. 8.4. (a) Positively charged
counterions swarm near a
large negatively charged
plate. (b) The density ρ of
positive charges decreases as
a function of distance from
the plate z, having a value ρ_0
where the potential ψ
vanishes. (c) For a negative
plate, the electrostatic
potential ψ rises with z from
its value at the plate
boundary.

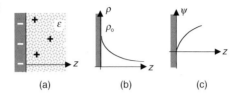

(a)   (b)   (c)

plate. By symmetry, the parallel components from all regions of the plate cancel, leaving the electric field pointing in a direction normal to the plate, the same as the area element $d\mathbf{A}$. To evaluate the integrals in Eq. (8.1) for this situation, we choose a mathematical surface like a pancake or tabletop: two identical flat sheets lie parallel to the plate and are joined along their perimeter by a surface perpendicular to the plate. The area of this mathematical pancake is much smaller than the plate itself, and it does not approach the plate boundary. The plate slices through the middle of the pancake, parallel to its two faces, such that all sections of a face are the same distance from the plate and experience the same electric field E (ignoring direction). Thus, each face of the pancake contributes EA to the integral on the left-hand side of Eq. (8.1), but no contribution arises from the rim of the pancake where $\mathbf{E}$ is perpendicular to $d\mathbf{A}$. The left-hand side of Eq. (8.1) is then $2EA$, and the charge enclosed on the right-hand side is $\sigma_s A$. Solving the equality, the magnitude of the electric field must be $E = |\sigma_s| / 2\varepsilon$, independent of the distance from the plate. As the electric field is the (negative) derivative of the potential, a constant value for E implies that the magnitude of ψ must grow linearly with z as $\sigma_s z / 2\varepsilon$. As usual, the location where the potential vanishes can be chosen arbitrarily.

Now, let's add counterions to one side of the plate as in Fig. 8.4(a), taking the plate to be negative and the counterions positive. At low temperature, the counterions cluster near the plate, their density ρ falling with distance z from the plate as in Fig. 8.4(b). As the temperature increases, the tail of the distribution extends further into the medium on the right. We choose the origin of the z-axis to lie at the plate boundary, where we fix $\psi = 0$. If the field is independent of direction parallel to the plate, the only non-vanishing part of $\nabla^2 \psi$ is $d^2\psi/dz^2$ and Eq. (8.5) reads

$$d^2\psi/dz^2 = -(q\rho_0/\varepsilon) \exp(-q\psi/k_B T). \tag{8.6}$$

It is useful to incorporate the constants $q/k_B T$ into the potential to render it unitless through the definition

$$\Psi(z) \equiv q\psi(z)/k_B T, \tag{8.7}$$

such that Eq. (8.6) becomes

$$d^2\Psi/dz^2 = -\rho_0(q^2/\varepsilon k_B T) e^{-\Psi}. \tag{8.8}$$

The combination $q^2/\varepsilon k_B T$ in this equation has units of [*length*] and appears in the Bjerrum length ℓ_B:

$$\ell_B \equiv q^2/4\pi\varepsilon k_B T. \tag{8.9}$$

Through q and ε, this quantity depends upon the properties of the counterions and the medium, not on the charge density of the plate. For instance, the Bjerrum length of a single electron charge in air at room temperature is 58 nm, dropping to 0.7 nm in water where the permittivity is eighty times that of air. We introduce yet another parameter K, which has the units of [*length*]$^{-1}$, by

$$K^2 \equiv 2\pi\ell_B \rho_o, \tag{8.10}$$

to further simplify the appearance of the Poisson–Boltzmann equation:

$$d^2\Psi/dz^2 = -2K^2 e^{-\Psi}. \tag{8.11}$$

To solve Eq. (8.11), we first multiply both sides by $d\Psi/dz$ and then use the following two relations

$$(d\Psi/dz)\cdot(d^2\Psi/dz^2) = (1/2)\, d[d\Psi/dz]^2/dz \tag{8.12}$$

$$e^{-\Psi}(d\Psi/dz) = -\, de^{-\Psi}/dz, \tag{8.13}$$

to rewrite the equation as

$$d[d\Psi/dz]^2/dz = 4K^2\, de^{-\Psi}/dz. \tag{8.14}$$

Integrating this expression over z and taking the square root of the result gives

$$d\Psi/dz = 2Ke^{-\Psi/2}, \tag{8.15}$$

where the positive root is chosen to correspond to Fig. 8.4(c). The integration constant that should appear in Eq. (8.15) has been set equal to zero permitting the electric field, which is proportional to $d\Psi/dz$, to vanish at the large values of Ψ expected as $z \to \infty$. Eq. (8.15) can be integrated easily by rewriting it as

$$\int e^{\Psi/2}\, d\Psi = 2K \int dz, \tag{8.16}$$

which gives

$$e^{\Psi/2} = K(z + \chi), \tag{8.17}$$

where the integration constants have been rolled into the factor χ.

We have now established that the functional form of $e^{-\Psi}$ is $[K(z + \chi)]^{-2}$, allowing us to generate an expression for the charge density from $\rho = \rho_o e^{-\Psi}$:

$$\rho(z) = \rho_o/[K(z + \chi)]^2 = 1/[2\pi\ell_B (z + \chi)^2]. \tag{8.18}$$

The integration constant is fixed by the value of the surface charge density σ_s. For the system to be electrically neutral, the integral of the positive charge density $\int q\rho \, dz$ must equal σ_s in magnitude; in symbols

$$-\sigma_s=(q/2\pi\ell_B)\int dz/(z+\chi)^2, \tag{8.19}$$

where the integral covers $0\leq z\leq\infty$. The solution to this equation requires the distance χ to be

$$\chi=q/(-2\pi\ell_B\sigma_s)=2\varepsilon k_B T/(-q\sigma_s), \tag{8.20}$$

which is positive because q and σ_s have opposite signs here. There are several features to note about the form of the counterion density. First, the density falls like the square of the distance from the plate, with half of the counterions residing within a distance χ of the plate. Second, from Eq. (8.20), the width of the distribution grows linearly with temperature, as the counterions venture further into the surrounding medium. Other properties of the solution are explored in Problem 8.9, confirming the expected behavior of the potential compared to that of a charged plate in the absence of counterions.

Charged plate in a salt bath

Both the interior of a cell and the environment surrounding it contain various organic compounds as well as ions released from salts such as NaCl, which we will refer to as "bulk" salts or ions. The effect that a negatively charged bilayer has on these bulk ions depends upon their charge: the density of positive ions is enhanced near the plate while negative mobile ions are depleted. The distributions are shown schematically in Fig. 8.5, where $\rho_+(z)$ and $\rho_-(z)$ are the number densities of positive and negative ions carrying charges $+q$ and $-q$, respectively. Only the case where the ions are monovalent ($q=e$) will be considered here. This situation is somewhat more complicated than the previous problem of a plate with counterions, but it can be approached with a suitably generalized version of the Poisson-Boltzmann equation.

Averaged over distances of tens of nanometers or more, the overall concentration of bulk salts generally exceeds that of the counterions. As a numerical example (see Problem 8.1), consider a negatively charged plate with the substantial surface charge density $\sigma_s=-0.3$ C/m², surrounded by water with permittivity 80 ε_o. If each counterion carries just one electron charge, their density may be in the molar range right at the interface with the plate, although they are largely confined to a region (χ) just 0.1–0.2 nm deep. Because the counterion density ρ drops like the distance squared according to Eq. (8.18), then even at a distance of 10 nm, ρ has fallen by a factor of 10^4. For comparison, the concentration

Fig. 8.5. Schematic representation of densities ρ_+ and ρ_- of positive and negative ions, respectively, as a function of distance z from a negatively charged plate (shaded region). The densities approach a common value ρ_s at large distance.

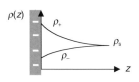

of ions in the cell's environment is about 0.2 M, and the salt concentration in the oceans themselves is 0.6 M. Thus, as far as the cell is concerned, the overall abundance of positive or negative ions in solution is dominated by that of the bulk salts. Now, in order to express ρ_+ and ρ_- according to the Boltzmann factor, a choice must be made for the position where the potential ψ vanishes. Given that $\rho_\pm$ approach their bulk value ρ_s as $z \to \infty$, a natural choice is $\psi \to 0$ as $z \to \infty$, and the Boltzmann factors are

$$\rho_+(z) = \rho_s \exp(-q\psi(z)/k_B T) \text{ and } \rho_-(z) = \rho_s \exp(+q\psi(z)/k_B T), \quad (8.21)$$

leading to a net charge density of

$$\rho_{ch}(z) = q\rho_s [\exp(-q\psi(z)/k_B T) - \exp(+q\psi(z)/k_B T)]. \quad (8.22)$$

This expression may be substituted into Eq. (8.3) to yield

$$\nabla^2 \psi = -(q\rho_s/\varepsilon) [\exp(-q\psi(z)/k_B T) - \exp(+q\psi(z)/k_B T)], \quad (8.23)$$

which can be generalized still further to include multiple ionic species.

The first steps in solving Eq. (8.23) are parallel to Eq. (8.6). To make the expression less visually cumbersome, the potential $\psi(z)$ is replaced by the dimensionless function $\Psi(z) \equiv q\psi(z)/k_B T$ defined in Eq. (8.7). The Bjerrum length appears unchanged at $\ell_B \equiv q^2/4\pi\varepsilon k_B T$, but we introduce a second length scale by

$$\ell_D^{-2} \equiv 8\pi\ell_B \rho_s. \quad (8.24)$$

Defined as the Debye screening length, ℓ_D depends only on the properties of the bulk medium, including its ionic content. Thus, the Poisson-Boltzmann equation for this situation becomes

$$d^2\Psi/dz^2 = -(1/2\ell_D^2)(e^{-\Psi} - e^{+\Psi}), \quad (8.25)$$

where we again assume that Ψ depends only on the distance from the plate, z. Now, the exponentials can be replaced by the hyperbolic function $\sinh\Psi$, yielding

$$d^2\Psi/dz^2 = +\sinh\Psi/\ell_D^2. \quad (8.26)$$

As with Eq. (8.11), we multiply both sides of Eq. (8.26) by $d\Psi/dz$, and then use Eq. (8.12) and $(d\Psi/dz)\cdot\sinh\Psi = d[\cosh\Psi]/dz$ to leave us with

$$(d\Psi/dz)^2 = (2\cosh\Psi - 2)/\ell_D^2, \quad (8.27)$$

where the additive factor of 2 from the integration constant permits $d\Psi/dz$ to vanish with Ψ at large distances. The identity $2\cosh\Psi - 2 = 4\sinh^2(\Psi/2)$ permits this expression to be simplified to

$$d\Psi/dz = \pm 2\sinh(\Psi/2)/\ell_D. \quad (8.28)$$

The potential is obtained by integrating this expression, as demonstrated in the problem set (or see Section 12.15 of Israelachvili, 1991).

Knowing the potential, the ion densities can be calculated from Eq. (8.21), although the resulting expressions may appear somewhat ungainly, given the exact solution for $\Psi(z)$ derived in Problem 8.10. However, if the potential is small, then we can replace $\sinh\Psi$ by Ψ, leading to the Debye–Hückel approximation to Eq. (8.26):

$$d^2\Psi/dz^2 \cong \Psi/\ell_D^2. \tag{8.29}$$

This has the physical solution for the potential

$$\Psi(z) = \Psi_o \exp(-z/\ell_D), \tag{8.30}$$

which could also be obtained directly from the solution for $\Psi(z)$ in Problem 8.10 if Ψ_o is small. We see that ℓ_D characterizes the "screening" of the charged plate by the salt. Now, ℓ_D depends on the properties of the bulk medium, leaving us to obtain Ψ_o from the surface charge density σ_s, which is given by

$$\sigma_s = -q \int (\rho_+ - \rho_-)dz = +2q\rho_s \int \sinh\Psi\, dz, \tag{8.31}$$

if the system is electrically neutral. The second equality follows from Eq. (8.21); again, σ_s and q have opposite signs here. The integration variable can be changed from dz to $d\Psi$ using Eq. (8.28), leading to

$$\sigma_s = 2q\Psi_o\ell_D\rho_s = \varepsilon\psi_o/\ell_D, \tag{8.32}$$

after invoking $\sinh\Psi \cong \Psi$ for small Ψ and recalling its definition from Eq. (8.7).

What is the magnitude of the Debye length in a typical cellular environment? As stated above, the Bjerrum length of monovalent ions ($q = e$) in water is just 0.7 nm. Taking this value along with 0.2 M salt (for a number density $\rho_s = 1.2 \times 10^{26}\ \text{M}^{-3}$), Eq. (8.24) yields $\ell_D = 0.7$ nm as well. In other words, the electric potential associated with a charged bilayer dies off rapidly over a distance of a few nanometers in a typical cellular electrolyte. Our results for the Debye length can be generalized to include several types of ions with charges q_i through the definition

$$\ell_D^{-2} = \Sigma_i \rho_{\infty,i} q_i^2/\varepsilon k_B T, \tag{8.33}$$

where $\rho_{\infty,i}$ is the number density of ion type i in bulk. This relationship and other aspects of multicomponent systems are treated in Chapter 12 of Israelachvili (1991).

In summary, we see that the presence of a negative charge on the lipid bilayer influences the distribution of mobile ions in its vicinity. Negative mobile ions are depleted near the bilayer while positive ions, both those that neutralize the bilayer and others from dissociated salts, are enhanced. For the typical cellular environment, the effects of a charged bilayer on mobile ions extend for several nanometers, as characterized by the Bjerrum and Debye lengths. In the following section, we bring two charged plates together and investigate the van der Waals force between them, as well as their interaction in an electrolyte.

8.3 van der Waals and electrostatic interactions

The components of the cell boundary – including the plasma membrane and extracellular structures – form a complex material of inhomogeneous elasticity and charge density. One aspect of this boundary region was treated in Section 8.2 by examining the distribution of mobile ions near a single charged plate. We now extend our analysis to consider two rigid plates interacting by:

- the van der Waals force between electrically neutral materials
- the electrostatic force between charged objects in an electrolyte.

Our approach is to determine mathematically the strength of these interactions as a function of separation between the plates, and then uncover their domains of importance in the cell.

van der Waals forces

What is often labeled the van der Waals force between neutral atoms or molecules arises from a number of effects, including:

- the attractive interaction between electric dipole moments (Keesom)
- the attraction between permanent electric dipoles and induced dipoles in a neighboring molecule (Debye)
- the attraction between fluctuating dipole moments created by instantaneous movement of the electrons in an atom (London),

where the name in brackets is the person generally credited with calculating the effect (for a review, see Mohanty and Ninham, 1976). The potential energy associated with each of these angle-averaged contributions is proportional to r^{-6}, so they are often collectively written as

$$V_{mol}(r) = -C_{vdw}r^{-6}, \tag{8.34}$$

where r is the separation between molecules and C_{vdw} is a constant. The subscript *mol* serves as a reminder that Eq. (8.34) applies to molecules. The complete van der Waals potential between molecules incorporates a short-range repulsive term (proportional to $+r^{-12}$) complementary to the long-range attractive term $(-C_{vdw}r^{-6})$.

Fig. 8.6. Geometry of two
rigid molecular slabs
separated across a vacuum by
a gap of width D_s. The slabs
are semi-infinite in that they
extend to infinity in directions
parallel to the gap and their
thickness is much greater than
their separation.

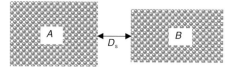

Fig. 8.6. Geometry of two rigid molecular slabs separated across a vacuum by a gap of width D_s. The slabs are semi-infinite in that they extend to infinity in directions parallel to the gap and their thickness is much greater than their separation.

The interaction energy between aggregates such as sheets and spheres can be obtained by integrating V_{mol}, as performed for a variety of surface geometries in Chapter 10 of Israelachvili (1991). The calculations are easy enough to perform under the assumption that the force between extended objects is equal to a pairwise sum over individual molecules, neglecting correlations. The approach gives simple functions which capture a good fraction of the applicable physics, although the interpretation of macroscopic observables in terms of C_{vdw} is not always direct. Of interest for the cell boundary is the interaction energy per unit area between two rigid slabs separated by a distance D_s, as shown in Fig. 8.6. This situation is treated in Problem 8.12 for a molecular potential energy of the form $V_{mol}(r) = -C_{vdw}/r^n$ under the assumption of pairwise addition of forces. We show that the energy per unit area on one slab (B) due to its interaction with the other (A) is

$$V_{slab}(D_s)/A = -2\pi\rho^2 C_{vdw}/(n-2)(n-3)(n-4)D_s^{n-4}, \tag{8.35}$$

where ρ is the molecular density of the medium (i.e., the number of molecules per unit volume). The slabs are much thicker than the gap between them, and they extend to infinity in directions parallel to the gap. Substituting $n=6$ of the van der Waals interaction into Eq. (8.35) gives

$$V_{slab}(D_s)/A = -\pi\rho^2 C_{vdw}/12D_s^2, \tag{8.36}$$

demonstrating that the energy density decays like the square of the separation. If D_s is not small compared to the thickness d_{sh} of the slab, Eq. (8.36) must be modified. As shown in Problem 8.13, the energy density of a thin sheet interacting with a semi-infinite slab is $-\pi C_{vdw}\rho^2 d_{sh}/6D_s^3$, while that of two thin, rigid sheets is $-\pi C_{vdw}\rho^2 d_{sh}^2/2D_s^4$, from which we see that the energy density decreases faster with distance as the sheets become thinner: D_s^{-2}, D_s^{-3}, and D_s^{-4} for the situations considered (see also Section 5.3 of Safran, 1994).

Equation (8.36) includes interactions between molecules on separate slabs, but not between molecules within the same slab, nor does it allow for the presence of a medium in the gap between A and B. If the materials are dissimilar, then ρ^2 in Eq. (8.35) is replaced by the product of densities $\rho_A\rho_B$ from each medium. The combination $\pi^2 C_{vdw}\rho_A\rho_B$ is collectively referred to as the Hamaker constant (often denoted by the symbol A, the same as our notation for area), having a value of about 10^{-19} J or $25k_BT$ for many condensed phases interacting across a vacuum. The presence of an electrolyte solution between charged plates screens their interaction and reduces the energy density, as described further in Section 8.2. For the same reasons, the van der Waals force is also screened, although the reduction in the energy is not completely exponential in the distance, and the screening length is half the Debye length ℓ_D (see Section 11.8 of Israelachvili, 1991); for instance, the effective Hamaker constant in an electrolyte may be reduced by a factor of ten for distances of a few nanometers. Finite temperature and other aspects of the van der Waals interaction in continuous media can be found in Lifshitz (1956) and Ninham et al. (1970), as explained in Chapter 11 of Israelachvili (1991) and Section 5.4 of Safran (1994) (see also Mohanty and Ninham, 1976).

The van der Waals interaction between plates results in a pressure P, which can be obtained from the change in the energy density $V(D_s)/A$ by

$$P = -d(V(D_s)/A)/dD_s, \tag{8.37}$$

where $P<0$ corresponds to attraction. As applied to Eq. (8.36), this yields

$$P = -\pi\rho^2 C_{vdw}/6D_s^3, \tag{8.38}$$

for two slabs; the attractive pressure decreases with D_s more rapidly than this for sheets whose thickness is much less than their separation (see Problem 8.13). To estimate the importance of the van der Waals interaction between bilayers, we consider two blocks of hydrocarbons, which we represent as aggregates of methyl groups with $C_{vdw} \sim 0.5 \times 10^{-77}$ J·m^6 and $\rho = 3.3 \times 10^{28}$ m^{-3} (from Table 11.1 of Israelachvili, 1991). With these values, the pressure from Eq. (8.38) is displayed in Fig. 8.7; for example, at $D_s = 5$ nm the pressure is 0.23×10^5 J/m$^3 = 0.23$ atm. We now contrast this interaction with the electrostatic force between charged plates.

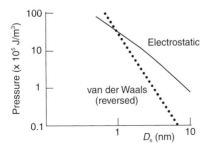

Fig. 8.7. Estimated pressure between two rigid plates as a function of their separation D_s, ignoring screening effects from the intervening medium. The van der Waals pressure (reversed in sign) is from Eq. (8.38) with parameters $C_{vdw} = 0.5 \times 10^{-77}$ J·m^6 and $\rho = 3.3 \times 10^{28}$ m^{-3}. The electrostatic pressure is a numerical solution of Eq. (8.56) with $\sigma_s = -0.1$ C/m^2, $q = e$ and $\varepsilon = 80\varepsilon_o$; screening reduces the calculated pressure.

Fig. 8.8. (a) Mobile (positive)
counterions spread through
the region between two
negatively charge plates. The
counterion density ρ has a
minimum (b) and the
potential ψ has a maximum
(c) at the midplane ($z=0$),
where their derivatives vanish.
The vertical bars in (b) and (c)
indicate the positions of the
plate.

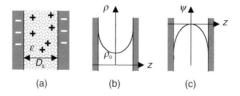

(a)   (b)   (c)

Fig. 8.8. (a) Mobile (positive) counterions spread through the region between two negatively charge plates. The counterion density ρ has a minimum (b) and the potential ψ has a maximum (c) at the midplane ($z=0$), where their derivatives vanish. The vertical bars in (b) and (c) indicate the positions of the plate.

Charged plates with counterions

Described in Section 8.2, phospholipids with serine head-groups are present in significant concentrations in some bilayers, awarding them a negative surface charge density that may range up to 0.3 C/m^2 in magnitude. As a first step towards understanding the forces between charged membranes, we calculate the electric field **E** between two rigid charged plates, as in Fig. 8.8(a) but in the absence of positive counterions. The plates, which extend to infinity, are separated by a distance D_s; they carry a charge density of σ_s and are immersed in a medium with permittivity ε, which equals the product of the dielectric constant and ε_o, the permittivity of free space. As established in Section 8.2, and most introductory physics textbooks, the magnitude of the electric field E from *one* plate is $E=\sigma_s/2\varepsilon$. For two parallel plates of the same charge density, *including the sign*, E vanishes between the plates and is equal to σ_s/ε outside the plates. That is, the fields from each plate cancel because they point in opposite directions between the plates. This is the reverse of a parallel plate capacitor, where the plates have opposite charges; there, the fields cancel outside the plates, but add between them, yielding $E=\sigma_s/\varepsilon$.

Now, let us introduce some (positive) counterions into the space between the plates, as illustrated in Fig. 8.8(a). If a single positive ion were placed between the plates, it would experience no net force because the electric field vanishes. However, a group of positive ions, initially spread throughout the space between the plates, would be driven towards the plates by their mutual repulsion. At finite temperature, the ions have thermal energy, allowing them to wander away from the plates into the medium. The form of their density distribution $\rho(z)$ reflects a competition between energy and entropy: entropy favors a uniform distribution with the ions exploring all available configuration space, while energy favors the ions piled up against the plates to minimize their mutual repulsion. This cloud of counterions at the plate is referred to as the electric double layer.

Our next step is to obtain the potential $\psi(z)$ in Eq. (8.5), subject to constraints imposed by symmetry and the value of the surface charge density. The mirror symmetry of the system in Fig. 8.8 means that the potential and charge distribution must be symmetric about the midplane ($z=0$), where we choose $\psi(0)=0$. As seen in Figs. 8.8(b) and (c), the counterion density has a minimum and the potential has an extremum (depending on its sign) at the midplane: $d\rho/dz=0$ and $d\psi/dz=0$. Imposing the condition of overall electrical neutrality means that the integral of the counterion charge density $\rho_{ch}=q\rho$ from $z=0$ to $z=D_s/2$ must equal σ_s in magnitude:

$$\sigma_s=-q\int_0^{D_s/2} \rho\,dz, \qquad (8.39)$$

where the minus sign arises because σ_s and $q\rho$ have opposite signs. Poisson's equation can be used to replace $q\rho$ by $-\varepsilon\nabla^2\psi$, which is equal to $-\varepsilon d^2\psi/dz^2$ because the function depends only on coordinate z. The integral in Eq. (8.39) is thus

$$(-1)^2\,\varepsilon\int_0^{D_s/2} (d^2\psi/dz^2)\,dz=+\varepsilon\,(d\psi/dz)_{D_s/2}, \qquad (8.40)$$

where we have used the condition $(d\psi/dz)_0=0$ imposed by symmetry. Combined, Eqs. (8.39) and (8.40) yield

$$(d\psi/dz)_{Ds/2}=\sigma_s/\varepsilon, \qquad (8.41)$$

which includes the correct signs. Being equal to $|(d\psi/dz)_{D_s/2}|$, the magnitude of the electric field at the plate is $|\sigma_s|/\varepsilon$.

Armed with an expression for $(d\psi/dz)_{D_s/2}$, we can determine $\psi(z)$ from the Poisson–Boltzmann equation, Eq. (8.5), which, in its one-dimensional form, reads

$$d^2\psi/dz^2=-\,(q\rho_o/\varepsilon)\,\exp(-q\psi/k_BT). \qquad (8.42)$$

As in Section 8.2, we replace $\psi(z)$ by the dimensionless function

$$\Psi(z)=q\psi(z)/k_BT, \qquad (8.43)$$

so that Eq. (8.42) becomes

$$d^2\Psi/dz^2=-\,(q^2\rho_o/\varepsilon k_BT)\,\exp(-\Psi)=-2K^2\exp(-\Psi), \qquad (8.44)$$

where

$$K^2=q^2\rho_o/(2\varepsilon k_BT)=2\pi\ell_B\rho_o, \qquad (8.45)$$

and where ℓ_B is the Bjerrum length $q^2/4\pi\varepsilon k_BT$. Although symbolically identical, ρ_o is evaluated at different locations in Eqs. (8.10) and (8.45). Eq. (8.44) has the solution

$$\Psi(z)=\ln(\,\cos^2[Kz]\,), \qquad (8.46)$$

as can be verified by first demonstrating

$$d\Psi/dz = -2K \tan(Kz), \tag{8.47}$$

from which it follows that

$$d^2\Psi/dz^2 = -2K^2/\cos^2(Kz). \tag{8.48}$$

The boundary condition $(d\Psi/dz)_{D_s/2} = q\sigma_s/\varepsilon k_B T$ from Eq. (8.41) fixes the value for K from Eq. (8.47):

$$-2K \tan(KD_s/2) = q\sigma_s/\varepsilon k_B T. \tag{8.49}$$

Note that $1/K$ has the units of [*length*]. The solution for $\Psi(z)$ in Eq. (8.46) vanishes at $z = 0$ and becomes negative for $|z| > 0$, as expected for negatively charged plates in Fig. 8.8(c).

Expressing Eq. (8.46) as $\exp[\Psi(z)] = \cos^2(Kz)$ permits the counterion density profile to be extracted easily from Eq. (8.4), namely

$$\rho(z) = \rho_o/\cos^2(Kz) = \rho_o + \rho_o \tan^2(Kz), \tag{8.50}$$

where the second equality follows from the trigonometric identity $1 + \tan^2\theta = 1/\cos^2\theta$. From Eq. (8.49), the value of $\tan^2(KD_s/2)$ at the plate is $[q\sigma_s/(2\varepsilon k_B T)]^2/K^2$, which simplifies to just $\sigma_s^2/(2\rho_o \varepsilon k_B T)$ when Eq. (8.45) is used for the definition of K^2. Thus, the counterion density at the plate obeys the particularly simple expression

$$\rho(D_s/2) = \rho_o + \sigma_s^2/(2\varepsilon k_B T). \tag{8.51}$$

In other words, the counterion density is lowest at $z = 0$, from which it rises to $\rho(D_s/2)$ at the plates. From Eq. (8.51), the smallest value of $\rho(D_s/2)$ is $\sigma_s^2/(2\varepsilon k_B T)$ for a given charge density σ_s. For instance, the magnitude of σ_s could be as large as 0.3 C/m² for one charge per lipid in a bilayer, yielding $\rho(D_s/2) = 1.6 \times 10^{28}$ m⁻³ = 26 M for water with $\varepsilon = 80\varepsilon_o$. Eq. (8.51) demonstrates that the counterion density at the plates declines with increasing temperature, as entropy encourages the counterions to explore new territory away from the plate boundaries.

The pressure P between charged plates in the absence of salts has the appealing form (see Section 12.7 of Israelachvili, 1991)

$$P = \rho_o k_B T = 2\varepsilon K^2 (k_B T/q)^2. \tag{8.52}$$

Except that ρ_o is the density of ions at the midplane, the first equality in this expression looks like the ideal gas law. Under what conditions do the counterions fill the gap between the plates and physically behave like a gas? The ions should spread away from the plates when the surface charge density is small, or the temperature is large. In this limit, K is small according to the right-hand side of Eq. (8.49), which then can be solved to yield

$$K^2 = -q\sigma_s/\varepsilon k_B T D_s. \tag{8.53}$$

Using $\rho_0 = 2\varepsilon k_B T K^2 / q^2$ from Eq. (8.45), this region of K corresponds to

$$\rho_0 = -2\sigma_s / qD_s, \qquad (8.54)$$

which is just the density expected if the counterions are spread evenly across the gap. The same expression for ρ_0 applies at small gap width, where the product KD_s in Eq. (8.49) is proportional to $\sqrt{D_s}$. Under these conditions (large T; small σ_s or D_s), the system has the same pressure as an ideal gas of counterions

$$P = -2\sigma_s k_B T / qD_s, \text{ (electrostatics, ideal gas limit)} \qquad (8.55)$$

which is inversely proportional to D_s. As usual, σ_s and q have opposite signs, making $P > 0$ and repulsive.

If the charges are more concentrated at the plates, Eq. (8.49) is easy enough to solve numerically by writing it as

$$y \tan y = -\zeta, \qquad (8.56)$$

where $y = KD_s / 2$ and $\zeta = qD_s\sigma_s / 4\varepsilon k_B T$, a constant of the configuration. At large values of ζ, the electrostatic interaction is approximately

$$P = \pi k_B T / 2\ell_B D_s^2, \text{ (electrostatics, large } D_s, \text{ no screening)} \qquad (8.57)$$

an expression which depends only on the bulk properties of the medium (see Problem 8.14). Displayed in Fig. 8.7 is a sample calculation for $\sigma_s = -0.1$ C/m² in water; here, $\zeta = 1.4$ at $D_s = 1$ nm, so the solution to Eq. (8.56) approaches the ideal gas result and the large-D_s expression at its respective limits. Interested readers may try a similar calculation themselves in Problem 8.4. The figure demonstrates that the repulsive electrostatic pressure *without screening* dominates the attractive van der Waals pressure for $D_s \geq 1$ nm as it must: the electrostatic pressure decays like $1/D_s^2$ while the van der Waals declines more rapidly as $1/D_s^3$.

Charged plates in an electrolyte

The electrostatic pressure between two plates separated by an electrolyte solution may be considerably less than that predicted by Eq. (8.57). The ion content of the medium for this case is illustrated in Fig. 8.9(a), and represented as the density function $\rho(z)$ in Fig. 8.9(b); for negatively charged plates, the density of positively charged ions is elevated near the plates while the negatively charged ions are suppressed. The presence of the bulk ions requires a modification of Eq. (8.52) for the electrostatic pressure between plates. The general case is treated in Section 12.17 of Israelachvili (1991); here, we treat only a monovalent electrolyte with charges $\pm e$. For this situation, the counterion density at the midplane ρ_0 in Eq. (8.52) must be replaced with the difference in the total ion

Fig. 8.9. (a) Ions from dissolved salts (black disks) may augment counterions between charged plates. (b) The number density of positive ions tends to be enhanced near a negatively charged plate, while negative ions are reduced. Vertical bars indicate the positions of the plates.

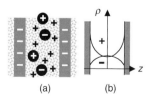

(a)   (b)

density at the midplane when the plates are separated by a distance D_s compared to when they are infinitely separated:

$$\rho_o \rightarrow [\textit{ionic density at separation } D_s]_m$$
$$- [\textit{ionic density at infinite separation}]_m$$

where the subscript indicates that the density is evaluated at the midplane. For each species of ion, the difference in densities at the midplane is $\rho_s [\exp(-q\psi_m/k_BT) - 1]$, where ψ_m is the potential evaluated at the midplane. Invoking the dimensionless potential $\Psi \equiv q\psi/k_BT$ from Eq. (8.43), the replacement for ρ_o becomes

$$\rho_o \rightarrow \rho_s \{[\exp(-\Psi_m) - 1] + [\exp(+\Psi_m) - 1]\},$$

for a monovalent electrolyte of positive and negative ions, permitting the pressure to be written as

$$P = 4\rho_s k_B T \sinh^2(\Psi_m/2). \tag{8.58}$$

As expected, the pressure goes to zero as the potential vanishes.

The potential Ψ_m receives contributions from both plates, each a distance $D/2$ from the midplane. We now make the assumption that Ψ_m is small, and can be regarded as a sum of independent contributions from each plate, which we show in Problem 8.10 to be

$$\Psi(z) = 2 \ln[(1+\alpha)/(1-\alpha)], \tag{8.59}$$

where $\alpha = \tanh(\Psi_o/4) \exp(-z/\ell_D)$ and ℓ_D is the Debye screening length $\ell_D^{-2} = 8\pi\ell_B\rho_s$. For small values of α, this individual plate potential becomes $4 \cdot \tanh(\Psi_o/4) \cdot \exp(-z/\ell_D)$ leading to a pressure of

$$P \cong 64\rho_s k_B T \tanh^2(\Psi_o/4) \exp(-D_s/\ell_D)$$
$$\text{(screened electrostatics),} \tag{8.60}$$

when the individual potentials are evaluated at $z = D_s/2$. The potential Ψ_o applies at the plates, not the midplane, and is given by $\Psi_o = q\sigma_s\ell_D/\varepsilon k_B T$ at low values of ψ_o according to Eq. (8.32). Thus, we are left with

$$P = (2\sigma_s^2/\varepsilon) \cdot \exp(-D_s/\ell_D) \text{ (screened electrostatics, small } \psi_o), \tag{8.61}$$

after eliminating ρ_s with Eq. (3.24). We see from Eqs. (8.60) and (8.61) that the presence of the electrolyte screens the interaction and suppresses the repulsive pressure: the power-law decay in Eqs. (8.55) and (8.57) becomes an exponential decay in an electrolyte.

Combined interactions

Given the negative charge on phophatidylserine, the force between lipid bilayers may include both electrostatic and van der Waals components.

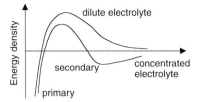

Fig. 8.10. van der Waals attraction dominates the interaction between charged plates at short distances; the behavior of the energy density at longer distances depends on the screening of the charges by the electrolyte solution.

At short distances, the van der Waals contribution dominates because of its stronger power-law dependence on $1/D_s$. At longer distances, the electrostatic interaction dominates if the electrolyte is dilute, resulting in a large Debye length. This is shown as the upper curve in Fig. 8.10. However, in a concentrated electrolyte with a small Debye length, the electrostatic interaction is rapidly extinguished with increasing distance, opening up the possibility that the electrostatic interaction is important only at intermediate separation, as illustrated by the lower curve in Fig. 8.10. In this situation, there are two minima in the potential energy density: a global (primary) minimum at short distance and a local (secondary) minimum at intermediate distance. The combined electrostatic and van der Waals interaction, and its dependence on electrolyte concentration, forms the basis for the DLVO theory of colloids (Derjaguin and Landau, 1941; Verway and Overbeek, 1948; for reviews, see Derjaguin, 1949, or Sec. 12.18 of Israelachvili, 1991). If the temperature is low enough, a system may become trapped in the secondary minimum, the thermal fluctuations in its energy being insufficient to carry it into the global minimum on a reasonable time frame. Thus, individual colloidal particles may be stabilized in an electrolyte, rather than adhere to each other at close contact.

8.4 Entropic repulsion of sheets and polymers

Treating membranes as flat rigid sheets, as we have done in Sections 8.2 and 8.3, facilitates comparison of various membrane forces in a standard geometrical context. Yet we stress in Chapters 5 and 6 that biomembranes are flexible, such that thermal fluctuations in their energy result in visible undulations at room temperature (see Brochard and Lennon, 1975; Harbich *et al.*, 1976). These undulations generate a repulsive pressure from their entropic resistance to compression. Further, the extracellular matrix provides the cell with a fuzzy coat that also may inhibit the close approach of two cell envelopes. The transverse length scale of the undulations is not trivial: the mean displacement of a membrane from a planar configuration is given by $k_B T A / 4 \pi^3 \kappa_b$ for a single gently undulating membrane, according to Eq. (6.82). For instance, taking a patch of membrane area A to be 1 μm^2 and the

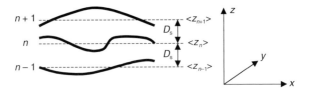

bending modulus κ_b to be 20 k_BT, we expect the mean displacement to be 20 nm, although the surface orientation may change only slowly with distance along the membrane. In this section, we analyze the free energy of an undulating membrane stack and then briefly introduce the reader to the properties of polymer brushes.

As a model system, consider a membrane stack lying parallel to the xy-plane, as displayed in Fig. 8.11. The location of a given membrane with label n is described by the function $z_n(x,y)$ with respect to the xy-plane. The mean separation between successive membranes is defined as D_s, or

$$\langle z_{n+1} \rangle - \langle z_n \rangle = D_s, \tag{8.62}$$

where the average is taken over the xy-coordinates of an individual surface. Following Chapter 6, we introduce a height function $h_n(x,y)$ which is the displacement in the z-direction of a location on the membrane from its mean position

$$h_n(x,y) = z_n(x,y) - nD_s, \tag{8.63}$$

whose mean value vanishes

$$\langle h_n \rangle = 0, \tag{8.64}$$

for equally spaced membranes.

The free energy of our model system must capture the membrane bending resistance and steric repulsion between neighboring sheets. Omitting the Gaussian rigidity as in Chapter 5, the simplest expression for the bending energy density of a single membrane is given by Eq. (5.17) as $(\kappa_b/2) \cdot (C_1 + C_2)^2$ in the absence of spontaneous curvature. For gentle undulations, the sum of the principal curvatures C_1 and C_2 can be replaced by $h_{xx} + h_{yy}$ according to Eq. (6.19), where the subscripts on the function $h(x,y)$ refer to derivatives (i.e., $h_{xx} = \partial^2 h / \partial x^2$, etc.). The entropic contribution to the free energy density depends upon the local value of $h_{n+1} - h_n$, which is the difference between the membrane separation $(h_{n+1} - h_n + D_s)$ and its mean value D_s at a given (x,y) location. In its simplest form, this repulsive energy is quadratic in $h_{n+1} - h_n$, leading to a model for the free energy per unit area in the xy-plane of

$$\mathcal{F} = (\kappa_b/2) \sum_n (h_{n,xx} + h_{n,yy})^2 + (B/2) \sum_n (h_{n+1} - h_n)^2, \tag{8.65}$$

where the sum is over all membranes in the stack. The term B represents the stack's (entropic) compression resistance but, unlike the volume compression modulus, it has units of $[energy] \cdot [length]^4$ and depends upon the separation D_s. This model energy was proposed by Helfrich (1978).

As in the single membrane situation studied in Chapter 6, Eq. (8.65) must be integrated over the membrane surfaces, and averaged over thermal fluctuations in their configurations, although the problem here is made complex by the presence of the stack. Using two slightly different approaches, Helfrich (1978) performed the thermal average by converting $h_n(x,y)$ to Fourier components in the z-direction as well as the xy-directions of Section 6.3. The derivation is somewhat lengthy for our context; instead, we quote and then analyze the result of the transformation, following closely the notation and approximations in Section 6.6 of Safran (1994). The Fourier components along the xy-axes introduce the two-dimensional wavevector $\mathbf{q}$ familiar from Section 6.3, and a second vector Q along the z-axis. The resulting free energy per unit volume has the form

$$\Delta \mathcal{F}_{\mathrm{v}} = (k_{\mathrm{B}}T/16\pi^3) \int_{-\pi/D_s}^{\pi/D_s} dQ \int d\mathbf{q} \, \ln[1 + B(Q)/\kappa_{\mathrm{b}}q^4], \tag{8.66}$$

where the function $B(Q) \equiv 2B(1 - \cos QD_s)$ depends only on Q. This expression is the difference between the free energy of the stack and the membranes in isolation, vanishing with B, as appropriate.

It is not difficult to perform the integrals in Eq. (8.66). The $\mathbf{q}$ angle can be removed immediately such that $\int d\mathbf{q} = \pi \int dq^2$. By changing variables to $t^2 = \kappa_{\mathrm{b}}q^4/B(Q)$, the integral over $\mathbf{q}$ in Eq. (8.66) becomes

$$\pi(B(Q)/\kappa_{\mathrm{b}})^{1/2} \int_0^\infty dt \, \ln(1 + 1/t^2) = \pi^2(B(Q)/\kappa_{\mathrm{b}})^{1/2}, \tag{8.67}$$

yielding a free energy density of

$$\Delta \mathcal{F}_{\mathrm{v}} = (k_{\mathrm{B}}T/16\pi) \int_{-\pi/D_s}^{\pi/D_s} dQ \, [2B(1 - \cos QD_s)/\kappa_{\mathrm{b}}]^{1/2}. \tag{8.68}$$

Collecting numerical constants and changing variables once again, this integral has the form

$$\Delta \mathcal{F}_{\mathrm{v}} = (k_{\mathrm{B}}T/16\pi D_s) \cdot (2B/\kappa_{\mathrm{b}})^{1/2} \int_{-\pi}^{\pi} (1 - \cos\theta)^{1/2} d\theta, \tag{8.69}$$

which can be solved by the trigonometric substitution $(1-\cos\theta)^{1/2} = \sqrt{2}\sin(\theta/2)$. Thus,

$$\Delta\mathcal{F}_{\mathrm{v}} = (k_B T/2\pi D_s)\cdot(B/\kappa_b)^{1/2}. \tag{8.70}$$

Helfrich (1978) uses a self-consistency argument to determine the functional dependence of B on D_s. Recall that for Hookean springs obeying $V(x) = k_{\mathrm{sp}}x^2/2$, the spring constant is equal to the second derivative of the potential $V(x)$ with respect to the displacement from equilibrium x. In our case, the displacement variable is D_s and the relationship between B and the derivative of the energy is slightly modified to read

$$B = D_s\, \partial^2\Delta\mathcal{F}_{\mathrm{v}}/\partial D_s^{\,2}. \tag{8.71}$$

Helfrich proposed that B is a polynomial function of D_s, namely

$$B = \alpha D_s^{\,n}, \tag{8.72}$$

where α and n are constants to be determined. After some algebra, the right-hand side of Eq. (8.71) has the form

$$D_s\, \partial^2\Delta\mathcal{F}_{\mathrm{v}}/\partial D_s^{\,2} = (k_B T/2\pi)\cdot(\alpha/\kappa_b)^{1/2}(n/2-1)(n/2-2)D_s^{\,n/2-2}, \tag{8.73}$$

which can be equated with Eq. (8.72) to yield $n = -4$ and $\alpha = (6k_B T/\pi)^2/\kappa_b$, or

$$B(D_s) = (6k_B T/\pi)^2/\kappa_b D_s^{\,4}. \tag{8.74}$$

Replacing B in Eq. (8.70) with this solution gives a free energy per unit area $\Delta\mathcal{F}$ of

$$\Delta\mathcal{F} = D_s\Delta\mathcal{F}_{\mathrm{v}} = (3/\pi^2)\cdot(k_B T)^2/\kappa_b D_s^{\,2}. \tag{8.75}$$

This expression, with a slightly different numerical prefactor, was obtained by Helfrich (1978); the argument has been extended by Evans and Parsegian (1986) to small separations. Multimembrane systems also have been investigated by computer simulation, which confirm the functional form of Eq. (8.75), but not the numerical prefactor. In light of this, it is more appropriate to write the free energy density as

$$\Delta\mathcal{F} = c_{\mathrm{fl}}\, (k_B T)^2/\kappa_b D_s^{\,2}, \tag{8.76}$$

where the constant c_{fl} is close to 0.1 (Gompper and Kroll, 1989; Janke and Kleinert, 1987; Netz and Lipowsky, 1995). The dynamics of fluctuating membranes are not covered in this book; pioneering work on this problem can be found in Milner and Safran (1987). Lastly, we mention that steric interactions between *polymerized* membranes (as opposed to *fluid* bilayers) have been studied via computer simulation by

Leibler and Maggs (1989), who find that the energy density decreases like D_s^{-3}.

The behavior of the free energy per unit area according to Eq. (8.75) is much as we might expect. Being entropic, the free energy increases with temperature from a vanishing contribution at $T=0$ where undulations are absent if $\kappa_b \neq 0$. In addition, the free energy falls as the mean spacing D_s rises, because the large excursions needed for adjacent membranes to collide at large D_s are uncommon. This dependence on D_s is the same as that displayed by the van der Waals expression for the attraction between two rigid slabs. Choosing $\pi^2 C_{vdw} \rho^2 = 10 \; k_B T$, the attractive van der Waals energy density is $-0.3 \; k_B T / D_s^2$ according to Eq. (8.36). For the entropic contribution, even if κ_b is as low as $10 \; k_B T$, the entropic contribution to the energy per unit area is $+0.03 \; k_B T / D_s^2$, which is certainly lower, but not by many orders of magnitude. Now, the interaction between slabs overestimates the interaction between thin sheets at large D_s: the energy density for sheets falls much faster like D_s^{-4}. Thus, there may be domains of D_s where the repulsive force from undulations dominates over other contributors.

Experimental evidence for the importance of membrane undulations comes from at least two sources: (i) measurements of the pressure between membranes and (ii) secondary effects of undulations on membrane adhesion. An example of item (i) is the observation by Evans and Parsegian (1986) of an enhanced repulsive stress between electrically neutral egg lecithin membranes at $D_s \sim 2$–3 nm, which they attribute to undulations (data from Parsegian et al., 1979). The physics behind item (ii) is described in Section 6.5: lateral tension suppresses undulations, thus increasing the possibility that two membranes may mutually adhere at close contact. Helfrich and coworkers proposed that this phenomenon underlies the onset of adhesion they observed for egg lecithin membranes under tension (Servuss and Helfrich, 1989; for earlier measurements, see Helfrich and Servuss, 1984).

Over time, our stack of model membranes separates because of entropic pressure, which can be obtained from Eq. (8.76) by the usual relation $P = -\partial \Delta \mathcal{F} / \partial D_s$. This exfoliation can be prevented by applying an external pressure P_{ext}, which generates an equilibrium spacing D_{eq} of

$$D_{eq}^{\;3} = 2 c_{fl} \, (k_B T)^2 / \kappa_b P_{ext}, \qquad (8.77)$$

found by balancing the internal and applied pressures. The presence of a strong attractive force also may offset the entropic pressure between membranes. At low temperature, the attraction may be sufficiently strong to bind the membranes, while at high temperature, the entropic pressure leads to unbinding (Lipowsky and Leibler, 1986). To estimate the magnitude of this effect, we consider just two membranes subject to an attractive potential energy density of the step-function form

$$
\begin{aligned}
V(z) &= \infty & z &< 0 \\
V(z) &= -V_o & 0 &\leq z \leq w \\
V(z) &= 0 & z &\geq w,
\end{aligned}
\tag{8.78}
$$

where V_o and w are constants. Lipowsky (1994) provides a guide to membrane potentials capturing more physics than the simple form of Eq. (8.78). Mathematically, the behavior of two membranes of bending rigidity κ_1 and κ_2 interacting by $V(z)$ is the same as that of a single membrane of rigidity $\kappa = \kappa_1 \kappa_2 / (\kappa_1 + \kappa_2)$ interacting with a rigid wall via the same potential. Although the system is trapped by the potential to lie within the distance w at low temperature, the binding becomes ever less effective as the temperature rises. The approximate temperature scale T^* at which the system escapes from the potential well can be obtained by equating V_o with $\Delta \mathcal{F}$ in Eq. (8.76)

$$
k_B T^* = (\kappa_b w^2 V_o / 2 c_{fl})^{1/2},
\tag{8.79}
$$

for two identical membranes with bending modulus κ_b (such that the relevant bending rigidity is $\kappa_b/2$).

A variety of theoretical approaches and simulation studies confirm the existence of a distinct unbinding transition, although the transition is continuous, in the language of phase transitions, and not discontinuous as one might expect from a comparison of energies as in Eq. (8.79) (Lipowsky and Leibler, 1986). As determined in simulation studies (Lipowsky and Zielinska, 1989), Eq. (8.79) sets the order of magnitude for the transition temperature, but the exact value depends on the short length scale of the simulation (i.e., the graininess of the discretized membrane). Further simulations have examined the unbinding transition of three or more membranes (Cook-Röder and Lipowsky, 1992; Netz and Lipowsky, 1993). The first experimental observation of a temperature-dependent unbinding transition was made by Mutz and Helfrich (1989) for digalactosyl-diacylglycerol (DGDG) in 0.1 M NaCl aqueous solution. We return to the general topic of membrane adhesion in Section 8.5.

The furry coat of polymers attached to the exterior of the plasma membrane also impedes cell adhesion. As displayed in Figs. 8.1 and 8.12(a), some attached polymers are relatively stiff polysaccharides (typical persistence length of 10 nm) and appear twig-like on the bilayer length scale of 4 nm. Other polymers are much more flexible, with persistence lengths of a few ångstroms, forming shapes imaginatively described as mushrooms or brushes in Fig. 8.12(b), at low or high surface density, respectively. The size of an unhindered polymer config-

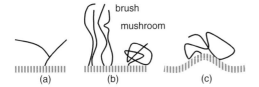

brush

mushroom

(a)   (b)   (c)

Fig. 8.12. Stiff polysaccharides appear like twigs on the bilayer length scale (a), while more flexible polymers resemble mushrooms or brushes, depending on their mean spacing (b). The entropic pressure of an attached polymer may cause the neighboring section of a flexible membrane to bend (c).

uration like a mushroom, as measured by its end-to-end distance r_{ee} or radius of gyration R_g (see Problem 2.12), increases slowly with its contour length L_c, where L_c is the total length of the polymer chain, including branches. From Table 2.1, the radius of gyration of a branched polymer or an ideal chain (mathematically permitted to intersect itself) is proportional to $L_c^{1/2}$ in three dimensions, although the proportionality constant is inequivalent for these two polymer types. For instance, the end to end displacement of an ideal chain obeys

$$\langle r_{ee}^2 \rangle = 2\xi_p L_c - 2\xi_p^2 [1 - \exp(-L_c/\xi_p)],\tag{8.80}$$

from Section 2.3, where ξ_p is the polymer persistence length; as an example, $\langle r_{ee}^2 \rangle^{1/2}$ is just 10 nm for a polymer with $L_c = 50$ nm and $\xi_p = 1$ nm. However, if the polymer has strong internal attraction, it may adopt condensed configurations whose size is proportional to $L_c^{1/3}$, more dense than ideal chains.

The effective volume of a polymer flopping about in solution, based upon Eq. (8.80), is much larger than its true physical volume, such that there is entropic repulsion when the fuzzy coats on two cells are brought into contact. Such a repulsive interaction is well known in colloids, where polymer coatings may keep particles from adhering under some conditions (see Napper, 1983, or Van de Ven, 1989, for a guide to the extensive literature on this subject). Depending as it does on polymer density and solvent conditions, the functional form of the repulsive pressure is rather complex and is not presented here. The reader is directed to Chapter 14 of Israelachvili (1991) for an introduction to this field; entry points to more recent literature can be found in Milner (1991), Ross and Pincus (1992), Lai and Binder (1992), Grest and Murat (1993), Yeung *et al.* (1993) and Soga *et al.* (1995). Experimental measurements of the effects of grafted polymers on membrane adhesion can be found in Evans *et al.* (1996).

Lastly, we mention the effect that a polymeric mushroom has on the membrane to which it is attached. Just as a random chain has a compression resistance arising from its entropy, an attached chain may exert a steric pressure on its membrane base, as illustrated in Fig. 8.12(c). This pressure forces the membrane away from the attachment point, giving it an induced curvature. The magnitude of the curvature is reviewed in

Lipowsky *et al.* (1998); being entropic, the induced curvature increases with temperature as $k_B T / \kappa_b$, where κ_b is the bending modulus. Of course, the presence of attractive interactions between polymer and membrane could modify the curvature, or even reverse its sign.

8.5 Adhesion

As determined in Sections 8.2–8.4, the pressure arising from several types of interactions between membranes decreases like an inverse power of D_s, the intermembrane separation. Crudely speaking, the pressure is 10^{-1} to 1 atm at $D_s \sim 2$ nm, built up from van der Waals (attractive), electrostatic (repulsive) and undulation (repulsive) terms. These contributions rise to 10^1 to 10^3 atm as D_s drops to 0.5 nm, where they are augmented by the steric repulsion between the extracellular matrix and other surface features. Finally, when the membranes are in very close contact, non-covalent binding is possible between specific molecules. We now describe these short-range couplings for biomembranes, before summarizing measurements of membrane adhesion energies and determining their effect on the shapes of adhering cells.

Site-specific interactions

Protein binding sites in immature cells, such as are present in an embryo, tend to be diffuse and may not display a locally organized structure in the plasma membrane. Organized adhesive structures appear as the cell matures, in some cases acting as junctions or switchyards for components of the cytoskeleton. In the epithelial cells shown schematically in Figs. 1.9 and A.2, for instance, desmosomes link bundles of intermediate filaments on adjacent cells, while hemidesmosomes join such filaments to the basal lamina. These structures provide local reinforcement to a membrane such that adhesion proteins are not simply ripped from the bilayer when the cell is subjected to stress. The largest families of proteins involved in cell adhesion are cadherins, integrins and selectins, with molecular masses around 80, 250 and 80 kDa, respectively, with some variation. The configurations adopted by proteins and their bound ligands (a generic term referring to any molecule capable of binding to a specified receiving molecule) fall into several categories. Simple binding geometries for pairs of identical or non-identical proteins are displayed in Figs. 8.13(a) and (b), corresponding to homophilic or heterophilic binding, respectively. For example, two desmosomes are bonded through their cadherin proteins as in (a), while the integrin proteins in a hemidesmosome are bonded to ligands on the extracellular matrix as in (b) (reviewed in Hynes, 1992). Another binding arrange-

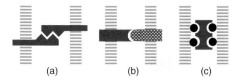

(a)   (b)   (c)

Fig. 8.13. Sample
configurations of bound
membranes. (a), (b) Direct
binding between identical
(homophilic) and non-
identical (heterophilic)
membrane proteins. (c)
Extracellular protein linked
to four small ligands (disks)
attached to neighboring
membranes. Bilayers are
indicated by striped bars.

ment involves an extracellular protein attached to ligands resident on different cells, as in Fig. 8.13(c). Although less common than the other mechanisms, this is the structure of the biotin–avidin complex exten- sively used for cellular studies (see Meier and Fahrenholz, 1996). Avidin is a tetrameric glycoprotein of molecular mass 68 kDa; as expected, it has four sites for receiving the vitamin biotin, a small molecule of molecular mass 0.244 kDa.

The free energies of association have been measured for a number of protein–ligand pairs, starting at as low as 5 $k_B T$ per bond (for a review, see Weber, 1975). Binding is particularly strong for biotin–streptavidin at 32 $k_B T$, although other ligands bind to streptavidin with as little as 8 $k_B T$ (Weber *et al.*, 1992). The free energy for binding biotin to avidin is even higher at 35 $k_B T$ (see Green, 1975; the force–distance relationship of this binding is investigated in Wong *et al.*, 1997). However, even these strong associations are still weaker than the typical energy of a covalent bond at 180 $k_B T$ for a C–H bond, or 150 $k_B T$ for C–C.

Measurements also have been made of the force required to break an adhesive bond, using an atomic force microscope or a synthetic vesicle as a force transducer. As emphasized by Bell (1978), any constant force ultimately causes a bond to break at non-zero temperature; it's only a question of the time scale (see also Evans and Ritchie, 1997). This phe- nomenon is introduced in the context of membrane rupture under a lateral tension in Section 5.5; according to Fig. 5.21, the membrane con- figuration with the lowest free energy for any tension contains a hole, although the time required to cross an energy barrier to reach this rup- tured state may be exceedingly long for low tensions. Two studies have confirmed the relation between applied force and bond lifetime for adhesion molecules. Merkel *et al.* (1999) have shown that the force required to break a biotin–streptavidin bond increases with the loading rate of the force: the faster the force is applied, the stronger the bond appears to be. The effect is dramatic, with the apparent rupture force growing from 5 to 170 pN when the loading rate increases by six orders of magnitude. Shao and Hochmuth (1999) have measured the related problem of removing an integrin or selectin from a neutrophil surface, observing that the time taken to extract the protein decays exponentially with the applied force; i.e., the protein could be extracted by forces in the 20–100 pN range, but took the least time with the strongest force.

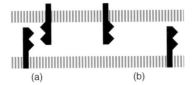

Fig. 8.14. Although some pairs of proteins on neighboring membranes may form an adhesive bond easily (a), others may have to diffuse through their membranes or reorient themselves to do so (b). Bilayers are indicated by the two striped bars.

(a)   (b)

Bearing this interpretational question in mind, the typical force required to break an adhesive bond on laboratory time-scales is quoted at 50–150 pN (for example, Florin *et al.*, 1994; Dammer *et al.*, 1995); similar values are observed for the removal of proteins from a bilayer (Waugh and Bauserman, 1995; Shao and Hochmuth, 1999).

The formation of specific ligand–receptor bonds introduces additional time-scales to the adhesion process beyond the dynamics of two membranes riding down a potential gradient that depends only on their separation D_s. As illustrated in Fig. 8.14, some interacting molecules may be close enough to easily form a bond (configuration (a)), while others may have to laterally diffuse along the bilayer to reach a suitable mate (b). Further, the participants may have to rotate in order to present the correct orientation for binding. Fundamental aspects of diffusion and bond formation have been incorporated into a set of rate equations developed by Bell (1978). This formalism involves several of the same diffusion concepts discussed in Section 9.2 on filament growth rates; namely, there are capture rate constants for the encounter of receptor–ligand pairs and dissociation constants for the breakup of bound complexes. The resulting rate equations interpret the binding process as a series of steps (e.g., initial association followed by local association to form a bound pair) and permit estimates of the rate constants on the basis of lateral diffusion constants and interaction distances (see Problem 8.6). The reaction rates have been measured for a number of bound systems; for recent examples, see Swift *et al.* (1998) or McKiernan *et al.* (1997).

Adhesion energies

A number of techniques have been employed to measure the adhesion energy density W_{ad} of pure bilayers, synthetic vesicles and simple cells. For instance, if two vesicles adhere as in Fig. 8.15(a), W_{ad} may be determined from the angle ϕ between the flat adhesion disk and the adjacent curved region (Bailey *et al.*, 1990). In the micromanipulation approach of Fig. 8.15(b), one vesicle is kept taut (*B*) while its contact area with a second flaccid vesicle (*A*) is controlled by the application of suction pressure, permitting W_{ad} to be extracted from the resulting dependence

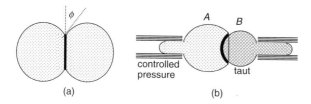

Fig. 8.15. (a) Two identical vesicles adhere, making an angle ϕ between the adhesion disk and the curved membrane. (b) Held in place by micropipettes, a flaccid vesicle (*A*) spreads onto a pressurized vesicle (*B*) in the shape of a sphere.

of pressure on contact area (Evans and Needham, 1988). Yet another approach uses interference patterns to obtain the height profile of a cell or vesicle near its contact region with a substrate as in Fig. 8.16(b), yielding a contact angle as in Fig. 8.15(a) (Rädler and Sackmann, 1993; Rädler *et al.*, 1995). Contact angles have also been measured for partially bound stacks of membranes (Servuss and Helfrich, 1989).

The adhesion energies of a number of pure bilayer systems are known experimentally, starting from the measurements by LeNeveu *et al.* (1975) on dimyristoyl phosphatidyl choline (DMPC); under many experimental conditions (see, however, Marra and Israelachvili, 1985), W_{ad} for DMPC lies close to 10^{-5} J/m². In 0.1 M NaCl buffer, SOPC vesicles (1-stearoyl-2-oleoyl phosphatidyl choline) display $W_{ad} = 1.3 \times 10^{-5}$ J/m², while DGDG bilayers (digalactosyl diacyl glycerol) are ten times stickier at 2.2×10^{-4} J/m² (Evans and Metcalfe, 1984; Evans and Needham, 1988). Introducing a polymer into the medium surrounding the SOPC vesicles raises W_{ad} linearly with concentration of the polymer; for instance, W_{ad} increases by a factor of ten in the presence of dextran at 10% by volume, in agreement with a mean field calculation of adhesive forces in polymeric solutions. In addition to pure bilayer vesicles, the adhesion of complete cells may be examined by interference microscopy; for example, W_{ad} for mutants of *Dictyostelium* has been found to lie in the range 0.6×10^{-5} to 2.2×10^{-5} J/m² (Simson *et al.*, 1998). Applied to synthetic model cells with surface receptors and a polymer glycocalyx, interference microscopy has established that the adhering region may phase separate, with local domains of strong contact separated by weakly associated membrane patches undergoing undulations (Albersdörfer *et al.*, 1997). The relationship between W_{ad} and other interfacial energies is established in Chapter 15 of Israelachvili (1991); note that the measured values of W_{ad} are several orders of magnitude lower than the surface tension of water with common alkanes, for example.

Let's compare the adhesion energy scales for a typical cell with a 10 μm² contact area. Consider two contributions:

- non-specific adhesion of a pure bilayer with $W_{ad} = 10^{-5}$ J/m², resulting in an adhesion energy of 10^{-16} J

- specific protein binding at 15 $k_B T$ per bond for 100 bonds/μm^2 (i.e., 10^4 copies over a surface area of 10^2 μm^2), yielding an energy of 0.6 $\times 10^{-16}$ J.

These two contributions are comparable in magnitude, and larger than the total membrane bending energy of 10^{-18} J for a spherical shell $(8\pi\kappa_b)$ at $\kappa_b = 10$ $k_B T$. Thus, a cell's shape may be strongly deformed upon adhering to a substrate; the rapid time scale for this change is estimated in Problem 8.8.

Cell shape with adhesion

The shape of an adhering cell in Fig. 8.16(b) bears some resemblance to the liquid drop on a solid substrate of Fig. 8.16(a), although we caution that the bilayer has more complex elastic properties than the interface between structureless fluids. The adhering droplet serves as a good introduction to adhering cells, so we first derive the classic equation for its boundary, taking the medium above the substrate to be a vapor with the same composition as the droplet. We define the surface tension between the vapor and the solid substrate or liquid drop to be γ_S and γ_L, respectively, while the energy density for the substrate and droplet interface is γ_{SL}. Assuming that the shape of the droplet is a truncated sphere of radius R (proven in Section 4.2 of Safran, 1994), then the area of the curved region is $A_{curve} = 2\pi R^2 (1 - \cos\theta)$ while that of the flat adhesion region is $A_{flat} = \pi R^2 \sin^2\theta$, where θ is the contact angle shown in Fig. 8.16(a). Hence, the total interfacial energy from the adhering droplet is

$$E = (\gamma_{SL} - \gamma_S)A_{flat} + \gamma_L A_{curve}, \qquad (8.81)$$

where the γ_S term arises from the removal of substrate area in contact with the vapor. The droplet shape that minimizes this energy is characterized by a contact angle θ_o satisfying $\partial E / \partial A_{flat} = 0$. One only needs $\partial A_{curve} / \partial A_{flat} = \cos\theta$ from Problem 8.15 to prove

$$\cos\theta_o = (\gamma_S - \gamma_{SL}) / \gamma_L, \qquad (8.82)$$

a result known as the Young equation (Young, 1805); the Young–Dupré equation, which involves a different parametrization for the energies, is discussed in Chapter 15 of Israelachvili (1991).

The value of the contact angle clearly depends on the relative magnitudes of the interfacial energies. For instance, taking water on wax as a typical situation, the relevant energy densities are $\gamma_L = 0.073$, $\gamma_S = 0.025$ and $\gamma_{SL} = 0.051$ J/m^2, yielding $\theta_o = 111°$. This confirms our experience that a water droplet forms a "bead" on a wax surface, rather than coating it ($\theta_o \ll \pi/2$) as in Fig. 8.16(a). The quantitative reason for this

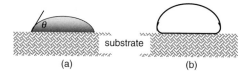

behavior is that $\gamma_{SL} > \gamma_S$: more energy is needed to create new area at a wax–water interface than at a wax–vapor interface. For less polar liquids than water, γ_{SL} may be much lower and consequently $\theta_0 < \pi/2$.

The presence of a lipid bilayer surrounding a fluid interior may strongly influence the appearance of a cell adhering to a substrate. A bilayer resists area expansion through its area compression modulus K_A (although the deformation energy has a different dependence on area change than does a liquid with surface tension γ), and also resists bending through the rigidities κ_b and κ_G. The sharp change in orientation of the droplet interface at the boundary of Fig. 8.16(a) would involve severe bending of a membrane at a large energy cost; consequently, an ideal bilayer must meet the adhesion zone smoothly with a contact angle of $\theta_0 = \pi$ when represented as a mathematically continuous surface. In such a representation, the principal curvatures at the contact point are $(2W_{ad}/\kappa_b)^{1/2}$ along a meridian of the vesicle and zero along the line of contact (Seifert and Lipowsky, 1990). Compared with an unconstrained cell or vesicle, the shape equations of an adhering cell have only a somewhat more complex mathematical form; however, the cell shape depends on more variables than κ_b and the reduced volume of the spontaneous curvature model of Sections 7.2 and 7.3, and must recognize

- the work of adhesion W_{ad}
- K_A of the membrane
- conservation of enclosed volume, if applicable.

With such a large parameter space, it is not surprising that the phase diagram of adhering cell shapes is nowhere near complete. We mention two special situations of several investigated thus far

- Bailey *et al.* (1990) have analyzed two adhering cells to determine the contact angle including only W_{ad} and K_A, but ignoring κ_b; expressions were obtained for θ_0 under conditions of either fixed or variable volume.
- Seifert and Lipowsky (1990) have determined cell shape for $\kappa_b \neq 0$, assuming that the bilayer area is fixed but the enclosed volume is not. They established that, even at zero temperature, a minimum adhesion energy W_{ad} is required to bind a simple vesicle.

To gauge the importance of various contributions to the cell's energy, we estimate the minimum adhesion energy needed to bind a cell with a fixed surface area but unconstrained volume. Performed in Problem 8.8, the calculation yields a value for W_{ad} that is easily an order of magnitude less than W_{ad} found for some pure bilayer systems, indicating that the observed adhesion energy may be considerably larger than the change in bending energy for some cells (supporting the approach of Bailey *et al.*, 1990). Nevertheless, the phenomenon of unbinding at weak adhesion is expected to be valid at zero temperature, and also at finite temperature where the net adhesion energy may be reduced by membrane undulations (Lipowsky and Seifert, 1991).

The weakness of the adhesion energy W_{ad} at zero temperature is not the only reason for membranes to unbind. At $T>0$, steric repulsion between undulating membranes is present (Helfrich, 1978), and is predicted to cause unbinding at a well-defined transition (Lipowsky and Leibler, 1986; Lipowsky and Zielinska, 1989). In studies of egg lecithin membranes under mild tension, Servuss and Helfrich (1989) observe the apparent adhesion energy to increase linearly with lateral tension, reaching 0.6×10^{-6} J/m^2 at the highest tensions in the experiment, a magnitude not inconsistent with $W_{ad} > 10^{-5}$ J/m^2 extracted from bilayers stiffened by strong lateral tension (e.g. Evans and Needham, 1988, Bailey *et al.*, 1990). More recent studies with reflection interference microscopy, while confirming the presence of undulations, yield a quadratic dependence on lateral tension for the related situation of a membrane adhering to a solid substrate (Rädler *et al.*, 1995). Further theoretical studies have shown that adhesion arising from the suppression of undulations may occur only under restricted conditions (Seifert, 1995). We also mention that the effect of membrane undulations on the formation of local molecular bonds has been examined by a number of researchers (Bruinsma *et al.*, 1994; Zuckermann and Bruinsma, 1995; Lipowsky, 1996; Weikl *et al.*, 2000). As with many topics in this text, we explore mainly equilibrium aspects of cell adhesion; the reader is directed to Dembo *et al.* (1988) for an introduction to the dynamics of adhesion and to Foty *et al.* (1994) for its effect on cell movement.

8.6 Summary

With their complex molecular structures and charge distributions, biological membranes experience forces from a variety of physical sources. In this chapter, we determine the pressure between membranes for several idealized models in order to identify their effects and assess their importance. Of the various contributors to intermembrane forces, the bulk of our attention is devoted to van der Waals, electrostatic and steric interactions from undulations. We first discuss the properties of charged membranes. Now, electrostatics in a cellular environment is more complex than that encountered in freshman physics: charged membranes and filaments may be surrounded by mobile counterions, as well as other ionic species. Further, the zero-temperature behavior of charges must be modified to accommodate the ambient temperature of cells, where entropy encourages ions to explore their surroundings in spite of the energy cost in doing so.

A good starting point for understanding such phenomena is the Poisson–Boltzmann equation, which reads $\nabla^2\psi = -(q\rho_o/\varepsilon)\exp(-q\psi(\mathbf{r})/k_BT)$, where $\psi(\mathbf{r})$ is the electrostatic potential experienced by a charge q at position $\mathbf{r}$. The number density of charged objects is ρ_o at the location where the potential is made to vanish; in general, the number density is given by $\rho(\mathbf{r}) = \rho_o\exp(-q\psi(\mathbf{r})/k_BT)$. As determined by the Poisson–Boltzmann equation, the spatial distribution of counterions lying a distance z to one side (only) of a large charged plate is $\rho(z) = [2\pi\ell_B(z+\chi)^2]^{-1}$, where the Bjerrum length ℓ_B is a property of the medium through $\ell_B \equiv q^2/4\pi\varepsilon k_BT$. The length parameter χ is fixed by charge neutrality to be $\chi = 2\varepsilon k_BT/(-q\sigma_s)$, where σ_s is the surface charge density of the plate. In an electrolyte solution, the potential from a charged plate is reduced or *screened* by a factor $\exp(-z/\ell_D)$; for monovalent ions in solution at a number density ρ_s, the Debye screening length is $\ell_D^{-2} \equiv 8\pi\ell_B\rho_s$.

Both electrostatic and van der Waals forces may contribute to the interaction between membranes at separations D_s of a few nanometers or more. Unfortunately, the Poisson–Boltzmann equation must be solved numerically to determine the electrostatic pressure between charged plates, even in the absence of screening. However, there are two useful approximations to the exact result: when counterions are spread uniformly throughout the gap between the plates, the pressure is given by $P = -2\sigma_s k_BT/qD_s$, while at large separations, it becomes $P = \pi k_BT/2\ell_BD_s^2$. Noting that q and σ_s have opposite signs here, the pressure is always repulsive $(P>0)$. In the presence of an electrolyte, the electrostatic pressure is sharply reduced by Debye screening, as described above for a single charged plate. When the electrostatic potential is weak, the pressure decays as $P = (2\sigma_s^2/\varepsilon)\cdot\exp(-D_s/\ell_D)$; for instance, the Debye length of a 0.2 M solution of monovalent salt (e.g. NaCl) in water is 0.7 nm, showing that the pressure declines substantially over 5–10 nm.

The attractive part of the van der Waals potential for a pair of molecules falls rapidly with distance as $-C_{vdw}/r^6$, where C_{vdw} is commonly in the neighborhood of 5×10^{-78} J·m⁶ for hydrocarbons. This potential can be integrated easily for a variety of geometries to yield the energy density of bulk materials under the assumption that the pairwise molecular potentials add independently. For instance, semi-infinite slabs experience an attractive pressure of $P = -(\pi^2 \rho^2 C_{vdw})/6\pi D_s^3$; the pressure between a slab and a sheet falls more rapidly at D_s^{-4}, changing to D_s^{-5} for two sheets. The combination $\pi^2 \rho^2 C_{vdw}$ is called the Hamaker constant, having a value in the region of 10^{-20} J $\cong 2$–$3\, k_B T$ for common organics separated by water.

In the absence of an electrolyte, the electrostatic pressure falls more slowly with D_s than does the van der Waals pressure. Depending on numerical prefactors, there is a value of D_s below which the van der Waals dominates and above which electrostatics is more important, considering only these two contributions. This behavior may change in the presence of electrolytes because of screening, and it is possible that the electrostatic term decays so rapidly as to be relevant only at intermediate distances. With van der Waals attraction at both short and long separations, the electrostatic component may then just serve as an energy barrier between the global energy minimum at small separations, and a local minimum somewhat further away. This effect, which underlies the DLVO theory of colloid stability, allows systems at low temperatures to be trapped in the local minimum, rather than condense into the global one.

Other interactions between membranes include solvation forces at short distances, and steric interactions at intermediate and long distances. The fuzzy glycocalyx coating the cell often provides a repulsive force as the entropy of the polymeric fur resists compression when cells approach one another. Thermal undulations of the membrane boundary at long wavelengths generate an entropic pressure of $2c_{fl}(k_B T)^2/\kappa_b D_s^3$, where c_{fl} is a constant equal to about 0.1. For large separations, this pressure may be competitive with the attractive van der Waals pressure between membranes and consequently reduce cell adhesion; in some situations, placing the membranes under lateral tension to reduce their undulations is observed to encourage adhesion.

In close proximity, the association of protein receptors and ligands on neighboring membranes can also contribute to their adhesion. Although the binding energy of individual molecular pairs may range up to $35\, k_B T$, the bonds have less than one-quarter the strength of conventional covalent bonds. At typical receptor densities, these site-specific interactions may contribute up to about 10^{-5} J/m^2 to the short-distance attraction of membranes, which is competitive with an adhesion energy density W_{ad} of $(1$–$2)\cdot 10^{-5}$ J/m^2 observed for pure bilayers. Adhesion energies in this range are sufficient to deform the boundary of a cell, permitting it to bind to other cells or a rigid substrate at low temperature. However, thermal undulations of membranes at higher temperature reduce their binding strength, perhaps forcing them to become unbound if the adhesion energy W_{ad} is otherwise weak.

8.7 Problems

Biological applications

8.1 Consider a negatively charged plate with surface charge density σ_s $=0.3$ C/m², immersed in water ($\varepsilon = 80\varepsilon_o$) which contains counterions on one side of the plate only. The counterions have charge $q = +e$ and the electrolyte is at $T = 300$ K.

(a) Find numerical values for ℓ_B, χ, and $\rho(z = 0)$ (quote ρ in M).
(b) Make a semi-logarithmic plot of ρ in the domain $0 \le z \le 5$ nm.
(c) At what value of z does ρ fall below 0.2 M?
(d) What is the magnitude of the electric field at $z = 1$ nm?

8.2 To one side of a single plate with surface charge density σ_s lies a monovalent electrolyte solution with $\rho_s = 0.2$ M. The potential at the plate is $\psi_o = -0.0125$ volts.

(a) Find ℓ_B and ℓ_D for this system at $T = 300$ K.
(b) Calculate σ_s in the Debye approximation.
(c) Make a semilogarithmic plot of $\psi(z)$ for $0 \le z \le 5$ nm using both the exact expression from Problem 8.10 and the Debye approxima-tion.

8.3 Using the potential energy density from Problem 8.13(b), derive the van der Waals pressure for two parallel sheets, each of thickness $d_{sh} = 4$ nm, separated by a distance D_s. Numerically compare your expression for the pressure to Eq. (8.38) for two slabs over the range $10 \le D_s \le 50$ nm. Assume $C_{vdw} = 0.5 \times 10^{-77}$ J·m⁶ and $\rho = 3.3 \times 10^{28}$ m⁻³.

8.4 Two parallel plates, separated by a distance D_s, form the boundaries of an aqueous medium containing only counterions, not dissolved salt. Their surface charge density is -0.2 C/m², and the medium is at 300 K.

(a) Calculate the electrostatic pressure between the plates for $0.2 \le D_s$ ≤ 5 nm.
(b) Compare your result at $D_s = 0.2$ nm with the ideal gas expression for P.
(c) Compare your result at $D_s = 5$ nm with the large-D_s expression for P.

8.5 Plot (logarithmically) the magnitudes of the following contributions to the pressure between membranes over the range $2 \leq D_s \leq 10$ nm:

- van der Waals attraction between two semi-infinite slabs (small D_s approximation)
- van der Waals attraction between two sheets of thickness 4 nm, using the potential from Prob. 8.13 to derive the corresponding pressure (large D_s approximation)
- repulsion from undulations for $\kappa_b = 25 \, k_B T$.

At what value of D_s does the repulsion from undulations match the van der Waals attraction? Use $c_{fl} = 0.1$, $\pi^2 C_{vdw} \rho^2 = 2 \, k_B T$ at $k_B T = 4 \times 10^{-21}$ J.

8.6 Proteins A and B move in parallel planes, producing a bound state C when they approach each other within a distance R_{AB} according to rate constants k_+ and k_-:

$$A + B \underset{k_-}{\overset{k_+}{\rightleftharpoons}} C$$

The applicable rate equations can be found in most introductory chemistry texts. If governed by diffusion, the rate constants are given by (see Bell, 1978)

$$k_+ = 2\pi(D_A + D_B) \qquad k_- = 2(D_A + D_B)/R_{AB}^2,$$

where D_A and D_B are the appropriate diffusion constants. Assume for proteins in a bilayer that $D_A = D_B = 10^{-14}$ m²/s and $R_{AB} = 0.75$ nm.

(a) Find an expression for the equilibrium constant $K_{eq} = [C]/[A] \cdot [B]$ in terms of R_{AB} from the condition $d[C]/dt = 0$, where $[...]$ represents concentration (here, in two dimensions). Evaluate your expression for our specific example.
(b) If the initial concentrations are $[A] = [B] = 5 \times 10^{14}$ m^{-2} and $[C] = 0$, what is the initial value of $d[C]/dt$? What is this production rate for a 1 μm² membrane patch?
(c) If the total number of copies of each of A and B in a 1 μm² patch is 500, which of the following is the most appropriate for the membrane adhesion time by the formation of the C bound state: 10^{-2}, 1, 10^2 s?

8.7 In bulk, cells may adhere to many of their neighbors, not just the pair-wise associations described in this chapter. Consider a cell with the unlikely shape of a rounded cube, each flat face of dimension $4R \times 4R$, for an overall width of $6R$. The principal curvatures are 0 and R^{-1} along the edges, and R^{-1} at the corners.

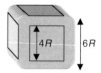

(a) Taking into account only its bending resistance κ_b, what is the deformation energy $\Delta E/\kappa_b$ of this shape compared to a sphere (refer to Table 7.1)?

(b) Evaluate the bending energy ΔE and the adhesion energy per cell for $\kappa_b = 10^{-19}$ J, $W_{ad} = 10^{-5}$ J/m² and $R = 3$ μm.

8.8 A vesicle with the shape of an axially symmetric pancake adheres to a flat substrate. The vesicle has a fixed area $A = 100$ μm² and radius parameter $R = 4$ μm, but its volume is unconstrained.

(a) What is the value of r?

(b) Find the energy $\Delta E/\kappa_b$ required to deform a spherical vesicle into this shape, using results from Table 7.1.

(c) What adhesion energy density W_{ad} is required to achieve this deformation if $\kappa_b = 20\, k_B T$? Quote your answer in J/m².

Formal development and extensions

8.9 The electrostatic potential $\psi(z)$ and counterion distribution $\rho(z)$ are determined in Section 8.2 for a large plate with surface charge density σ_s adjacent to a bath of counterions, each with charge q, in a medium of permittivity ε.

(a) Find $\rho(z=0)$ in terms of σ and ε.

(b) What fraction of the counterions are present in the region $0 \le z \le \chi$?

(c) Compare the behavior of the potential with that of a plate in the absence of counterions, particularly near $z = 0$ and $z \to \infty$.

8.10 Show that the dimensionless potential $\Psi(z)$ for a single charged plate in a salt medium is

$$\Psi(z)/2 = \ln([1+\alpha]/[1-\alpha]),$$

where $\alpha = \tanh(\Psi_0/4) \cdot \exp(-z/\ell_D)$ and $\Psi_0 = \Psi(z=0)$. Start from $d\Psi/dz = \pm 2\sinh(\Psi/2)/\ell_D$, as established in Section 8.2, and be cautious with the choice of sign. You may need to use $\int dx/\sinh x = \ln \tanh(x/2)$ from Section 2.423 of Gradshteyn and Ryzhik (1980).

8.11 Suppose that the potential energy between two molecules separated by a distance r has the power-law form $V(r)=-C_{vdw}/r^n$, where C_{vdw} and n are constants.

Show that a point molecule a distance D_s from a semi-infinite flat slab of material (infinite in the x- and y-directions, extending to $-\infty$ along the z-axis) experiences a potential energy given by

$$V_{point}(D_s)=-2\pi\rho C_{vdw}/(n-2)(n-3)D_s^{n-3} \quad (\text{for } n>3)$$

where ρ is the number density of molecules (units of $[length]^{-3}$). Assume that the potential is additive (*Hint: perform the integration using cylindrical coordinates.*)

8.12 Consider two molecules interacting with a potential energy $V(r)$ $=-C_{vdw}/r^n$, as in Problem 8.11, where r is the separation between molecules. Show that the energy of the

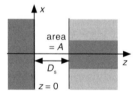

molecules in the slab to the right (of the diagram) arising from their interaction with the molecules in the semi-infinite slab on the left is

$$V_{slab}(D_s)/A=-2\pi\rho^2 C_{vdw}/(n-2)(n-3)(n-4)D_s^{n-4}.$$

Both materials have the same molecular density ρ and are semi-infinite in extent. Assume that the potential energy of the molecules is additive and use the result from Problem 8.11.

8.13 Molecules interacting according to the van der Waals potential $-C_{vdw}/r^6$ are present in the sheets and slabs shown in the diagram below. In (a), a sheet of thickness d_{sh} interacts with a semi-infinite slab separated by a distance D_s, while in (b), two identical

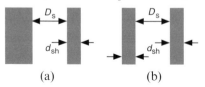

(a)   (b)

sheets are a distance D_s apart. Starting with the result from Problem 8.11 for the point + slab combination, show that the energy per unit area of the sheet in configuration (a) is $V(D_s)/A = -\pi C_{vdw}\rho^2 d_{sh}/6D_s^3$. Repeating Problem 8.11 for the point + sheet geometry, establish that the energy per unit area of a sheet in configuration (b) is $V(D_s)/A = -\pi C_{vdw}\rho^2 d_{sh}^2/2D_s^4$. In both cases, assume $D_s \gg d_{sh}$.

8.14 Show that, in the limit of large separation D_s, the electrostatic pressure between two charged plates bounding a medium containing counterions (but no salt) is given by

$$P = \pi k_B T / 2\ell_B D_s^2.$$

You may start your proof from $-2K\tan(KD_s/2) = q\sigma_s / \varepsilon k_B T$.

8.15 Assume that the shape of a liquid droplet adhering to a planar substrate is a section of a sphere with radius R. The areas of the curved and flat regions of the droplet are defined as A_{curve} and A_{flat}, respectively, and the boundary has a contact angle θ.

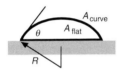

(a) Show that the areas and volume of the droplet are:

$$A_{flat} = \pi R^2 \sin^2\theta \qquad A_{curve} = 2\pi R^2(1-\cos\theta)$$
$$V = (\pi R^3/3)\cdot(1-\cos\theta)^2 \cdot (2+\cos\theta).$$

(b) Establish that $\partial A_{curve}/\partial A_{flat} = \cos\theta$ at fixed V by evaluating $\partial A_{curve}/\partial\theta$, etc. (where R is a function of θ, according to your results from part (a)).

Chapter 9 *more movement rather than assembly)*

Dynamic filaments

Cells are more than just passive objects responding to external stresses: they can actively change shape or move with respect to their environment. A very familiar example of cellular shape change is the contraction of our muscle cells. Less familiar, but very important to our health, is the locomotion of cells such as macrophages, which work their way through our tissues to capture and remove hostile cells and material. Another example of cell movement is the rotation of flagella (Latin plural for the noun *whip*) which extend from some cells and provide them with propulsion in a fluid medium. Flagella can be seen at both ends of the bacterium pictured in Fig. 9.1. Structurally related to flagella are cilia (Latin plural for *eyelash*), which occur on the surfaces of some cells and wave in synchrony like tall grass in the wind, creating currents in their fluid environment. What microscopic mechanisms underlie a cell's movement or *motility*? *use one — go into them*

In Section 9.1, we review several aspects of cell movement and trace their origin to the molecular level. Dynamic biofilaments alone can be responsible for certain shape changes because they can generate a force by increasing their length, as described in Section 9.2. However, contractile forces in muscles, or transport along a microtubule, arise from specialized motor proteins capable of crawling along a biofilament. A number of models have been proposed for the operation of such molecular motors, and these are explained in Section 9.3. Finally, we summarize the magnitude of the motile forces, and their effect on cell shape, in Section 9.4.

Fig. 9.1. Transmission electron micrograph of a platinum-shadowed replica of *Aquaspirillum metamorphum*, showing the flagella at each end of the bacterium. The arrow indicates the direction from which platinum was deposited on the bacterium to make the image. The bar is 500 nm in length (courtesy of Dr Terry Beveridge, University of Guelph).

Fig. 9.2. Actin and tubulin filaments are asymmetric, and growth through polymerization occurs more rapidly at one end than the other (called the plus and minus ends, respectively). Depolymerization also can occur at either end.

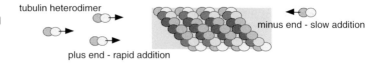

tubulin heterodimer

minus end - slow addition

plus end - rapid addition

9.1 Movement in the cell

Actin and tubulin are dynamic polymers: their fundamental protein building blocks (G-actin or the tubulin heterodimer) can both polymerize and depolymerize, depending on the conditions, changing the length of the filament in the process. Further, each end of the filament grows and shrinks at a different rate: the rapidly growing end is called the plus end, while the slowly growing end is minus. This is illustrated in Fig. 9.2 for tubulin; it has been argued that the α-tubulin units of the heterodimer face towards the plus end of the filament (Oakley, 1994). For actin, electron microscopy shows the physical appearance of each filament end to be different: the plus end resembles the feathered end of an arrow (hence the "barbed" end) while the minus end looks like the arrowhead (or "pointed" end).

Once a unit has attached to the end of the filament, a further compositional change occurs by hydrolysis. In the case of tubulin, each α- and β-unit contains GTP (guanosine triphosphate) binding sites; GTP in the β-unit hydrolyzes to GDP (guanosine diphosphate) shortly after polymerization. Similarly, actin monomers contain ATP (adenosine triphosphate), which hydrolyzes to ADP (adenosine diphosphate) after polymerization. In both cases, the effect of the hydrolysis is to weaken the polymeric bonds, making depolymerization easier. Depending on the environment, growth of the filament may be so rapid that hydrolysis lags behind the addition of new units, resulting in a region near the end of the filament where hydrolysis has yet to occur. For tubulin, this region is called the GTP cap, as illustrated in Fig. 9.3, and is *more* stable against depolymerization than the bulk filament itself. The growth and shrinkage rates of these filaments are discussed at length in Section 9.2.

The growth of dynamic filaments such as microtubules affects cell shape and motility. As shown in Fig. 9.4(a), the centrosome of most animal cells provides a nucleation region (or microtubule organizing center, MTOC) from which literally hundreds of microtubules radiate

Fig. 9.3. When tubulin heterodimers are added to a microtubule faster than the rate of GTP hydrolysis, the filament acquires a GTP-rich cap.

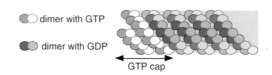

dimer with GTP

dimer with GDP

GTP cap

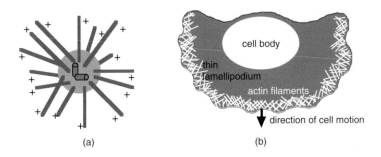

(a)   (b)

Fig. 9.4. (a) Long microtubules radiate from the cell's centrosome (fuzzy region around two cylindrical centrioles), their plus ends extending towards the cell boundary (after Alberts *et al.*, 1994; reproduced with permission; © 1994 by Routledge Inc., part of The Taylor and Francis Group). (b) Actin filaments occur at high density within several microns of the leading edge of a moving cell on a substrate; in fish scale keratocytes, the filaments often meet at right angles (Small *et al.*, 1995). Seen in cross section, the thin lamellipodium hugs the substrate and only the cell body has an appreciable thickness.

at any given instant. The filaments grow and shrink continuously, with their rapidly growing plus ends extending out towards the cell periphery. Thus, the filaments can exert a collective pressure on the plasma membrane in some cases, and, as a result, can push the nucleation region towards the center of the cell. During cell division, the centrosome replicates, providing two centers for microtubule nucleation and growth. Imagine two copies of Fig. 9.4(a) placed beside each other. Some microtubules from each centrosome approach each other plus-to-plus, meeting near the midplane of the dividing cell; these can link to form long bridges between the centrosomes. Other filaments can attach to the cell's chromosomes lying between the centrosomes. In one mechanism for cell division, the bridging microtubules force the centrosomes apart, dragging sister chromatids along with them by means of the attached microtubules. Further aspects of microtubule organization are described by Dustin (1978).

Filament growth also plays a role in cell locomotion. For example, fibroblasts and other cells can move along a substrate, adhering to it by spreading a sheet-like structure (the lamellipodium) across the surface, as illustrated schematically in Fig. 9.4(b). Other types of cells create extensions called pseudopods (literally false feet) to enable them to crudely walk over a surface. Observations of fluorescently labelled filaments show that the leading edge of such cells is actin-rich, with filament growth occuring at the cell boundary. In a study of keratocytes, the actin filaments move back through the cell body at a speed roughly equal to that of the cell (up to 0.1 μm/s), such that a given position on a filament remains roughly stationary with respect to the substrate (Theriot and Mitchison, 1991; see also Cao *et al.*, 1993, Lin and Forscher, 1993; actin growth rates of 0.05–0.1 μm/s are observed for neuronal growth cones by Forscher and Smith, 1988). At a speed of 0.1 μm/s, newly incorporated actin can be carried along a filament in the lamellipodium of these cells in about 20–30 s.

Actin and microtubules need not exist as unconnected entities in the cell. We have already described in Section 4.1 a number of actin-binding

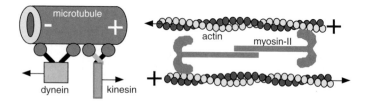

Fig. 9.5. (a) Kinesin and dynein walk towards the plus and minus ends of a microtubule, respectively. (b) Myosin walks towards the plus end of actin; here, the minus ends of the actin filaments are pulled towards each other (after Alberts *et al.*, 1994; reproduced with permission; © 1994 by Routledge Inc., part of The Taylor and Francis Group).

proteins (ABPs) capable of linking actin filaments into networks, and there also exist microtubule-associated proteins (MAPs) for fabricating bundles of microtubules. Further, specialized proteins can slide along actin and tubulin filaments, the energy for their motion being provided by ATP hydrolysis. Although their specific amino-acid sequence varies with the cell type, these so-called motor proteins can be grouped into three families: myosins associate with actin, while kinesins and dyneins associate with microtubules. Samples of movements possible with motor proteins are shown in Fig. 9.5.

Neither actin filaments nor microtubules are symmetric end-to-end, and the motor proteins have a preferred direction to their movement.

- Microtubules: kinesins generally walk towards the plus end, cytoplasmic dyneins walk towards the minus end
- Actin: myosins walk towards the plus end.

Given that the plus ends of microtubules radiate from the centrosome, kinesin and dynein together provide a mechanism for transport in either direction from the nucleus with modest speeds at cellular length scales. Motors on axonal microtubules can walk at speeds up to 2–5 μm/s per second towards the end of the axon, a rate which permits a chemical cargo of neurotransmitter, for example, to be transported in 2–6 days from a production site in the brain to the end of a motor neuron a meter away.

Actin filaments and bundles of myosin may be organized into highly cooperative structures in our muscles. Figure 9.6 is a schematic representation of skeletal muscle, which is the common muscle of our biceps etc. (the other types of muscle cells are described in Appendix A). The

Fig. 9.6. Arrangement of actin filaments and myosin bundles in a skeletal muscle cell. The myosin bundles contain more than one hundred individual myosin-II filaments, as shown in the inset (the average lateral separation between myosin bundles is 40–50 nm). Myosin "walks" along actin towards its plus end, causing the muscle to contract. The Z-discs define the boundary of a single sarcomere about 2.2 μm long (after Alberts *et al.*, 1994; reproduced with permission; © 1994 by Routledge Inc., part of The Taylor and Francis Group).

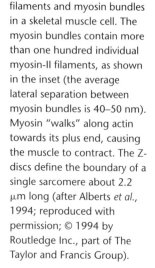

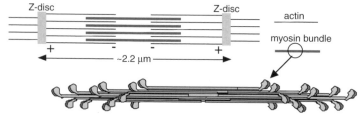

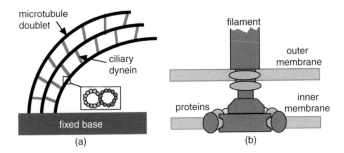

(a)   (b)

Fig. 9.7. (a) The bending of a cilium results from the motor protein ciliary dynein walking along a neighboring doublet of microtubules. If the cross-linking structure at the base of this diagram were removed, the filaments would slide past one another without bending. Insert shows a cross section of a microtubule doublet. (b) Highly schematic representation of the motor region of a (Gram-negative) bacterial flagellum (based upon material drawn from Francis *et al.*, 1992, and Kubori *et al.*, 1992; © 1992 by Academic Press).

thick filaments are bundles of myosin with a mean diameter of perhaps 30 nm, separated from adjacent myosin bundles by 40–50 nm. Walking towards the plus end of the actin, as shown in Fig. 9.5, mysosin pulls the minus ends of the filament towards one another, contracting the muscle along the horizontal direction in Fig. 9.6.

Molecular motors attached to parallel microtubules generate the whip-like motion of cilia and some flagella. The core of these thin structures is the axoneme, a bundle of microtubules about 225 nm in diameter, consisting of two microtubules in its center, surrounded by nine microtubule doublets. Each doublet has a single microtubule (13 protofilaments) to which is attached an incomplete second tubule (11 protofilaments), forming a structure whose cross section looks like the letters CO. The doublets are cross-linked at various points to form a bundle, and are also linked by a form of dynein called ciliary dynein (as opposed to cytoplasmic dynein described above) at 24 nm intervals. Because the microtubules are cross linked, the effect of the dynein motors attempting to walk along the filaments causes them to bend, as demonstrated in Fig. 9.7(a). The period of the wave motion is 0.1–0.2 s.

Flagella attached to sperm cells undulate transversely, although their motion is not identical to that of cilia. In contrast, the flagella of the bacterium in Fig. 9.1 rotate about their axis, driven by a much more complicated mechanism than that of sliding filaments. A schematic drawing of the rotary mechanism for a Gram-negative bacterium is shown in Fig. 9.7(b), although the drawing does not do justice to its complexity. Just as with a conventional electric motor, there is a rotor surrounded by a stator, in this case consisting of proteins. Seals must be present as well, as the motor passes through two membranes of the bacterium. These flagella are capable of 150 rotations per second, and provide a means of propelling the bacterium (see Chapter 6 of Berg, 1983, and Berg, 2000). We now decribe several molecular aspects of motility, before returning in Sec. 9.4 to the forces and torques generated within the cell.

Fig. 9.8. The capture (or release) rates at each end of a filament need not be equal if there is a chemical change of free monomer A to monomer B upon polymerization, as occurs for actin and tubulin. In the simplest polymerization process, A and B are identical.

9.2 Polymerization of actin and tubulin

Actin and tubulin filaments arise from the polymerization of subunits, which undergo a chemical change upon incorporation into a filament. Actin monomers contain an ATP molecule that hydrolyzes to ADP shortly after polymerization; similarly, the β-unit of the tubulin heterodimer contains a GTP molecule that hydrolyzes to GDP after polymerization. As the subunits in the filament are chemically inequivalent to free monomers, the filaments are capable of two phenomena – treadmilling and dynamic instability – not present in the simplest form of polymerization where free and bound subunits are equivalent. Our pedagogical approach to understanding this complex behavior is to present a simple model for polymerization first, followed by discussions of treadmilling and dynamic instability.

Simple polymerization

Consider the time rate of change, dn/dt, of the number of monomers n in a single filament. In principle, monomers can be both captured and released by a filament, as illustrated in Fig. 9.8, the two processes generally having different rates. The rate equation for the capture of monomers by a single filament has the form

$$dn/dt = +k_{on}[M] \quad \text{(capture)}, \tag{9.1}$$

where $[M]$ is the concentration of free monomer in solution and k_{on} is the capture rate constant, with units of $[concentration \cdot time]^{-1}$. That is, the number of monomers captured per unit time is proportional to the number of monomers available for capture. In contrast, the release rate from a single filament need not depend upon the free monomer concentration, and may be governed by the simple expression

$$dn/dt = -k_{off} \quad \text{(release)}, \tag{9.2}$$

where k_{off} has units of $[time]^{-1}$. Taking both processes into account, we have

$$dn/dt = +k_{on}[M] - k_{off}, \tag{9.3}$$

a form proposed by Oosawa and Asakura (1975). Note that Eq. (9.3) applies only once a nucleation site for filament growth is available.

The linear dependence of the elongation rate on monomer concentration is observed experimentally for both actin and tubulin filaments, although their behavior is slightly more subtle that Eq. (9.3) because of hydrolysis. The two rate constants k_{on} and k_{off} are easily extracted from a plot of dn/dt against [M], as shown in Fig. 9.9; k_{on} is the slope of the line and $-k_{off}$ is the y-intercept. Negative values of dn/dt mean that the filament is shrinking, with the result that the minimum concentration for filament growth [M]$_c$ (often called the *critical concentration*) is

$$[M]_c = k_{off}/k_{on}, \tag{9.4}$$

arising when dn/dt in Eq. (9.3) vanishes.

Effects of hydrolysis

The protein subunits of actin filaments and microtubules undergo more complex polymerization reactions than the simple situation summarized in Fig. 9.9. First, the subunits are not symmetric and add to each other with a preferred orientation, giving rise to an oriented filament as well. Second, the free subunits carry a triphosphate nucleotide (ATP for actin, GTP for tubulin) which is hydrolyzed to a diphosphate (ADP for actin, GDP for tubulin) after polymerization. Being chemically inequivalent, the ends of the filament polymerize at different rates, with the faster-growing (slower-growing) end refered to as the plus (minus) end. The various combinations of filament end (plus or minus), nucleotide state (di- or triphosphate) and reaction process (capture or release) mean that a minimum of $2^3 = 8$ rate constants are needed to describe the simplest generalization of Eq. (9.3), as illustrated in Fig. 9.10. Accommodating the hydrolysis of ATP and factors such as local chemical composition of the filament requires even more terms in the rate equations (Pantaloni *et al.*, 1985).

The rate constants are obtained by observing the time-evolution of filaments by a variety of methods (for a review, see Chaps. 3 and 7 of Amos and Amos, 1991). Usually, a mechanism must be devised for identifying the plus and minus ends of the filament in order to isolate the plus and minus rate constants. For example, growth from a particular nucleating center may allow the orientation of the filament to be distinguished (e.g., Bergen and Borisy, 1980; Bonder *et al.*, 1983; Pollard, 1986). Alternatively, a specific protein might be used to cap one end of the filament and restrict further changes to it (see Korn *et al.*, 1987); in

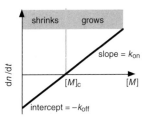

Fig. 9.9. Filament growth rate dn/dt as a function of free monomer concentration [M] expected from Eq. (9.3). The slope of the line is k_{on} and the y-intercept is $-k_{off}$. Filaments grow only if [M] exceeds the critical concentration [M]$_c$ defined by Eq. (9.4).

Fig. 9.10. The subunits of actin filaments and microtubules are asymmetric and contain a triphosphate nucleotide (T) that hydrolyzes after polymerization to a diphosphate (D). The capture and release of these subunits is described by a minimum of eight inequivalent rate constants. Note the stylized difference in conformations of the free and polymerized subunits.

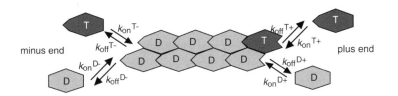

actin filaments, for example, phalloidin inhibits disassembly, while cytochalasin inhibits assembly at the plus end. The lengths of the filaments themselves can be obtained directly through electron microscopy (Bergen and Borisy, 1980; Bonder *et al.*, 1983; Mitchison and Kirschner, 1984; Pollard, 1986) and optical microscopy (Horio and Hotani, 1986; Walker *et al.*, 1988; Verde *et al.*, 1992) or indirectly via light scattering or spectroscopically (Lal *et al.*, 1984).

The dependence of the rate constants on the nucleotide composition of the monomer is obtained by observing filaments in pure solutions of ATP- or ADP-monomers; the rates also may depend strongly on other solutes. Sample values for the rate constants obtained for both actin and tubulin are given in Table 9.1. The first column on the left specifies the monomer in solution, while the headings above the data indicate the filament end at which the reaction occurs. The actin results quoted here are similar to those found with electron microscopy by Bonder *et al.* (1983), but are larger than other measurements summarized in Korn *et al.* (1987). The quoted microtubule rates are somewhat larger than Bergen and Borisy (1980). The data establish that:

- for both triphosphate and diphosphate monomer, the capture and release rates are almost always larger at the plus end than at the minus end
- the capture rates of the triphosphate are larger than the diphosphate at both ends.

In addition, the rate constants depend on the composition of the solvent; for example, altering the salt content of a solution at fixed monomer concentration may change some ks by more than a factor of two for actin (see collection of measurements in Lal *et al.*, 1984, and Pollard, 1986). The ratio k_{off}/k_{on} of the rate constants gives the critical concentrations shown in the two columns at the right, as expected from Eq. (9.4). Other determinations of the critical concentrations for ATP-actin (Lal *et al.*, 1984) are generally higher than the range quoted in the table.

Microtubules appear to grow only in the presence of GTP-tubulin (Walker *et al.*, 1988). Once incorporated into the microtubule and hydrolyzed, GDP-tubulin is capable of very rapid disassembly: k_{off} is one or two orders of magnitude larger for tubulin than it is for actin. Other studies of microtubules using dark field microscopy (Horio and Hotani, 1986) and electron microscopy (Mitcheson and Kirschner, 1984) also yield values of hundreds of dimers per second for k_{off}. Note that the critical concentrations $[M]_c$ for GTP-tubulin are the same at each filament end to within experimental uncertainties. We return to the dynamics of microtubule growth after describing the phenomenon of treadmilling observed for actin.

Table 9.1. *Samples of measured values for rate constants of actin filaments (Pollard, 1986) and microtubules (Walker et al., 1988); units of* $(\mu M \cdot s)^{-1}$ *for* k_{on}, s^{-1} *for* k_{off} *and* μM *for* $[M]_c$

monomer in solution	k_{on}^+	k_{off}^+	k_{on}^-	k_{off}^-	$[M]_c^+$	$[M]_c^-$
	(plus end)		(minus end)			
actin						
ATP-actin	11.6±1.2	1.4±0.8	1.3±0.2	0.8±0.3	0.12±0.07	0.6±0.17
ADP-actin	3.8	7.2	0.16	0.27	1.9	1.7
microtubules						
growing (GTP)	8.9±0.3	44±14	4.3±0.3	23±9	4.9±1.6	5.3±2.1
rapid disassembly	0	733±23	0	915±72	not applicable	

Notes:

The actin measurements vary by a factor of two or more with the solute concentration (for a more complete compilation, see Sheterline et al., 1998). The growth mode of microtubules is measured with GTP-tubulin, while rapidly retreating microtubules are presumably GDP-tubulin.

Fig. 9.11. (a) If $[M]_c^+=[M]_c^-$, both filament ends grow or shrink simultaneously. (b) If $[M]_c^+ \neq [M]_c^-$, there is a region where one end grows while the other shrinks. The vertical line indicates the steady-state concentration $[M]_{ss}$ where the filament length is constant.

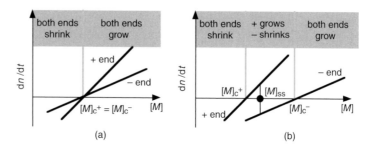

(a)

(b)

Treadmilling

The fact that actin and tubulin subunits undergo hydrolysis leads to an interesting dynamical mode called treadmilling (Wegner, 1976; Bonder *et al.*, 1983). We can see how this arises by inspecting the rate equations for each end of the filament, having the approximate form

$$dn^+/dt = k_{on}^+ [M] - k_{off}^+ \tag{9.5a}$$

$$dn^-/dt = k_{on}^- [M] - k_{off}^- \tag{9.5b}$$

where $+/-$ refer to the filament end. The triphosphate version of free actin and tubulin is copiously present *in vivo*, and clearly dominates the rate constants in Table 9.1; hence we take $[M]$ to be the concentration of free triphosphate proteins. Viewed independently, each of these equations may lead to a different critical concentration

$$[M]_c^+ = k_{off}^+ / k_{on}^+ \quad [M]_c^- = k_{off}^- / k_{on}^-. \tag{9.6}$$

Examples of the critical concentrations for solutions containing ATP-actin and GTP-tubulin are shown on the two right-hand columns of Table 9.1. In the special case $[M]_c^+ = [M]_c^-$, which is approximately obeyed by tubulin, the rate of growth or shrinkage has the schematic form in Fig. 9.11(a): at a given free monomer concentration, both ends grow or both ends shrink simultaneously, although the *rates* of growth or shrinkage may be different.

The dynamics are more intriguing for the general situation $[M]_c^+ \neq [M]_c^-$ displayed in Fig. 9.11(b). At high or low values of $[M]$, both ends grow or both ends shrink simultaneously, as in Fig. 9.11(a). However, in the intermediate concentration range $[M]_c^+ < [M] < [M]_c^-$, the plus end grows at the same time as the minus end shrinks. A special case occurs when the two rates have the same magnitude (but opposite sign): the total filament length remains the same although monomers are constantly moving through it. Setting $dn^+/dt = -dn^-/dt$ in Eqs. (9.5), this steady-state dynamics occurs at a concentration $[M]_{ss}$ given by

$$[M]_{ss} = (k_{off}^+ + k_{off}^-)/(k_{on}^+ + k_{on}^-). \tag{9.7}$$

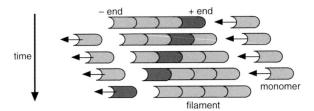

Fig. 9.12. Time evolution of a filament undergoing treadmilling. The dark monomer is seen to move along the filament during succesive additions, while the length of the filament remains constant. Whether the filament remains stationary depends upon its mechanical environment.

Here, we have assumed that there is a source of chemical energy to phosphorylate, as needed, the diphosphate nucleotide carried by the protein monomeric unit; this means that the system reaches a steady state, but not an equilibrium state.

The behavior of the filament in the steady state condition is called treadmilling, as illustrated in Fig. 9.12. Inspection of Table 9.1 tells us that treadmilling should not be observed for microtubules since the critical concentrations at the plus and minus ends of the filament are the same; that is, $[M]_c^+ = [M]_c^-$ and the situation in Fig. 9.11(a) applies. However, $[M]_c^-$ is noticeably larger than $[M]_c^+$ for actin filaments, and treadmilling should occur. If we use the observed rate constants in Table 9.1 for ATP-actin solutions, Eq. (9.7) predicts treadmilling is present at a steady state actin concentration of 0.17 μM, with considerable uncertainty. A direct measure of the steady state actin concentration under not dissimilar solution conditions yields 0.16 μM (Wegner, 1982). At treadmilling, the growth rate from Eqs. (9.5) is

$$\mathrm{d}n^+/\mathrm{d}t = -\mathrm{d}n - /\mathrm{d}t = (k_{on}^+ k_{off}^- - k_{on}^- k_{off}^+)/(k_{on}^+ + k_{on}^-), \qquad (9.8)$$

corresponding to $\mathrm{d}n^+/\mathrm{d}t = 0.6$ monomers per second for $[M]_{ss} = 0.17$ μM.

Dynamic instability

Rate equations such as Eq. (9.3) generally represent average behavior, perhaps an average over a number of systems, or perhaps a time average over a single system. Clearly, in a molecular picture, the capture or release of a monomer at a filament end is stochastic, such that $\mathrm{d}n/\mathrm{d}t$ varies with time. Over how long a time interval must the rate be measured in order to appear relatively constant? Observations of a single filament using optical microscopy, such as the plot shown in Fig. 9.13, indicate that the growth rate is uniform when sampled at time intervals of fractions of a second. However, for microtubules, growth is interrupted at seemingly random locations where dramatic shortening of the filament occurs. That is, growth does not appear to continue forever (unbounded), in spite of the presence of tubulin monomer; rather, the

Fig. 9.13. Growth of the plus and minus ends of a single microtubule as seen with optical microscopy (redrawn with permission from Horio and Hotani, 1986; © 1986 by Macmillan Magazines Limited).

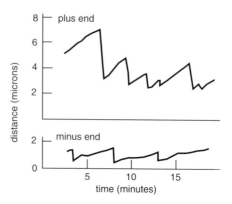

filaments alternately lengthen and shrink. The figure shows that disassembly events occur at both the plus and minus ends of the filament. This intermittent disassembly of the microtubule, referred to as dynamic instability, can be suppressed by the presence of a number of microtubule associated proteins (or MAPS; see Horio and Hotani, 1986).

Time-scales for the growth and retreat phases were first measured by Horio and Hotani (1986) and more systematically by Walker *et al.* (1988). The frequency of disassembly is found to decrease linearly with monomer concentration above $[M]_c$, becoming very small above $[M] \sim$ 15 μM. Similarly, the frequency of recovery during a disassembly event rises linearly with monomer concentration for $[M] > [M]_c \sim 5$ μM. That is, disassembly events occur less frequently and are of shorter duration as the concentration of monomers rises.

It is generally thought that the instability is associated with the hydrolysis of GTP at the microtubule tip (Mitchison and Kirschner, 1984; see also Carlier and Pantaloni, 1981). As illustrated in Fig. 9.3, the GTP-bearing tubulin hydrolyzes to GDP-tubulin once the molecule is polymerized. If the growth rate of the tip is fast enough, the hydrolysis reaction lags behind monomer capture and a GTP-rich cap builds up at the filament tip. From Table 9.1, GTP has a much lower off-rate k_{off} than GDP-tubulin, so the GTP cap is more stable than the main body of the filament. However, if growth slows for whatever reason (e.g. fluctuations in monomer concentration) the cap may be shortened or lost, permitting the filament to disassemble. Thus, this picture predicts that at high monomer concentration $[M]$

- growth is rapid, the GTP cap is long, and disassembly is infrequent
- disassembly is more likely to be reversed because $k_{on}[M]$ is large.

These model expectations are in general agreement with measurements of growth and shrinkage reversal by Walker *et al.* (1988).

A simple set of rate equations that include a parametrization of the random switching between growth and shrinkage epochs has been solved numerically or analytically for some special cases (Verde *et al.*, 1992; Dogterom and Leibler, 1993). This model exhibits a phase transition between unbounded filament growth and the bounded behavior of Fig. 9.13 when

$$f_{-+} dn^+/dt = f_{+-} dn^-/dt, \tag{9.9}$$

where f_{+-} is the frequency of changing from growth to shrinkage, and f_{-+} is the inverse. Although the rate equations do not include a microscopic prediction for parameters such as f_{-+}, the model permits a data analysis whose internal consistency can be checked through observables such as the mean filament length.

Dynamic instability is an efficient growth strategy for microtubules. Among their other roles, microtubules function as probes, which may seek out the cell boundary or organize components of the cell. In its "seeking phase", a microtubule should grow only at a modest rate as it extends through the cell; but once its job is complete (even if ending in failure), there is no need for the microtubule to retreat gracefully. Rapid disassembly releases the contents of the microtubule to the cytoplasm and permits it, or nearby microtubules, to start afresh in a different direction. In some sense, the process is like fly fishing, where placing a fishing lure in the water is a fast process, but retrieving the lure is done slowly to increase its likelihood of attracting a fish.

Benchmarks for k_{on} and k_{off}

We have used the measured values of k_{on} and k_{off} to understand dynamic behavior such as treadmilling and catastrophic disassembly, without attempting to interpret their numerical values within a molecular context (for a review of microscopic approaches to filament growth, see Bayley *et al.*, 1994). Molecular diffusion is one component of the overall reaction mechanism and is easily estimated, as we now show. Diffusion-based benchmarks for k_{on} and k_{off} permit us better to visualize the process of monomer capture and release in filament formation.

The disassembly process could be viewed as sequential if the average monomer randomly diffuses away from the filament before the next monomer is released. If the mean time between release is much smaller than this diffusion time scale, the disassembly is, in a sense, more cooperative. Now, the mean square displacement $\langle r^2 \rangle$ of a randomly diffusing particle in three dimensions increases linearly with time t according to

$$\langle r^2 \rangle = 6Dt, \tag{9.10}$$

where D is the diffusion constant. This equation is just another version of the random walk problem from Chapter 2, where the contour length has been replaced by the number of time steps. The diffusion constant for a monomer depends on its size and the viscosity η of the medium; the Stokes–Einstein relation (see Chapter VI of Landau and Lifshitz, 1987)

$$D = k_B T / 6\pi\eta R, \tag{9.11}$$

for spherical particles of radius R, provides a starting point for estimating the applicable D. The viscosity of the cytoplasm should be higher than water ($\eta = 10^{-3}$ kg/m·s) but less than glycerine ($\eta = 830 \times 10^{-3}$ kg/m·s), such that Eq. (9.11) predicts D in the range 6×10^{-14} to 5×10^{-11} m²/s for proteins with a 4 nm radius. For an experimental comparison, lactate dehydrogenase (from dogfish) with $R \sim 4$ nm has $D = 5 \times 10^{-11}$ m²/s in water (see Chapter 7 of Creighton, 1993). We choose $D \sim 10^{-12}$ m²/s to be representative of cellular conditions, noting that D for actin monomers in endothelial cells is observed at $3–5 \times 10^{-12}$ m²/s (McGrath et al., 1998).

Let's use the filament diameter to set $\langle r^2 \rangle$ of the diffusion process, recognizing that Eq. (9.10) averages over a distribution of paths, including those where the particle has not moved very far. The corresponding time scale for a protein to diffuse over a distance $\langle r^2 \rangle^{1/2} = 25$ nm is 10^{-4} s with our choice of D in Eq. (9.10). Thus, if each successive monomer unbinds once its former neighbor has moved away by a filament diameter, the monomers can be released at a rate k_{off} of order 10^4 s^{-1} for microtubules. The off-rates of tubulin in Table 9.1 are an order of magnitude lower than this benchmark, indicating that disassembly is moderately fast on a molecular length scale but not so fast as to be cooperative. Actin monomers, in contrast, are released much more slowly, presumably reflecting their tighter binding to the filament.

Diffusion also provides us with an estimate for k_{on}. The growing tip of a filament depletes monomers from its neighborhood, such that the local concentration of free monomers is less than the bulk concentration $[M]_\infty$ some distance away. As a result, free monomers flow into the depleted region, driven by the concentration gradient d$[M]$/dr, where r is the distance between the monomer and the capture site. The flux of monomers (the number per unit area per unit time) is simply $-D$d$[M]$/dr from Fick's first law of diffusion, where the minus sign indicates that the flow is in the opposite direction to the concentration gradient (see Section 12.5 of Reif, 1965). The gradient can be integrated to give a local concentration profile, rising smoothly to $[M]_\infty$ from zero at $r = R$, where all monomers are captured. From the number of monomers passing into the capture region, one obtains (see problem set)

$$k_{on} = 4\pi DR, \tag{9.12}$$

for a stationary reaction region (see Section 6.7 of Laidler, 1987, or Section 6.3 of Weston and Schwarz, 1972; for other biological applications, see Berg and von Hippel, 1985). The expression for k_{on} depends upon the trapping mechanism; e.g., for the bimolecular reaction $A + B \rightarrow AB$ of two mobile species, D is replaced by the sum of the individual diffusion constants $D_A + D_B$. Let us apply Eq. (9.12) to microtubules by taking $D = 10^{-12}$ $(m^2 \cdot s)^{-1}$, as usual, and R to be the sum of the microtubule radius (12 nm) and tubulin size (about 6 nm). With these values, Eq. (9.12) gives $k_{on} = 2.3 \times 10^{-20}$ $m^3/s = 138$ $(\mu M \cdot s)^{-1}$, about an order of magnitude larger than the results in Table 9.1. Given the uncertainty in D and the omission of rotational diffusion when the monomer docks at the filament, this estimate is appropriately larger than the observed values. In other words, the diffusion benchmarks for k_{on} and k_{off} provide a rough value for the capture rate and establish that disassembly is more random than cooperative along the length of the filament.

9.3 Molecular motors

The specialized proteins capable of walking or crawling along a cellular filament are referred to as motor proteins, of which there are separate families for actin and tubulin. The principal motor proteins associated with actin are myosin-I and myosin-II, having the distinct structures shown in Fig. 9.14 (the myosin family has more than two members, by the way). Myosin-I has a single globular "head" of mass ~80 kDa, containing the actin-binding site, and a tail of mass ~50 kDa, giving a typical overall mass in the 150 kDa range. The tail is attached to another structure such as a vesicle, which acquires movement as the myosin head crawls along an actin filament. Myosin-II is a dimer with two actin-binding heads attached to a flexible, but not floppy, tail (see problem set). The head region, at 95 kDa, is augmented by two light chains to bring the total mass to ~150 kDa, while each component of the tail has a mass of ~110 kD. For both myosins, the linear dimension of the heads is about 15–20 nm. In skeletal muscle, more than one hundred myosin-II

Fig. 9.14. Schematic representation of the motor proteins myosin-I and myosin-II, showing their actin-binding sites. These proteins walk on actin filaments towards their plus end.

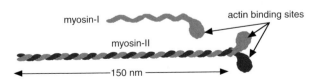

myosin-I

myosin-II

actin binding sites

←————————150 nm————————→

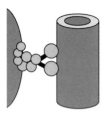

Fig. 9.15. Schematic representation of cytoplasmic dynein showing its attachment to a microtubule (right) and a section of a vesicle (left). It has been suggested that the heads may be linked to microtubules by thin stalks about 10 nm long (see Chapter 8 of Amos and Amos, 1991). Dynein usually walks towards the minus end of a microtubule and kinesin towards the plus end.

Fig. 9.16. Postulated mechanism for the movement of a myosin head (redrawn with permission, after Rayment et al., 1993b; © 1993 by the American Association for the Advancement of Science). Steps (i)–(v) are described in the text; in (iii) and (v) the initial configuration during the step is indicated by the light shaded region and the final configuration is dark. The tail of the myosin molecule, indicated by the wiggly line at the bottom of each frame, would be buried in a myosin bundle in Fig. 9.6. The symbols T, D and P represent ATP, ADP and phosphate, respectively.

dimers can form a bipolar bundle with a length in the micron range, as illustrated in Fig. 9.6. Myosin motors walk towards the plus end of actin filaments.

Structurally, the kinesin dimer has the same general appearance as myosin-II, except that the tail region is shorter (60–70 nm at ~50 kDa per strand) and the heads are smaller (10–15 nm at 45 kDa). Each strand of the tail terminates in several intermediate length chains, bringing the overall mass to about 180 kDa for each element of the dimer. Both heads of kinesin attach to a single microtubule and slide parallel to a protofilament, with the tail perpendicular to the axis of the microtubule, as shown in Fig. 9.5. Having a mass in the range 1200–1300 kDa, cytoplasmic dynein is much heavier than the other motor proteins and is structurally different, as seen in Fig. 9.15. The globular heads are spherical with diameters of 9–12 nm; cytoplasmic dynein is two-headed whereas ciliary dynein may have between one and three heads. The orientation of cytoplasmic dynein as it links a microtubule to a cellular component like a vesicle is displayed in Fig. 9.15 as well. A more detailed exposition of these proteins can be found in Amos and Amos (1991).

The energy source of the motors is ATP, which induces configurational changes during its capture, hydrolysis and release. A proposed mechanism for the structural changes of the myosin motor, based upon a determination of its head geometry from X-ray scattering, is shown in Fig. 9.16 (Rayment et al., 1993a, b). Five configurations are shown in the figure:

(i)   tight binding of the myosin head to the actin filament in the "rigor" position (so named from *rigor mortis*)
(ii)   release of the filament upon capture of ATP; the head and filament remain in close proximity because of neighboring connection sites
(iii) configurational change to the cocked position during hydrolysis of ATP; as a result, the head moves along the length of the filament
(iv)   weak binding of the head to the filament in a new position accompanying release of a phosphate group

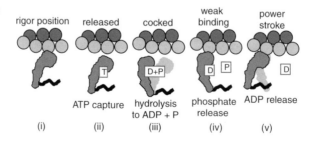

(v) configurational change during the power stroke, initiated by the release of ADP.

In the configuration changes of steps (iii) and (v), the position of the head in the previous step is indicated by the light shaded region, while its position after the change is indicated by the dark region with a black perimeter. Typical displacement of the head is 5 nm, occuring with a repetition rate which may be well above a cycle per second. The evidence in support of the directed motion model of Fig. 9.16 has been reviewed by Howard (1995); as we now describe, an alternative approach involves biased random motion, although this mechanism may not be relevant for today's cells.

Thermal ratchets

A simple, but inefficient, means of generating motion is a thermal ratchet, which makes use of a spatially asymmetric potential that oscillates with time, as sketched out in Fig. 9.17 (for general properties of thermal ratchets, see Ajdari and Prost, 1992; Doering and Gadoua, 1992; Magnasco, 1993; Bier and Astumian, 1993; Prost *et al.*, 1994). The potential $V(x)$ is taken to be periodic in the figure, with a peak-to-trough difference in part (a) that is large compared with $k_B T$; the potential is alternately turned on and off in the sequence (a)–(c). In part (a), a particle lies near the minimum of the potential, where it has a sharply peaked probability distribution. However, when the potential is turned off (b), the particle undergoes a random walk, such that the probability of finding it in a given location is described by a Gaussian distribution centered at the original potential minimum with a dispersion that increases with time. In part (c), the potential is turned back on again, and the particle tumbles towards the nearest potential minimum, which may not be its original position. Because the potential is asymmetric, the probability of moving to the right may be different from that of

Fig. 9.17. Probability distribution (top row) of a particle subject to an asymmetric potential (bottom row) that is alternately switched on and off from (a) to (c). Initially very sharply peaked at the potential minimum, the probability distribution broadens when the potential is off (b). Restoration of the potential in (c) results in a greater probability that the particle will move to the right than to the left.

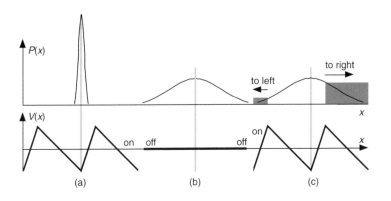

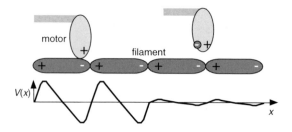

moving to the left. Thus, application of an oscillating asymmetric potential can generate a net movement in one direction, although the movement is not steady and may reverse direction from one step to the next. Clearly, if the potential is completely symmetric, a particle may jump between local potential minima, but with equal probability in either direction, resulting in no preferred direction of travel over long times. The form of the potential energy function is reminiscent of a mechanical ratchet, hence the model is referred to as a thermal or Brownian ratchet, after random or Brownian motion.

Why might this mechanism be important for cells (Astumian and Bier, 1994; Prost *et al.*, 1994)? Suppose that there is a charge distribution along a line of monomers in a filament as in Fig. 9.18. If it were charged, the head of a motor molecule could bind to this filament electrostatically; in the figure, the head on the left side carries a positive charge, and experiences the potential energy shown at the bottom. Upon capture of a charged molecule like ATP, the net charge on the head is reduced or eliminated, rendering the potential energy featureless and releasing the head from the filament. Lastly, when ATP is hydro-lyzed and ultimately released, the charge is restored and the head binds to the filament once again. As we now show, this mechanism is not 100% efficient for generating motion, and may not be the method of choice in today's cells. However, because it is operationally simple and may have been present earlier in evolution, we develop an approximate model for thermal ratchets that permits us to probe their efficiency and flux.

We invoke a number of approximations to get an analytically simple form for the thermal ratchet model. First, the rounded peaks and troughs of the potential in Fig. 9.18 are sharpened to give the sawtooth form in Fig. 9.19, where the periodic length is defined as b and the short-est peak-to-trough distance is αb. Second, the one-dimensional Gaussian probability distribution $\mathcal{P}(x)=(2\pi\sigma^2)^{-1/2}\exp(-x^2/2\sigma^2)$ is replaced by a triangular form $\mathcal{P}(x)=(1-|x|/w)/w$, such that $\int \mathcal{P}(x)\,dx = 1$. Equating the dispersions of the two distributions, namely $\langle x^2\rangle = \sigma^2$ and $\langle x^2\rangle = w^2/6$, gives the relation

$$w = \sqrt{6}\,\sigma \cong 2.5\,\sigma, \tag{9.13}$$

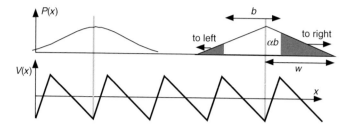

Fig. 9.19. In a simplified model for a thermal ratchet, the potential energy has a sharply peaked sawtooth form with period b and minimum peak-to-trough separation αb. We replace the Gaussian probability distribution by a triangular one with base $2w$ to calculate the probability of motion.

as one might intuitively expect from the shapes of the distributions. The accuracy of the triangle approximation is tested further in the problem set.

To begin our calculation of motion in the ratchet model, we refer to Fig. 9.19 and calculate the probability of a particle hopping to the right. If the potential is turned off for too short a time, a particle diffuses locally and the probability distribution is very narrow, such that w is less than αb. In such situations, a particle simply returns to its original position once the potential is turned on. For somewhat longer diffusion times with $w > \alpha b$, the particle may have moved so far that it seeks out the neighboring minimum once the potential is restored. Now, we will assume that the potential is turned on so fast that any particle moves rapidly down its potential gradient to a local energy minimum and does not diffuse into other regions of the system. The probability of movement to the right is then equal to the area of the right-hand shaded triangle in Fig. 9.19, or $\int P(x)\,dx$ integrated over $\alpha b \le x \le w$. With the form $P(x) = (1 - |x|/w)/w$, the probability of movement to the right is $\mathcal{P}_R = (1 - \alpha b/w)^2/2$, for $w > \alpha b$. This same argument can be repeated for motion to the left. If $w < (1-\alpha)b$, the distribution is sufficiently narrow that the probability of motion to the left is zero in the triangle approximation: $\mathcal{P}_L = 0$. However, once the distribution is wide enough, the probability of a hop to the left is $\mathcal{P}_L = [1 - (b - \alpha b)/w]^2/2$. The net probability of motion, $\mathcal{P}_{net}$, is the difference between $\mathcal{P}_R$ and $\mathcal{P}_L$

$$\mathcal{P}_{net} = \mathcal{P}_R - \mathcal{P}_L, \tag{9.14}$$

and can be summarized by:

$$= 0 \qquad\qquad 0 < w < \alpha b \tag{9.15a}$$

$$\mathcal{P}_{net} = (1 - \alpha b/w)^2/2 \qquad\qquad \alpha b < w < (1-\alpha)b \tag{9.15b}$$

$$= (b/w)\cdot(1 - 2b/w)\cdot(1 - 2\alpha) \quad (1-\alpha)b < w. \tag{9.15c}$$

Let's examine the properties of $\mathcal{P}_{net}$ as a function of α. When $\alpha = 1/2$, the potential is symmetric and there is equal probability of movement to the left or right; hence, $\mathcal{P}_{net} = 0$ in Eq. (9.15) (note that Eq. (9.15b) has

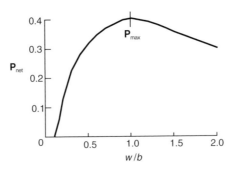

Fig. 9.20. The probability of net motion to the right $\mathcal{P}_{net}$ as a function of w/b for a sawtooth potential with $\alpha = 0.1$, according to Eqs. (9.15); $\mathcal{P}_{max} = 0.4$ at $w/b = 1$.

no domain at $\alpha = 1/2$). At the other extreme, when $\alpha = 0$, one wall of the potential is vertical and the probability distribution is split right through the center when the potential is turned on. This gives an upper bound on $\mathcal{P}_{net}$ of $1/2$, again as can be verified in Eqs. (9.15). At intermediate values of α, the maximum value of $\mathcal{P}_{net}$ is less than $1/2$, as shown for the case $\alpha = 0.1$ in Fig. 9.20. The probability rises steadily from zero once $w > \alpha b$, reaching a maximum value of $2/5$ at $w/b = 1$ before slowly falling towards zero at large w/b. By solving the condition $d\mathcal{P}_{net}/dw = 0$ at $w/b > \alpha b$, it is easy to show that the maximum probability for net motion $\mathcal{P}_{max}$ occurs at $w/b = 1$, and has the value

$$\mathcal{P}_{max} = (1 - 2\alpha)/2. \tag{9.16}$$

This expression yields $\mathcal{P}_{max} = 0.4$ for the parameter choice in Fig. 9.20.

We were forced to make a number of strong approximations in deriving Eqs. (9.15), and each of these approximations leads to an overestimate of $\mathcal{P}_{net}$ for molecular motors. For example, the physical potential does not have a sharp peak and is likely to appear bumpy in all regions because of the charge distribution and other effects. In addition, the time for switching the potential on and off is not instantaneous, but is at least on the order of the time to capture ATP or release ADP. Both of these effects give the motor molecule more time to diffuse forward or backward along the filament, thus reducing the probability of motion in a specific direction. More accurate evaluation of the properties of simple potentials has been performed by Astumian and Bier (1994) and Prost *et al.* (1994), who find that $\mathcal{P}_{max}$ may be 0.25 or much less, depending on the potential.

Let us estimate the flux of thermal ratchets to determine if they are even relevant for molecular motors in the cell. First, we use the diffusion equation in one dimension to estimate the time-scale for the step. The trajectory of a diffusing particle obeys the usual Gaussian distribution with $\sigma^2 = 2Dt$, where D is the diffusion constant and t is the time, such that $w = \sqrt{6}\,\sigma = (12Dt)^{1/2}$ according to Eq. (9.13). Now, in Section 9.2, our order of magnitude estimate for D was 10^{-12} to 10^{-14} m²s, indicating that t must be 5×10^{-6} to 5×10^{-4} s if we impose the condition for $\mathcal{P}_{max}$: $w=b=8$ nm. This time-scale is rather small: if the period for one cycle of the motion is twice this time, the motor could step at a rate of 10^3–10^5 steps per second. Of course, the local chemical environment of the motor may not be immediately ready for a new cycle at the end of the power stroke, increasing the time between completed steps; for instance, myosin is observed to move at less than 200 steps per second (Finer et al., 1994). At 10^3 attempts per second and 8 nm per step at a success rate of, say $\mathcal{P}_{net}=1/10$, the motor could advance at 800 nm/s. This is clearly in the experimental range ~0.5 µm/s observed for kinesin motors by Svoboda et al. (1993, 1994).

The primary drawback of the ratchet mechanism, for an advanced cell, is efficiency: it may take the hydrolysis of 10 ATP molecules to complete one step (if, for example, $\mathcal{P}_{net}=0.1$). Certainly, this efficiency is much lower than the directed motion of conformational changes such as Fig. 9.16, which could complete one step for every ATP hydrolysis. This question has been addressed experimentally by Svoboda et al. (1994) who measured the fluctuations, as a function of time, in the position of a kinesin motor sliding on a microtubule. Having established that the movement occurs in 8 nm steps (Svoboda et al., 1993), they show that the probability of completing a step per ATP hydrolysis is 0.5, which lies at the upper limit of the ratchet model developed here. Realistic ratchet models have efficiencies that range from 0.02 to 0.25 as considered by Astumian and Bier (1994) or Prost et al. (1994). The implications of these observations for two-step models are described by Svoboda et al. (1994). What we have demonstrated, then, is that the speeds of simple thermal ratchets are not unreasonable, but their efficiency is low, which may restrict their importance to earlier stages in the evolutionary development of motor mechanisms.

9.4 Forces from filaments

The previous two subsections are devoted to quantifying filament dynamics at the microscopic level: Section 9.2 analyzes the rates of growth and shrinkage of filaments in solution, while Section 9.3 examines the motion of molecular motors as they slide, dance or walk along a filament. The speeds observed for several types of motion associated with filaments are summarized in Table 9.2. The important feature to note is that the benchmark speed is 1 µm/s for many processes, although cellular movements such as slow axonal transport admittedly are much less. We have already established that these speeds could be achieved by a variety of mechanisms, although we have not ranked the importance,

Table 9.2. *Summary of speeds observed in various types of cellular movement. The monomer concentration in solution is denoted by [M]. Fast axonal transport involves kinesin or dynein moving along microtubules*

motion	typical speed (μm/s)	example
actin filament growth	10^{-2}–1	0.3 μm/s at $[M] = 10$ μM (Table 9.1)
actin-based cell crawling	10^{-2}–1	fibroblasts move at $\sim 10^{-2}$ μm/s
myosin on actin	10^{-2}–1	0.1–0.5 μm/s common in muscles
microtubule growth	up to 0.3	0.03 μm/s at $[M] = 10$ μM (Table 9.1)
microtubule shrinkage	0.4–0.6	0.5 μm/s (Table 9.1)
fast axonal transport	1–4	
slow axonal transport	10^{-3}–10^{-1}	

or even the relevance, of such mechanisms in actual cells. In this section, we place dynamic filaments in a cellular environment and explore their role in motility; our discussion is organized by filament type (actin first, then microtubules) and complexity of system (isolated filaments first, then composite mechanical systems).

Actin filaments

The growth rates of individual actin filaments depend on the concentration of actin monomer [M], which varies from cell to cell, and even within regions of a given cell. The total actin concentration of many cells falls in the 1–5 mg/ml range, although much of this actin may be bound in filaments rather than free in solution. Given a molecular mass of 42 000 Da, a 1 mg/ml actin solution is equivalent to 24 μM, for example. At this concentration, the plus end of an actin filament would grow at a rate of 280 monomers per second, according to the rate constants of Table 9.1. Each monomer adds 2.73 nm to the filament length (as can be verified from a mass per unit filament length of 16 000 Da/nm from Table 2.2), such that a rate of 300 monomers per second at the plus end corresponds to 0.8 μm/s in length. However, free actin concentrations in endothelial cells, for example, are 1/2 to 2/3 of total actin concentrations (McGrath *et al.*, 1998), pushing the expected growth rate down to about 0.3 μm/s. Thus, filament growth rates approaching 1 μm/s are achievable for actin, but only if the monomer concentration is rather large (see also Abraham *et al.*, 1999).

One mechanism for cell movement, shown in Fig. 9.4(b), involves the growth and disassembly of actin filaments at the leading edge of a lamellipodium. Is the rate of filament growth consistent with the movement of the cell? Although there are cells capable of crawling at a micron per second, typical speeds of crawling fibroblasts are closer to 10^{-2} μm/s, easily within the range of actin filament growth. Of course, the motion of a lamellipodium involves more than filament elongation at the cell boundary: materials such as monomeric actin and other proteins must be transported to the leading edge (where the plus-end of the filament resides) and each of these transport processes has a time frame. Several models have been proposed for the cooperative machinery of a crawling cell, and the reader is directed to the literature for more details (a sampling would include Dembo and Harlow, 1986; Dembo, 1989; Lee *et al.*, 1993; Evans, 1993; Lauffenburger and Horwitz, 1996; Mitchison and Cramer, 1996).

Does the behavior of actin filaments in lamellipodia arise from unassisted treadmilling? In isolated actin filaments, treadmilling is a relatively slow process: we showed from Eq. (9.8) that dn^+/dt is just 0.6 s^{-1} at steady state, or about 0.002 μm/s in filament length for the particular values of the rate constants in Table 9.1 (see also Selve and Wegner, 1986, and Korn *et al.*, 1987). This speed is rather low compared to most processes in Table 9.2. In some locomoting cells, actin filaments do have a roughly constant filament length even though the cell boundary may be moving at 0.1 μm/s; actin filament formation in neuronal growth cones also proceeds at 0.05–0.1 μm/s (Forscher and Smith, 1988). In such situations, the apparent treadmilling may be caused by actin-related proteins that sever the filament away from the leading edge of the cell, exposing regions of ADP-type actin that are more prone to disassembly.

Listeria monocytogenes and *Shigella flexneri* (Gram positive and negative, respectively) are two bacteria whose locomotion makes use of actin filaments within a host cell but *outside* the bacterium itself (Dabiri *et al.*, 1990; Theriot *et al.*, 1992; Goldberg and Theriot, 1995; and references therein). The surface of the bacterium contains a protein that helps organize free actin into a comet-like tail of cross-linked filaments pointing away from the direction of motion, as in Fig. 9.21. The plus end of the filaments is at the bacterium (Sanger *et al.*, 1992) and grows at rates up to 0.2–0.4 μm/s. Although composed of shorter actin segments, the tails themselves range up to about 10 μm in length and are stationary with respect to the host cell (Dabiri *et al.*, 1990; Theriot *et al.*, 1992). Newly incorporated monomers reside at their local position in

Fig. 9.21. Several bacteria are capable of sequestering actin monomers from a host cell into filaments with their plus ends at the bacterium. The filaments are stationary with respect to the host cell, and the capture of new monomers at the plus end generates a locomotive force on the bacterium.

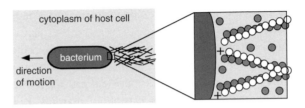

the tail for about 30–40 s on average, independent of the speed of the bacterium. Thus, the tail length increases with the speed of the bacterium: the separation between ends equals the product of the bacterial speed with the (constant) lifetime of the tail. Theoretical approaches to this motion have concentrated on mechanisms by which actin can insert at the plus end against the bacterial surface (Mogilner and Oster, 1996; for an earlier version, see Peskin *et al.*, 1993). Thermal fluctuations in the gap between the bacterium and filament end, arising in part from its bending motion (see Kroy and Frey, 1996), are sufficient to explain the observed filament growth rates.

Actin/myosin systems

The relative speed of actin and myosin falls in a range from 10^{-2} to 1 μm/s, depending on conditions in the cell. As discussed in Section 9.3, this range is anticipated by a variety of mechanical models, even the noisy motion of thermal ratchets. If the myosin head in Fig. 9.16 moves forward at about 5 nm per completed step (Veigel *et al.*, 1998), and if 10^1–10^2 steps can be completed per second (a range permitted by diffusion and also observed in optical trap experiments by Finer *et al.*, 1994) then the motor protein can advance at a rate of 0.05–0.5 μm/s. How is this speed related to the macroscopic speeds of our muscles? From Fig. 9.6, the relative motion of myosin on actin causes the two Z-discs (defining the boundary of the sarcomere) to advance towards each other at a speed twice that of the individual myosin head, i.e. 0.1–1 μm/s in this calculation. Given that the sarcomere in the figure has a length in its relaxed state of 2.2 μm, the muscle unit may contract by 5–45% per second. This admittedly broad relative change is consistent with the motion of the long muscles in our arms or legs; taking 30 cm as a typical muscle length, a 40% change corresponds to 12 cm per second, or 1 cm in a tenth of a second. Recalling the lever-like mechanical structure of our arms (i.e., our hand swings through perhaps 10 cm when our biceps contracts by 1 cm), the rapid contraction of our muscles has an order of magnitude of 10 cm per second.

The force generated by a single myosin motor has been measured by several techniques. Early experiments by Hill (1981) found a force of

8 pN (pN = 10^{-12} N) and more recent observations using optical traps by Finer *et al.* (1994) yields 3–4 pN. We can estimate the force generated in a single step knowing that the energy provided by the hydrolysis of an ATP molecule is about 20 $k_B T$. If 50% of the energy released results in a single 5 nm step of the motor, the resulting force must be $0.5 \cdot 20 k_B T / 5$ nm = 8 pN from [*work*] = [*force*]·[*distance*]. This estimate is consistent with the measured values.

Microtubules

The concentration of tubulin dimer in the cell, expressed as a mass per unit volume, is in the same general range as G-actin, although the corresponding molarity is a factor of two lower because of the difference in molecular mass – 42 kDa for G-actin and 100 kDa for tubulin dimers. The total tubulin concentration in a common cell like a fibroblast is 2 mg/ml, or 20 μM, of which half is free and half is bound in filaments, a similar ratio of free to bound monomer as found for actin. Thus, the typical concentration of free tubulin dimer is 10 μM. From the rate constants in Table 9.1, we expect the plus end of a microtubule to grow at ~50 units per second at 10 μM and to catastrophically disassemble at 800 s^{-1}, this latter value independent of tubulin concentration. At 13 protofilaments per microtubule, a single 8 nm long tubulin unit adds just 8/13 = 0.62 nm to a filament, so the growth and shrinkage speeds of a microtubule are 0.03 and 0.5 μm/s, respectively. Unlike actin, the crawling motion of cells may not require the presence of microtubules (Euteneuer and Schliwa, 1984).

Motor/microtubule systems

The motor proteins kinesin and dynein walk towards the plus and minus ends of microtubules, respectively, at a similar range of speeds, from 1 to 4 μm/s. Given that the polarity of microtubules in most cells has the plus end towards the cell boundary, the two molecular motors permit the transport of cargo-laden vesicles or organelles to and from the protein manufacturing sites near the cell's nucleus. Vesicle transport over long distances is particularly important for nerve cells, where chemical neurotransmitters produced in the cell body must be delivered to a synapse (anterograde motion) and waste products must be returned to the cell body for recycling (retrograde motion). Parenthetically, we note that the microtubules of the axonal highways are continuous only over distances of a few hundred microns on average, which is much less than the length of some axons. Both anterograde and retrograde motion have a fast component of a few microns per second; their cargo may range from small vesicles to mitochondria, which provide ATP to molecular motors as they work their way along axonal microtubules. In addition to this *fast axonal transport* in both directions, there is also *slow axonal transport* in the anterograde direction, with much lower speeds of 10^{-3} to 10^{-1} μm/s (axonal transport is treated in more detail in Amos and Amos, 1991). The speed of the fast component *in vivo* is consistent with, and perhaps faster than, the motion of molecular motors as we described in Section 9.3.

The force generated by a motor on a single microtubule has been measured by several independent experiments to be ~2.6 pN (Ashkin *et al.*, 1990), 1.9 ± 0.4 pN (Kuo and Sheetz, 1993), 4–5 pN (Hunt *et al.*, 1994) and 4–6 pN (Svoboda and Block, 1994). Our general expectation of the force, based on nucleotide hydrolysis as we discussed for myosin, is 5 pN for 20 $k_B T$ released per hydrolysis resulting in a step of 8 nm at 50% efficiency. This force range is consistent with the drag force on a vesicle, as described in the problem set.

Flagella

Many bacteria and complex single-celled organisms inhabit aqueous environments and have developed ways of locomoting in fluids. We described above the actin-based movement of *listeria*; however, a more common propulsion mechanism employs whiplike flagella, which can be seen extending from both ends of the bacterium displayed in Fig. 9.1. Flagella have a typical length of 10 μm, although examples ten times this length are known, such that their length is usually several times that of the main body of the bacterium. Both torque and thrust are generated by a flagellum as it rotates about its axis, driven by a rotary motor schematically illustrated in Fig. 9.7(b). Embedded in the bacterial membrane and cell well, the motor consists of several protein-based components, including:

- bushings to seal the cell membrane
- a circular stator, attached to the cell
- a rotor, with a typical radius of 15 nm, attached to the flagellum.

Early observations favored eight force-generation units in the stator (see references in Berg, 1995; but see also Berry and Berg, 1999). A curved segment or "hook" separates the motor from the main length of the filament, such that the filament is bent away from the normal to the membrane over a distance of tens of nanometers. Thus, the filament executes a helical motion as its rotates, and acts like a propeller.

As illustrated in Fig. 9.22(a), the flagellar helix moves through a fluid at an angle with respect to the symmetry axis of the bacterium. The

(a)   (b)

Fig. 9.22. (a) Rotating about the symmetry axis of a bacterium, a flagellum adopts a helical shape. (b) Looking along the axis towards the cell, the flagellum rotates counter clockwise to provide thrust, and the cell slowly rotates clockwise in response. Darker regions are closer to the viewer.

resistive force from the fluid can be resolved into two components: one generates a thrust along the symmetry axis while the other results in a torque around the axis. As shown in Fig. 9.22(b), the rotation of the flagellum is balanced by the slower counter-rotation of the cell proper. The figure displays the rotation of *E. coli*: normally the filament rotates counter clockwise (CCW) as viewed looking along the axis towards the cell. When the appropriate flagella rotate CCW cooperatively, the cell can move forward at speeds of 20 μm/s in a mode of motion called a "run". However, the motor has a switch permitting it to run in reverse, or clockwise (CW); flagella rotate CW independently, resulting in no net thrust on the cell such that it "tumbles", losing its orientation. For *E. coli* under common conditions, a typical run lasts 1 s, and a typical tumble lasts 0.1 s, although the duration of a given mode is exponentially distributed.

As a bacterium swims, its flagella may rotate at 100 revolutions per second (Hz) or more. Such rotational rates are comparable to an automobile engine, which runs comfortably at 30 Hz (or 2000 rpm) and reaches its operating limit at ~100 Hz; also like a car engine, flagellar motors fail catastrophically if driven too hard. A single flagellum attached to a fixed substrate can cause the cell body to rotate at 10 Hz, a lower rotation rate than the free flagellum because of viscous drag on the cell. Extensive measurements have been made of the torque generated by the *E. coli* flagellar motor (see Iwazawa *et al.*, 1993, Berg and Turner, 1993, and references therein; the behavior of motors forced to run in reverse has been revisited by Berry and Berg, 1999); the torque decreases monotonically to zero at several hundred revolutions per second, depending on conditions. As estimated in the problem set, the magnitude of the torque lies in the range 2–6×10^{-18} N·m in a number of cells investigated to date.

Flagella are not driven directly by ATP hydrolysis; rather, the motion is generated by protons running down a potential gradient across the membrane – electrical and pH gradients are both possible (see Khan and Macnab, 1980, Manson *et al.*, 1980, and references therein). Although early measurements were not inconsistent with thermal ratchet models, conceptually similar to those for linear motion in Section 9.3, more recent observations (Berry and Berg, 1999) are less supportive to these models and are not unfavorable to viewing the motor as a proton turbine (Berry, 1993; Elston and Oster, 1997).

9.5 Summary

Cells use a variety of mechanisms to change their shape, move through their environment and internally transport chemical cargo and cellular subunits. The simplest dynamic mechanism involves growth and shrinkage of the cytoskeletal filaments actin and tubulin. Filament growth can exert internal pressure on a cell boundary causing it to move, as in the lamellipodium of migrating cells, or can be harnessed to exert an external pressure on the wall of a bacterial invader like *listeria*. The monomers in these filaments have asymmetric shapes and, once captured, a triphosphate nucleotide they carry is hydrolyzed to a diphosphate. As a result, the ends of the filament are inequivalent, with the faster (slower) growing end referred to as the plus (minus) end. In many situations, the plus ends of the filaments extend towards the cell boundary. The overall length of both types of filaments changes at a concentration-dependent rate: under common cellular conditions, actin filaments lengthen at 10^{-2} to 1 μm/s while microtubules grow up to 0.3 μm/s and shrink at 0.4–0.6 μm/s.

The simplest representation of the change in the number of monomers n in a filament is $dn/dt = k_{on} [M] - k_{off}$, where $[M]$ is the concentration of free monomers and k_{on}, k_{off} are rate constants. Because these filaments are in dynamic equilibrium, net loss of monomer occurs if $[M]$ is below the critical concentration $[M]_c = k_{off}/k_{on}$. For actin filaments and microtubules, k_{on} is often in the range 5–10 $(\mu\text{M}\cdot\text{s})^{-1}$, depending on conditions; k_{off} is in the range 1–10 s^{-1} for actin but dramatically larger at ~ 800 s^{-1} for microtubules. These rates can be interpreted in terms of monomer diffusion; for example, in a very simple reaction model where a monomer is captured if it strays within a distance R of a filament tip, we expect $k_{on} \sim 4\pi D R$, where D is the diffusion constant of the monomer in solution. Cytoskeletal filaments have two interesting dynamical characteristics. With inequivalent values of $[M]_c$ at each end, actin filaments may grow at one end while they simultaneously shrink at the other; in fact, the two rates are equal at a unique monomer concentration, resulting in a filament of constant length, a phenomenon known as treadmilling. Microtubules, for which treadmilling is not an important process, are subject to dynamic instability where disassembly is so rapid that one tubulin dimer has barely diffused away from the filament before another is released.

More complex machinery for cell movement and transport is provided by special motor proteins capable of walking along actin fila-

Table 9.3. *General attributes of four motor proteins found in the cell. Omitted from the list is ciliary dynein, which displays a wider range of mass than the other proteins*

motor protein	associated filament	mass (kDa)	usual direction of travel
myosin-I	actin	~150	plus
myosin-II	actin	~2×260	plus
kinesin	microtubules	~2×180	plus
cytoplasmic dynein	microtubules	1200–1300	minus

ments or microtubules, summarized in Table 9.3. These motors have one or more globular-shaped "heads" about 5–10 nm in size, connected to a longer "tail" which can be attached to other motor proteins or to small vesicles or organelles. Present in both one- and two-headed varieties, the motor protein myosin forms a mechanical complex with actin, a widespread example being our muscle cells. Myosin commonly attains a speed of 10^{-2} to 1 μm/s as it works its way towards the plus end of an actin filament. Two distinct motor proteins travel along microtubules: kinesin generally walks towards the plus end while cytoplasmic dynein walks towards the minus end (there are a few examples of kinesin moving in the minus direction). Again, the typical speeds of these motors may range up to 1–4 μm/s. The mechanism by which the motor proteins travel along their respective filaments involves conformational changes of the "head" region, during the capture, hydrolysis and release of ATP and its reaction product ADP. The most energy-efficient mechanism for movement of the heads involves deliberate steps like a walk; it is less efficient for a motor to use a mechanism called a thermal ratchet, in which trial steps are made subject to a potential bias, not every step being successful. Current experiment leans towards the direct motion picture, although thermal ratchets may have been important in the evolutionary development of today's molecular motors. Lastly, we mention that motor proteins generate a force of a few piconewtons, consistent with the chemical energy released during the hydrolysis of ATP as the motor makes steps of 5–8 nm, depending upon the system.

The rotary motors driving bacterial flagella are the most complex source of cellular motion: many different proteins are used to fabricate the stator, rotor and bushings of this cellular engine. The analogue with an everyday mechanical engine is not inappropriate, as both engines have operating ranges of 100 Hz or more, and both fail catastrophically if overdriven. However, the torque produced by a flagellar motor is in the range 10^{-18} N·m, much less than an automobile engine $(1-3\times10^2$ N·m), but sufficient to provide a few piconewtons of thrust and propel a bacterium at up to a few tens of microns per second. Although their propulsion mechanism is not yet completely understood, flagellar motors have been described as protein turbines, driven by the flow of protons down a potential gradient across the cell boundary.

9.6 Problems

Biological applications

9.1 Find the time taken for the plus end of a microtubule to grow 5 μm from the centrosome of a hypothetical cell to its boundary. How long does it take for the filament to shrink to zero length if it undergoes rapid depolymerization upon reaching the boundary? Assume $[M]=10$ μM and take rate constants from Table 9.1 (a tubulin dimer is about 8 nm long).

9.2 Suppose that we have an ensemble of microtubules which are constrained to grow/shrink only at their plus ends. If their mean length is constant, what fraction of their time is spent growing and what fraction shrinking? Use rate constants from Table 9.1 and assume $[M]=15$ μM (tubulin dimers are about 8 nm long).

9.3 There is evidence that the hydrolysis of the GTP cap of a microtubule occurs at a rate constant k_{hydro} that is independent of monomer concentration, having units of $[time]^{-1}$, just like k_{off}. Let's evaluate the growth of the GTP cap for a specific tubulin concentration.

(a) Show that the number of tubulin dimers in the cap grows like

$$dn^+_{cap}/dt = dn^+/dt - k_{hydro},$$

assuming that the monomer solution is pure GTP-tubulin.
(b) Calculate dn^+/dt from Table 9.1 if $[M]=20$ μM.
(c) Assume, without experimental justification, that $k_{hydro}=30$ s^{-1}. Plot the filament length, and the cap length, as a function of time until the filament reaches 10 μm assuming that the negative end does not change with time (a tubulin dimer is about 8 nm long).

9.4 The "straight" tail of myosin-II consists of two parallel segments, each an α-helix of mass of 110 kDa, wound together into a coil of contour length 150 nm. Find the flexural rigidity κ_f and persistence length ξ_p of the tail using results from Fig. 2.20. Find $\langle r_{ee}^2 \rangle^{1/2}$ for the tail using the continuum filament model of Section 2.3.

9.5 Using Stokes' relation, Eq. (9.11), for the diffusion constant, determine the diffusion model prediction for k_{on} in the capture of a protein of radius R_p by a filament according to Eq. (9.12). Assume that the capture occurs when the centres of the protein and filament are separated by $3R_p$.

(a) What is the value for k_{on} if the viscosity of the medium is 10^{-1} kg/m·s, quote k_{on} in units of $(\mu M \cdot s)^{-1}$? Does k_{on} depend on R_p?
(b) Compare your calculation for k_{on} with that of actin from Table 9.1. Comment on the difference between these values.

9.6 The viscous drag force exerted by a stationary fluid on a spherical object of radius R is, according to Stokes' law,

$$F = 6\pi\eta v R,$$

where η is the viscosity of the medium and v is the object's speed. Let's apply this to a spherical vesicle moving at 0.5 μm/s in a medium with $\eta = 10^{-2}$ kg/m·s, which is ten times the viscosity of pure water.

(a) Plot the drag force as a function of vesicle radius for the range $20 \le R \le 200$ nm.
(b) Find the radius at which the drag force equals 2 pN, the force typical of a molecular motor.

9.7 What force is needed to propel a spherical bacterium of radius $R = 1$ μm at a speed v of 20 μm/s in water. Take the drag force to be $6\pi\eta Rv$ with $\eta = 10^{-3}$ kg/ m·s (quote your answer in pN). What power must be supplied to keep the cell moving (quote your answer in watts)?

9.8 In a viscous medium, an object rotates at an angular frequency ω when subject to a torque τ according to

$$\tau = \omega f_{drag}$$

where f_{drag} is the drag coefficient. For a sphere of radius R, f_{drag} is given by

$$f_{drag} = 8\pi\eta R^3,$$

where η is the viscosity of the medium. Find τ if the rotational speed is 10 revolutions per second, $R = 1$ μm, and $\eta = 10^{-3}$ kg/ m·s (water).

Formal development and extensions

9.9 The rate expressions of Fig. 9.11(b) are one example of the general case where $k_{on}^{+} \neq k_{on}^{-}$ and $k_{off}^{+} \neq k_{off}^{-}$:

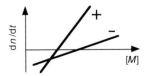

(a) What is the significance of the intersection point of the lines?
(b) What are the growth characteristics of a filament where the two lines are parallel?
(c) Suppose that the lines intersect when $dn/dt > 0$. What would be the properties of such a filament?

9.10 A filament, starting from a small seed, grows in a monomer solution of initial concentration M_0 in volume V. Suppose that no new monomer is added to the solution as the filament grows, and that the rate equation for the number of monomers in the filament is governed by the usual $dn/dt = k_{on}[M] - k_{off}$.

(a) Sketch $[M]$ as a function of time.
(b) What is the equation of this plot at $t \sim 0$?
(c) What is the asymptotic value of $[M]$ as $t \to \infty$?

9.11 Perform the following calculation to confirm how well the triangle distribution in Fig. 9.19 approximates the Gaussian distribution $\mathcal{P}(x) = (2\pi\sigma^2)^{-1/2} \exp(-x^2/2\sigma^2)$.

(a) Determine the functional form of $\mathcal{P}(x)$ in the triangle approximation (as parametrized by w) and evaluate the dispersion $\langle x^2 \rangle$. Equate this to the dispersion of the Gaussian distribution to establish $w = \sqrt{6}\,\sigma$.
(b) Compare $\mathcal{P}(0)$ and the full width at half maximum for both distributions, with $w = \sqrt{6}\,\sigma$ from part (a).

9.12 For the bimolecular reaction $A + B \to AB$, the diffusion constant in Eq. (9.12) is replaced by $D_A + D_B$ if both species are mobile (Chandrasekhar, 1943). If A and B are viewed as hard spheres of radius R_s, establish that

$$k_{on} = 8k_B T/3\eta$$

in the diffusion model.

9.13 Derive Eq. (9.12) for the capture of monomers in a solution. From Fick's law, the current of monomers (J, a number per unit time) inbound through a spherical shell of area A is $J = DA \, dc(r)/dr$, where $c(r)$ is the concentration of monomers at radius r (note the lack of a minus sign with this definition).

(a) Assuming that J is independent of r (why?), show that $c(r)$ must satisfy

$$c(r) = [M] - J/4\pi Dr,$$

where $[M]$ is the monomer concentration at infinite separation.
(b) Sketch $c(r)$ as a function of r. What is the physical significance of R, the radius at which the concentration vanishes?
(c) Use R to obtain an r-independent expression for J, from which you can deduce

$$k_{on} = 4\pi DR.$$

9.14 A spherical object of radius R moving in a stationary fluid of viscosity η experiences a drag force given approximately by Stokes' law

$$F = -6\pi\eta R \, \mathbf{v},$$

where $\mathbf{v}$ is the object's velocity and the minus sign signifies that $\mathbf{F}$ and $\mathbf{v}$ point in opposite directions. The velocity of an object subject only to drag steadily decreases from its initial value v_0.

(a) Starting from Newton's law, establish that the velocity decays with time t as

$$v(t) = v_0 \exp(- t/t_{visc}),$$

where the characteristic time $t_{visc} = m/6\pi\eta R$, m being the object's mass.
(b) Find t_{visc} for a cell with the density of water (10^3 kg/m^3) and radius $R = 3$ μm moving in a fluid with $\eta = 10^{-3}$ kg/m·s (water).
(c) By integrating the velocity, show that the cell comes to rest in a distance equal to $v_0 t_{visc}$. If the cell is initially moving at $v_0 = 10$ μm/s, find the distance over which it comes to a stop (after Chapter 6 of Berg, 1983).

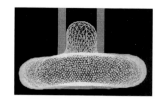

Chapter 10

Mechanical designs

Before launching into our concluding chapter, we catch our breath and survey the results obtained so far. In the opening chapter of the text, we identified the important structural elements of the cell as soft filaments and sheets, their flexibility arising from their nanometer thickness compared to their micron length. The mechanical properties of filaments in isolation (Chapter 2) and as components of networks (Chapters 3 and 4) were established in Part I, where we saw how the elasticity of soft filaments has its roots in both energy and entropy. The bending resistance of a biofilament may be sufficiently small that the filament noticeably changes shape in response to thermal fluctuations in its energy. Entropy is consequently reduced when a fluctuating filament is forced to straighten out at non-zero temperature and, of course, energy is increased when the intermolecular separation is driven away from its equilibrium value by tension; both of these effects contribute to the stretching resistance of a filament. The two chapters in Part II uncovered similar generic properties for flexible membranes: the energy required to bend a thin membrane may not be extravagant on the thermal energy scale $k_B T$, so that a biomembrane may display gentle undulations at ambient temperatures. Under lateral tension, energy increases with intermolecular separation and entropy is reduced when the undulations are suppressed; again, both of these effects lead to area compression resistance.

These two categories of materials – ropes and sheets – are used to construct cells in Part III. First, Chapter 7 demonstrated how cell shape is influenced by membrane bending or by the deformation of the cytoskeleton. Only the simplest systems were considered: vesicles bound by a fluid membrane and cells with a two-dimensional cytoskeleton or cell wall in addition to a membrane. Chapter 8 treated cells in contact, determining the adhesion energy between membranes and exploring its dependence on thermal undulations and molecular binding. Biochemically based forces made their appearance, at last, in Chapter 9, with the introduction of molecular motors. We now combine and extend these results to address several of the broad questions laid out in Chapter 1. First, in Section 10.1, we discuss strategies the cell uses to accommodate tension and compression, including forces generated during cell division. Then, in Section 10.2, we speculate on why cells have the size they do, and describe mechanical designs for cells with dimensions far away from microns.

10.1 Tension and compression

The filaments and sheets of a cell are subject to stress from a variety of sources. For example:

- the membrane and its associated networks may be under tension if the cell has an elevated osmotic pressure
- molecular motors may drag a vesicle along a filament or pull filaments past each other
- inequivalent elements of the cytoskeleton may bear differentially the compressive and tensile stresses of a deformation.

What are the magnitudes of these forces? Walking along a microtubule, a single molecular motor generates a force of about 5 pN; when multiplied by the powerstroke distance of 8 nm, the corresponding work of 4×10^{-20} J is consistent with the energy released from ATP hydrolysis (see Section 9.4; 1 pN $=10^{-12}$ N). A second example is the adhesion force between proteins on two membranes, which, at tens of piconewtons, may be an order of magnitude larger than the motor force. The stretching of a flexible filament provides a third yardstick; for instance, the effective spring constant of a spectrin tetramer is about $1-2\times10^{-5}$ J/m^2. As a result, it requires a piconewton to stretch this filament by 0.1 μm when a cell is moderately deformed.

To put these forces into perspective, consider first the viscous drag on a spherical cell of radius R moving with velocity v in a fluid according to Stokes' law, which has a magnitude $F=6\pi\eta Rv$, where η is the viscosity of the medium (see Problem 9.14). Taking $R=5$ μm, $v=1$ μm/s, and $\eta=10^{-3}$ kg/m·s (η of water) as representative values, the drag force on the cell is about a tenth of a piconewton. For a comparison at a macroscopic scale, we calculate the force required to bend a strand of hair. The flexural rigidity κ_f of a solid cylindrical filament of radius R is equal to $\pi YR^4/4$, where Y is the Young's modulus of the material (see Section 2.5); with $R=0.05$ mm and $Y=10^{10}$ J/m^3 (typical of biomaterials), we expect $\kappa_f=5\times10^{-8}$ J·m for a strand of hair. With one end of a filament of length L held fixed, the free end moves a distance $z=FL^3/(3\kappa_f)$ when subjected to a transverse force F (see Chapter 38 of Feynmann et al.,

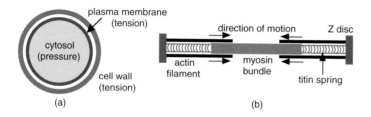

Fig. 10.1. (a) In a simple design for a pressurized cell, the membrane and cell wall may bear tensile stresses while the cytosol is under compression. (b) Schematic section of the actin–myosin filaments in a muscle cell; the spring representing the titin protein is not its true structure. Neither drawing is to scale.

1964). Thus, a force of 1.5×10^6 pN is required to move the free end of a 10 cm strand through a distance of 1 cm. In other words, even this imperceptibly small force on our finger tip is a million times larger than the typical force on a filament in the cell.

Newton's third law of mechanics tells us that a tensile stress on one component of a cell in equilibrium must be balanced by a compressive stress on another. Is there an optimal arrangement for the distribution of tension and compression? The simplest mechanical design for a cell would have the filaments and sheets bear any tensile stresses, while the cytosol is under hydrostatic compression. This strategy, shown in cross section in Fig. 10.1(a), is like the hot air balloon in Fig. 1.4(a). Now the plasma membrane alone cannot withstand a significant interior pressure: from Section 5.3, a typical bilayer ruptures at a lateral tension τ of about 0.02 J/m² on laboratory time scales. Using the relation $\tau = RP/2$ from Section 7.2 for a spherical shell of radius R, the rupture tension corresponds to a pressure difference P of just 8000 J/m³ = 0.08 atmospheres if $R = 5$ µm. This pressure is *much* less than the operational pressure of a typical bacterium or plant cell, both of which must augment their plasma membrane with a cell wall to prevent failure.

This design – fluids under compression and sheets under tension – can be extended to include filaments under tension, resulting in surprisingly complex biomachines. One example is the force-generating element of a skeletal muscle cell, introduced in Section 9.1 and redrawn in Fig. 10.1(b). Myosin motors, forming the thick bundle in the center of the figure, walk along reinforced actin filaments, pulling opposing filaments towards each other. To keep correctly oriented, the myosin bundle is attached to two mechanical springs in the form of the protein titin, each molecule extending all the way from the Z-disc to the midpoint of the bundle. All three elements – actin, myosin and titin – are subject only to tension in this configuration.

In the design of bridges and houses, one often sees rigid beams and bars carrying either tension or compression. The simple truss in Fig. 10.2(a) demonstrates how a vertical load is distributed across three beams in a triangle: the two thick elements are under compression while the thin element at the base is subject only to tension. The mast and

Fig. 10.2. (a) Three elements linked in a triangle bearing a vertical load; two bars are under compression while the rope is under tension. (b) A two-dimensional tensegrity structure of ropes and bars: no two compression-bearing bars are attached.

rigging of a boat, illustrated in Fig. 1.3, provide another example of a tension–compression couplet. As a design, these couplets may make efficient use of materials because the tensile elements, in general, need only a fraction of the cross sectional area of a compressive element to do their job properly. Space-filling structures built from components that individually bear only tension or only compression include the so-called tensegrities, a two-dimensional example of which is drawn in Fig. 10.2(b). Coined by R. Buckminster Fuller as *tensile-integrity structures* in 1962, tensegrities are intriguing in that rigid compressive elements are often linked only by floppy, tension-bearing ropes. The possibility that tensegrities can provide cells with rigidity at an economical cost of materials has been raised by Ingber (see Ingber, 1997, and references therein; Maniotis *et al.*, 1997). Certainly, the filaments of the cell do span a remarkable range of bending stiffness – a microtubule is a million times stiffer than a spectrin tetramer – and may be capable of forming a delicately balanced network if the cell could direct its assembly.

The importance of compression-bearing rods in the cell's architecture depends upon their buckling resistance; in Fig. 10.2(b), the two compressive elements will buckle if the tension sustained by the ropes is too great. Buckling occurs when a force applied longitudinally to a bar exceeds a specific threshold value, which depends upon the length of the bar and its rigidity. We calculate this buckling threshold in two steps. First, we describe the bending of a beam or rod in response to an applied torque (leading to Eq. (10.4)), then we apply this equation to the specific problem of buckling. The calculation follows that of Chapter 38 of Feynmann *et al.* (1964); a more general treatment can be found in Section 21 of Landau and Lifshitz (1986). Readers not interested in the derivation may skip to Eq. (10.11) to see the application to microtubules.

Suppose that we gently bend an otherwise straight bar by applying a torque about its ends. A small segment of the now-curved bar would look something like Fig. 10.3(a), where the top surface of the bar is stretched and the bottom surface is compressed – a situation we explored in Chapter 6 in the context of membranes. Near the middle of the bar (depending in part on its cross-sectional shape) lies what is called the neutral surface, within which there is no lateral strain with respect

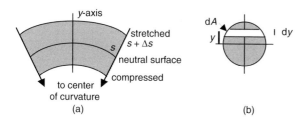

to the original shape. Let's assume that the bend is very gentle and that the neutral surface runs through the midplane of the bar. Measured from the neutral surface, the radius of curvature R is taken to be constant on the small segment in the figure.

The section has an arc length s along the neutral surface and a length $s + \Delta s$ at a vertical displacement y, where $\Delta s > 0$ when $y > 0$. Because the arcs in Fig. 10.3 have a common center of curvature, then by simple geometry $(s + \Delta s)/s = (R + y)/R$, or

$$\Delta s / s = y / R. \tag{10.1}$$

However, $\Delta s / s$ is the strain in the longitudinal direction (i.e., the change in the relative length; see Appendix D), telling us that the longitudinal strain at y is equal to y / R. The stress that produces this strain is the force per unit area at y, which we write as dF/dA, where dA is the unshaded region at coordinate y in the cross section displayed in Fig. 10.3(b). Stress and strain are related through Young's modulus Y by

$$[stress] = Y [strain] \tag{10.2}$$

which becomes $dF/dA = Yy/R$, or

$$dF = (yY/R)\, dA. \tag{10.3}$$

This element of force results in a torque around the mid-plane equal to $y\, dF$, which can be integrated to give the bending moment $\mathcal{M}$:

$$\mathcal{M} = \int y\, dF = (Y/R) \int y^2\, dA,$$

after substituting Eq. (10.3) for the force; equivalently

$$\mathcal{M} = Y\mathcal{I}/R. \tag{10.4}$$

The quantity $\mathcal{I}$, which made its debut in Section 2.3, is called the moment of inertia of the cross section, and has the form

$$\mathcal{I} = \int_{cross\ section} y^2\, dA, \tag{10.5}$$

where the integration is performed only over the cross section of the bar. For a cylinder of radius R, we showed in Section 2.2 that $\mathcal{I} = \pi R^4/4$; other shapes are considered in the problem set for Chapter 2.

Fig. 10.4. A rod subject to a sufficiently large compressive force *F* in its longitudinal direction will buckle. The deformed shape can be characterized by a function *h*(*x*).

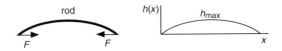

We now apply Eq. (10.4) to the buckling problem, specifically the forces applied to the bar in Fig. 10.4. The coordinate system is defined with $x=0$ at one end of the bar, whose contour length is L_c. Then, at any given height $h(x)$, the bending moment $\mathcal{M}$ arising from the force F applied to the ends of the bar is equal to

$$\mathcal{M}(x)=Fh(x),\qquad(10.6)$$

as expected from the definition of torque ($\mathbf{r} \times \mathbf{F}$). We replace the moment using Eq. (10.4) to obtain

$$Y\mathcal{I}/R(x)=Fh,\qquad(10.7)$$

where we emphasize that the radius of curvature R is a function of position. From Section 2.2, the radius of curvature at a position $\mathbf{r}$ is defined by $d^2\mathbf{r}/ds^2=\mathbf{n}/R$, where s is the arc length along the curve and $\mathbf{n}$ is a unit normal at $\mathbf{r}$. For gently curved surfaces, $d^2\mathbf{r}/ds^2$ can be replaced by d^2h/dx^2, so that $1/R=-d^2h/dx^2$ (the minus sign is needed because d^2h/dx^2 is negative for our bent rod). Thus, Eq. (10.7) becomes

$$d^2h/dx^2=-(F/Y\mathcal{I})h(x),\qquad(10.8)$$

which is a differential equation for $h(x)$.

From first-year mechanics courses, we recognize this equation as having the same functional form as simple harmonic motion of a spring $[d^2x/dt^2 \propto -x(t)]$, which we know has a sine or cosine function as its solution. For the specific situation in Fig. 10.4, the solution must be

$$h(x)=h_{max} \sin(\pi x/L_c),\qquad(10.9)$$

where h_{max} is the maximum displacement of the bend, occurring at $x=L_c/2$ in this approximation. As required, $h(0)=h(L_c)=0$. We can relate F to L_c by taking the second derivative of Eq. (10.9)

$$d^2h/dx^2=-(\pi/L_c)^2 h_{max} \sin(\pi x/L_c)=-(\pi/L_c)^2h(x).\qquad(10.10)$$

Comparing this with our differential equation (10.8) gives $F/Y\mathcal{I}=(\pi/L_c)^2$, or

$$F_{buckle}=\pi^2 Y\mathcal{I}/L_c^2=\pi^2\kappa_f/L_c^2.\qquad(10.11)$$

Now, this expression for the force is independent of h_{max}. What does this mean physically? If the applied force is less than F_{buckle}, the beam will not bend at all, simply compress. However, if $F>F_{buckle}$, the rod buckles as its ends are driven towards each other. To find out what happens at larger displacements, greater care must be taken with the expression for the curvature. This type of analysis can be applied to the buckling of membranes as well, and has been used to determine the bending rigidity of bilayers (Evans, 1983).

Just knowing that their persistence lengths are comparable to, or less than, cellular dimensions tells us that single actin and spectrin filaments do not behave like rigid rods in the cell. A microtubule, on the other hand, is gently curved, as its persistence length is ten to several hundred times the width of a typical cell (see Table 2.2). Can a microtubule withstand the typical forces in a cell without buckling? Taking its persistence length ξ_p to be 3 mm, in the mid-range of experimental observation, the flexural rigidity of a microtubule is $\kappa_f = k_B T \xi_p = 1.2 \times 10^{-23}$ J·m. Assuming 5 pN to be a commonly available force in the cell, Eq. (10.11) tells us that a microtubule will buckle if its length exceeds about 5 μm, not a very long filament compared to the width of some cells. If 10–20 μm long microtubules were required to withstand compressive forces in excess of 5 pN, they would have to be bundled, as they are in flagella, to provide extra rigidity.

Both of these expectations have been observed experimentally (Elbaum et al., 1996). When a single microtubule of sufficient length resides within a floppy phospholipid vesicle, the vesicle has the appearance of an American football, whose pointed ends demarcate the ends of the filament. As tension is applied to the membrane by aspiration, the microtubule ultimately buckles and the vesicle appears spherical. In a specific experiment, a microtubule of length 9.2 μm buckled at a force of 10 pN, consistent with our estimates above. When a long bundle of microtubules was present in the vesicle, the structure resembled the Greek letter ϕ, with the diagonal stroke representing the rigid bundle and the circle representing the bilayer. In other words, although they are not far from their buckling point, microtubules are capable of forming tension–compression couplets with membranes or other filaments. However, actin and spectrin are most likely restricted to be tension-bearing elements.

Cell division is one event in the life cycle of a cell where compressive forces may be borne by microtubules. The process by which a cell physically divides involves several steps and varies from one cell type to the next; for instance, a bacterium will have an easier time preparing its DNA for division than will a eucaryotic cell, whose nuclear envelope must be first disassembled and later reassembled after the chromosomes are separated. Sequestering the chromosomes into two separate nuclei is referred to as mitosis and is subdivided into five stages; the subsequent step of dividing the cytoplasm by forming two distinct cells is cytokinesis. Here, we describe only a few attributes of the mitosis and cytokinesis of animal

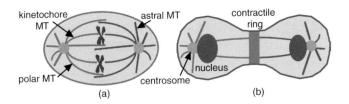

Fig. 10.5. (a) Having replicated, centrosomes (containing two centrioles in most animal cells) provide two nucleating sites for microtubules (MT), called astral, polar and kinetochore according to their function in separating sister chromatids, seen at the center. (b) Later in the cycle, an actin–myosin contractile ring assembles, ultimately forming a cleavage furrow as it shortens (after Alberts *et al.*, 1994; reproduced with permission; © 1994 by Routledge Inc., part of The Taylor and Francis Group).

cells, and refer the reader to standard texts for other aspects of the cell cycle such as DNA replication (e.g., Chapters 6, 8, and 18 of Alberts *et al.*, 1994). Fig. 10.5(a) illustrates the organization of the cytoskeleton in separating sister chromatids marshaled at the cell's midplane, after the nuclear membrane has been disassembled and two copies of the centrosome have moved to opposite ends of the cell, defining two spindle poles. Later still, as shown in Fig. 10.5(b), an actin-myosin contractile ring assembles just inside the plasma membrane, and cinches the cell into a dumbbell shape by shortening.

Microtubules are labeled according to their function or appearance during mitosis. Without considering the long preparation time involved in DNA replication, the onset of the visible part of mitosis is accompanied by a dramatic rise in the rate of growth and shrinkage of microtubules, as they seek out the attachment point (kinetochore) on a chromosome; indicated by the two small ellipses on the chromosome in Fig. 10.6(b), one kinetochore is present on each of the two sister chromatids and may be capable of attaching more than one microtubule, depending on cell type. Kinetochore microtubule is the label given to a filament that has successfully hit its target. If they are growing in the correct direction, some microtubules will snag, not a chromosome, but another microtubule radiating from the opposite centriole. Meeting plus end to plus, these microtubules may be linked by other proteins to form a dynamic bond, and are referred to as polar microtubules, running between the spindle poles defined by the centrosomes. Growing in the wrong direction, a third set of filaments has no luck at all in fishing for chromosomes, and are termed astral microtubules.

Several features of filament growth and molecular motors introduced in Chapter 9 are present during mitosis. The plus end of a microtubule (the rapidly growing end) is connected to the kinetochore of a chromatid in an arrangement that has been likened to an open collar, rather than a cap, as illustrated in Fig. 10.6(a); this structure permits new tubulin dimers to be added to, or lost from, the plus end according to dynamics established in Section 9.2. The minus end of the filament is buried in the centrosome, and loses tubulin with time, such that a tubulin dimer moves along a microtubule from plus to minus, as displayed in Fig. 10.6(b). A kinetochore microtubule can drag a chromatid

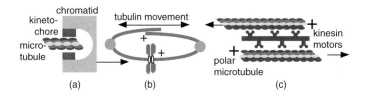

kineto-
chore

chromatid

micro-
tubule

tubulin movement

kinesin
motors

polar
microtubule

(a)   (b)   (c)

Fig. 10.6. (a) Cross section of a possible structure of a kinetochore collar surrounding the plus end of a microtubule. (b) Overall movement of newly incorporated tubulin towards the centrosome. (c) Possible arrangement of kinesin motors linking polar microtubules; the direction of their force is indicated by the arrows. Each of (a)–(c) is drawn to a different scale (after Alberts *et al.*, 1994; reproduced with permission; © 1994 by Routledge Inc., part of The Taylor and Francis Group).

towards a centrosome at a rate of about 2 μm per minute, corresponding to a loss of about 50 tubulin heterodimers per second, at 13 dimers for every 8 nm of filament length (Sawin and Mitchison, 1991; Mitchison and Salmon, 1992). About 60–80% of this tubulin loss occurs at the plus end and the remainder at the centrosome (Mitchison and Salmon, 1992). The loss rate at the plus end is not as catastrophic as permitted by the values quoted in Table 9.1, where the loss rate for rapid disassembly (k_{off}) is ~700 dimers per second. Section 9.2 also shows that the treadmilling of microtubules, where one end of the filament shrinks as the other grows, lies in a rather restrictive tubulin concentration regime, if it is present at all (see Fig. 9.11). The treadmilling rates extracted from experiment (Bergen and Borisey, 1980; Hotani and Horio, 1988) are of the order microns per hour, far slower than the migration of tubulin during mitosis.

Lest it lose mechanical contact, the collar around the microtubule at the kinetochore must pull itself along the filament towards the centrosome in advance of the plus end as it depolymerizes. The combination of collar movement and tubulin loss at the centrosome results in a tensile force of several hundred piconewtons applied to the attached chromatid; given that tens of microtubules are attached to a given kinetochore, the force per filament is much less, closer to 50 pN (Nicklas, 1983). Although the molecular basis of the collar movement is not fully resolved, it would require a small collection of molecular motors to produce this force, at about 5 pN per motor described in Section 9.5. Even this force is far more than what is needed to overcome the viscous drag on a chromatid, estimated at 10^{-1} pN (for a review, see Nicklas, 1988); further, the speed of a typical motor protein is microns per second, not the microns per minute of a chromatid. Like a sports car with a 5.0 litre engine stuck in city traffic, molecular motors can deliver more power and travel at higher speeds than are necessary for chromosome movement.

Pairs of opposing polar microtubules are thought to overlap at a junction complex containing the molecular motor kinesin. A candidate structure is displayed in Fig. 10.6(c), reminiscent of the actin–myosin complex in muscles of Fig. 10.1(b), where a motor pulls opposing filaments towards one another. Walking towards the plus end of the filament, kinesin motors generate compression on the microtubule, driving the spindle poles to move apart as observed in a later stage of mitosis. If the junction contains several kinesin motors, then the compressive force on the pair of microtubules will lie in the 10–20 pN range, which is not far from the threshold force that will cause buckling in a microtubule 5–10 μm long.

The mechanism for drawing the membrane together at the midplane is provided by the contractile ring, comprising some 20 actin filaments. The molecular motor myosin-II is thought to provide the driving force to shorten the ring, just as it draws actin filaments together in a muscle cell. The operation of the ring is somewhat like a belt, in that it cinches down the middle region of the dividing cell, ultimately forming a cleavage furrow where the cell splits. Unlike a belt, however, surplus actin is released from the ring as the neck narrows, and the ring disappears before the daughter cells separate, each carrying off their half of the cytoplasm and its organelles.

Experimental study of the formation and structure of the cytoskeleton has been aided by the development of synthetic model systems – a bilayer encapsulating a medium containing actin filaments or microtubules (Cortese *et al.*, 1989; Miyata and Hotani, 1992; Bärman *et al.*, 1992; Elbaum *et al.*, 1996). Some aspects of the microtubule-bearing vesicles were described above, so we briefly review just the actin systems. The strategy is to prepare a liposome (a bilayer vesicle) in a medium containing unpolymerized G-actin and perhaps other proteins. Once the liposome has closed into a seamless shell, polymerization of monomers can be induced by one of several means, such as increasing the temperature or altering the ion concentration. Shape changes of the liposome resulting from the synthetic cytoskeleton can be recorded microscopically. Crosslinking can be induced by means of the actin cross-linker filamin; the mean filament length can be controlled by introducing gelsolin as a nucleation site, giving mean filament lengths of 0.07–5.4 μm in one study (Cortese *et al.*, 1989). In some cases, F-actin is found to assemble into stiff bundles that force the vesicle into a highly prolate shape with the filaments running parallel to the axis of symmetry (Miyata and Hotani, 1992). The complete theoretical toolkit for the analysis of cell shape developed in Chapters 5–7 should be applicable to these model systems as quantitative measurements of their properties become available.

10.2 The largest and smallest cells

In part because of their evolutionary link, a common set of mechanical elements appears among modern terrestrial cells: bilayer-based membranes, several families of filaments, and, in some cases, interwoven mats external to the cell membrane. The composition of these elements is similar from cell to cell, although specialized conditions and functions may require additional chemical building blocks. Have terrestrial cells always had the same architecture and function as they do now? Are their molecular constituents the only ones capable of forming cells? Are there limits to the size and functionality of cells imposed by standard-issue bilayers and filaments? These questions are of fundamental importance as we try to understand the origin of terrestrial life and begin the search for life on other planets (see, for example, Jakosky, 1998; McKay *et al.*, 1996).

The Earth condensed about 4.5 billion years ago, but was inhospitable to life for at least half a billion years owing to its high surface temperature and a continual assault by meteors. Carbon-rich deposits in Greenland laid down 3.8 billion years ago have been interpreted as evidence for the rapid establishment of life on Earth once conditions were favorable; further, the oldest rocks containing recognizable fossilized cells date from 3.465 ± 0.005 billion years (Schopf, 1993; see Schopf, 1999, and de Duve, 1995, for overviews of the origin and evolution of life). These cells, having dimensions in the micron range, may have been preceded by pools of self-replicating polymers such as RNA, which could have formed from the simple organic molecules likely to have been present in the primitive atmosphere of the Earth (see, for example, Ferris *et al.*, 1996, and references therein). Yet, if individual strands of RNA can replicate in a suitable environment, why bother with the encapsulation provided by a membrane, particularly given the mechanical challenges posed by cell growth and division? It is argued that encapsulation accelerates evolution: a more efficient genetic macromolecule can take greater advantage of related molecules if they are sequestered within a cell, than if they are free to diffuse away in an uncontrolled environment.

Following the pioneering work on the production of organic compounds in a hydrogen-rich atmosphere (Miller, 1953), a number of scenarios for the origin of the cell's constituents have been advanced, including transport to Earth by cometary bombardment (see Delsemme, 1998). The primary chemical pathways for producing fundamental biomolecules have undoubtedly evolved with time, following the changing composition of the Earth's atmosphere and oceans. For example, the oxygen concentration in the atmosphere did not begin to accumulate until roughly two billion years ago, reaching its present abundance of 21% about a billion years ago (Chapter 6 of Delsemme, 1998). However, our approach here is not to catalogue the changing composition of the cell, but rather to explore the limits, if any, to cell size and functionality dictated by currently available materials assembled according to simple

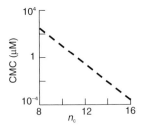

Fig. 10.7. Estimated critical micelle concentration c_{CMC} for double-chain phosphatidylcholines as a function of the number of carbon atoms n_c per individual acyl chain (according to the fit $c_{CMC} = 570 \exp(-1.8n_c)$ molar in Fig. 5.9).

blueprints. Cells operating outside of these rough bounds need modified construction. For example, we will place a bound on the maximum osmotic pressure sustainable by a bilayer according to cell size, demonstrating that micron-sized cells must modify their architecture to operate at several atmospheres pressure.

We begin our discussion of design limitations by considering biomembranes. The lipids present in a cell's bilayer possess a range of hydrocarbon chain lengths, 14–20 carbon atoms being among the most commonly observed (see Table 5.1). Why is this range the most favorable, and what are the attributes of membranes lying outside of this range? Let's review the mechanical properties of bilayers as a function of their thickness d_{bl} for conventional lipids before discussing the implications of bilayer thickness to overly large or small cells.

Aggregation

As discussed in Section 5.2, the concentration at which lipids form aggregates (the critical micelle concentration or CMC) drops steadily with increasing hydrocarbon chain length (Marsh, 1990, and Chapter 16 of Israelachvili, 1991). Shown in Fig. 10.7 is the fit to the CMC of dual-chain phosphatidylcholines as a function of the number of carbon atoms n_c per acyl chain in the region $8 < n_c < 16$, as determined from Fig. 5.9. A lower CMC means that the available lipids are used efficiently to form bilayers, an advantage for the cell, but offset by the disadvantage of having a longer synthetic pathway to produce the long chains. Now, the concentration benchmark for proteins and other compounds in the cell is 10^{-6} molar, and it would be beneficial to keep the concentrations of lipids at least an order of magnitude below this level. In analogy with a mediaeval town wall, it would be a waste of labor to produce an excess of blocks for the wall, only to have them littering the countryside or scattered throughout the town impeding foot traffic. One can see from the figure that dual-chain PC concentrations in the 10^{-7} to 10^{-8} molar region correspond to $n_c \sim 12$, at the low end of the distribution of chain lengths commonly found in the cell. In other words, the efficient usage of materials does not favor phospholipids with short chains.

Although lipids form bilayers with less waste as their chain length increases, other considerations argue against the presence of very long chains. For instance, the melting point of pure linear alkanes in the bulk increases by about 5°C for every CH_2-group added to the chain. Adding an extra five methylene groups to a 20-carbon chain would raise its melting point by about 25°C, which would be well above the operating temperature of a cell and cause the membrane to lose its fluidity. Now, there are ways of circumventing this catastrophe, such as introducing double bonds to increase the disorder of the chains, so the presence of long-chain lipids and thick bilayers is not ruled out. However, there is no compelling reason for the cell to adopt such bilayers in normal operating conditions, so the inventory of long-chain lipids is sparse.

Bilayer compression and bending resistance

After accounting for membrane undulations, the area compression modulus K_A of a lipid bilayer is roughly independent of bilayer thickness at just over 0.2 J/m², in agreement with models for K_A based upon surface tension (see Table 5.2). Thus, the compression modulus favors no particular membrane thickness, at least for conventional lipids. In contrast, the bending rigidity κ_b is proportional to $K_A d_{bl}^2$ for some systems and this dependence on d_{bl} may affect the properties of a cell in several ways. First, the membrane persistence length is proportional to $\exp(4\pi\kappa_b/3k_BT)$, from Eq. (6.50), implying that thin membranes fluctuate more than thick ones. However, the membrane would have to be very thin, perhaps just a nanometer across, in order that κ_b be driven down to the value of a few k_BT where the persistence length would be comparable to the cell size. Such thin membranes may be unfavorable on other grounds such as permeability or materials efficiency, so that κ_b is unlikely to be small enough to permit strong fluctuations on micron-sized membranes. Second, the energy required to bend a flat membrane into a closed shape increases linearly with κ_b; for example, the deformation energy of a sphere in the spontaneous curvature model is $8\pi\kappa_b + 4\pi\kappa_G$, where κ_G is the Gaussian bending rigidity. Increasing the bilayer thickness by 50% would roughly double κ_b and the bending energy. As discussed in Section 5.4, however, κ_b for cholesterol-rich membranes may be more than double that of pure bilayers, indicating that cell designs can accommodate $\kappa_b = 50\ \kappa_BT$. In summary, the compression and bending resistances of bilayers with $2 \leq d_{bl} \leq 6$ nm are structurally acceptable for terrestrial cells.

Bilayer rupture resistance

On laboratory time-scales, lipid bilayers rupture at strains of 2–5%, corresponding to a lateral tension of 2–5% of K_A, or 0.005–0.01 J/m² for K_A = 0.2 J/m² (Olbrich et al., 2000). Even at a fixed chain length, the rupture tension may vary by a factor of two according to the number of double bonds in the hydrocarbon chain. Now, the lateral tension τ in a membrane is determined by the pressure difference P across it and by the linear dimension of the cell. For cylinders and spheres, τ is equal to RP to within a factor of two, where R is the radius of the sphere or cylinder.

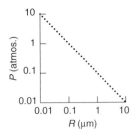

Fig. 10.8. Rough estimate of the maximal pressure sustainable for a bilayer-bounded cell as a function of radius R, assuming a rupture tension of 10^{-2} J/m^2.

Equating RP to the rupture tension yields a rough limit on the maximum internal pressure that a bilayer-bounded cell can maintain. This is plotted in Fig. 10.8 for a rupture tension of 10^{-2} J/m^2 on laboratory time-scales. For a very small cell with $R=0.15$ nm, the limiting pressure is about an atmosphere, which is at the low end of the range for common bacteria. However, the pressure drops rapidly with increasing radius, and is less than a tenth of an atmosphere for micron-size cells.

To survive at several atmospheres pressure, most bacteria need a cell wall whose thickness can be estimated from Young's modulus Y of peptidoglycan. Let us assume that the wall fails at 50% strain, i.e., at a lateral tension τ equal to $0.5K_A$. To within a factor of two, we can replace τ by RP for simple shapes and approximate K_A for solids by $K_V d$, where d is the wall thickness (see Problem 5.10). Thus, we expect d to be larger than $2PR/K_V$. For an isotropic material, K_V is close to Y (see Eq. (D.35)), which is about 3×10^7 J/m^3 for peptidoglycan. For example, if a cell with $R=0.5$ μm is to support a pressure of 1 atm $= 10^5$ J/m^3, we would estimate d to be greater than 3 nm, comparable to the observed wall thickness of Gram-negative bacteria. On the other hand, the wall must be 30–60 nm thick to withstand pressures of 10–20 atm, according to this estimate.

Large cells

The dimensions of the cell's filaments and sheets are appropriate for structures 1–10 μm across. What mechanical challenges are faced by large cells, for example, a sphere of radius 100 μm? First, the energy for deforming a membrane into a large cell should not be much different than a conventional cell; as we know, the deformation energy is independent of radius at $8\pi\kappa_b + 4\pi\kappa_G$ in the spontaneous curvature model. Next, the thickness of the plasma membrane need not increase to make the membrane stiffer and suppress undulations; even at a radius of 100 μm, the persistence length of a conventional bilayer is still much longer than the size of the cell, so that its surface undulations will not be unruly. However, if the cell were to operate at an elevated pressure, its cell wall would have to scale with the cell size according to our estimate of $2PR/K_V$ in the preceding paragraph. A novel feature of large cells would be the size of filaments used for compression resistance. In conventional cells, the persistence lengths of the filaments fall into families with 10^{-3}, 1 and 10^3 times the dimension of the cell (spectrin, actin and microtubules, respectively). To obtain a filament with a persistence length a thousand times larger than 100 μm, a cell might bundle together microtubules, or create a new filament with a somewhat larger radius, taking advantage of the R^4 scaling behavior of the persistence length of a cylindrical rod (see Section 2.2). In other words, there would have to be some adaptation, but not extensive revision, of design to scale up the cell size by a factor of one hundred.

Small cells

Let's now scale down the size of a cell and consider a spherical shape of radius 0.1 μm, a factor of ten or more smaller than a conventional cell. The energy to bend the membrane into a sphere is independent of radius, so a small cell gains little benefit here unless its membrane is thinner, and more flexible. In fact, there may be no energetic advantage for a really small membrane to close up into a sphere (Fromhertz, 1983). This possibility is discussed in Section 5.5, which compares the bending energy to the energy of a free boundary. There, Eq. (5.23) argues that vesicles are not favored at less than a radius $R_V{}^* = (2\kappa_b + \kappa_G)/\lambda$, where the edge energy λ is often measured at a few times 10^{-11} J/m. Taking $\kappa_b = \kappa_G = 10k_BT$, the minimal vesicle radius should be 12 nm from these considerations, although this value cannot be taken too seriously given the 4 nm thickness of the bilayer.

   With a persistence length 10^4 times the cell dimension, there is no need, and probably no space, for microtubules in our 0.1 μm cell: actin is probably stiff enough for most purposes. Further, because the lateral tension in a membrane is proportional to the cell radius, a very small cell could operate at an osmotic pressure of an atmosphere using just its plasma membrane without a cell wall. Without a wall or thick filaments, less genetic information need be encoded in the cell's DNA, another small blessing. The mechanical simplicity of small cells is attractive, but with volumes a factor of 1000 smaller than common bacteria, their biochemistry would be very limited. For instance, ribosomes used in protein production are about 20 nm across, and a suite of ribosomes would be hard-pressed to fit into a cell 200 nm in diameter.

Information storage

The storage of genetic information presents a challenge for many cells because of the length of their DNA. In double-stranded DNA, two linear polymers of sugar–base–phosphate monomers wrap around each other, as shown in Fig. B.6, their bonding provided by the interaction between the bases on opposite strands. A pair of bases, one from each strand, occupies 0.34 nm along the axis of the helix. As a rule of thumb, it takes about 1000 base-pairs to store the genetic code of a protein, at three base-pairs per amino acid of the protein, so about 300–400 nm of DNA are required per protein. Consequently, the contour length of

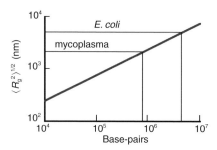

Fig. 10.9. Root-mean-square radius of gyration $\langle R_g^2\rangle^{1/2}$ for a random polymer with the persistence length of DNA. The number of base-pairs in mycoplasma and *E. coli* are indicated for comparison.

DNA required for 10^3 or 10^4 proteins (in addition to non-coding regions) can grossly exceed the dimension of a cell. For instance, the DNA of the bacterium *E. coli* is 1.6 mm long, a thousand times the diameter of the cell. Human DNA is even longer: the 24 distinct chromosomes (counting X and Y separately) range in contour length from 1.7 to 8.5 cm (see Chapter 8 of Alberts *et al.*, 1994).

Suppose that a cell did nothing special to package its DNA: how large a region would this molecular string occupy? To answer this question, let's characterize the dimension of the strand by its radius of gyration R_g, defined by $R_g^2 = N^{-1}\,\Sigma_{i=1,N}\,\mathbf{r}_i^2$, where $\mathbf{r}_i$ is the position of segment i on the chain as measured from the center-of-mass of the configuration. We showed previously that for ideal chains, $\langle R_g^2\rangle = \langle r_{ee}^2\rangle/6$ (Problem 2.12) and $\langle r_{ee}^2\rangle = 2L_c\xi_p$, (Eq. 2.31) where r_{ee} is the end-to-end displacement of a chain. Thus,

$$\langle R_g^2\rangle = L_c\xi_p/3, \tag{10.12}$$

for ideal chains. Using the DNA persistence length $\xi_p = 53$ nm, we plot in Fig. 10.9 the radius of gyration of ideal polymers as a function of their contour length L_c, expressed in terms of base pairs at 0.34 nm per base-pair. The DNA of mycoplasma has 800 000 base-pairs, or a contour length of 270 μm, which would coil up into a ball with $\langle R_g^2\rangle^{1/2} =$ 2.2 μm according to Eq. (10.12). Treated as a random linear polymer, the 4.7 million base pairs of *E. coli* would have a radius of gyration of about 5 μm from Fig. 10.9. In both examples, the jumble of DNA is wider than the cell itself, indicating that other mechanisms may be present for reducing the volume of the DNA ball.

Advanced cells have devised strategies for packaging their DNA by wrapping it around molecular barrels called histones, each barrel about 11 nm in diameter. This strategy is unnecessary for small cells with short DNA, and wouldn't be efficient in any event because of the space taken up by the histones. One mechanism that does not introduce extra packaging material is the phenomenon of supercoiling. This is familiar from a telephone cord, which twists around itself when overwound or under-wound compared to its form when manufactured. Supercoiling may produce straight sections of coiled coils, radiating from the axis of the DNA helix, thus reducing the effective size of the DNA ball consider-ably (see Chapter 23 of Lehninger *et al.*, 1993). High densities may also be achieved by polymers with strong self-attraction where R_g grows like $L_c^{1/3}$ (in three dimensions), which is more compact than the loose $L_c^{0.6}$ scaling of self-avoiding chains (see Section 2.3). This may be the situa-tion with a virus, for example, whose DNA is compressed into a dense ball just 30–40 nm across. The condensation of DNA into a more dense form has been induced in the laboratory through multivalent cations (reviewed in Bloomfield, 1991). Although it may be surprising that DNA, with one negative charge every 0.17 nm along its helix, could con-dense at all, the generation of attractive interactions between like-charged rods surrounded by counterions has been demonstrated theoretically (Grønbech-Jensen *et al.*, 1997; Ha and Liu, 1997), as recently reviewed in the DNA context by Gelbart *et al.* (2000).

10.3 Summary

Cells and their components are subjected to varying degrees of mechanical stress during their lifetime. Even in a quiescent environ-ment, most cells must generate internal forces to segregate copies of their DNA and divide the cell envelope during reproduction. What strategies are most appropriate for accommodating internal or exter-nal stresses? The simplest approach is to place the fluid interior of the cell under compression (as needed), against which the envelope bears a lateral tension. For cells several microns across, a pressure difference of just a tenth of an atmosphere can be sustained by a lipid bilayer, which must be reinforced by a wall tens of nanometers thick to with-stand the 10–25 atmosphere osmotic pressure of some bacteria. A more complex cell architecture involves the use of protein filaments having persistence lengths ξ_p a thousand times less (spectrin) or a thou-sand times more (microtubules) than the width of the cell itself. Now, a filament of contour length L_c experiencing a compressive force along its axis will buckle when the force exceeds $\pi^2 \kappa_f / L_c^2$, where $\kappa_f = k_B T \xi_p$ is the flexural rigidity. The typical compressive force an individual fila-ment must carry is a few piconewtons, which approaches the buckling point of a microtubule 5–10 μm long, and would easily buckle a single actin or spectrin filament of that length. Thus, many of the cell's fila-ments are limited to tension-bearing roles (unless they are rather short) and only microtubules, individually or in bundles, usually take part in compressive activities such as forcing the centrosomes apart during cell division.

The architecture of today's cells results from many contributing factors, such as the availability of construction materials, the desired functionality of the cell, and its ease of reproduction. What limits, if any, do conventional biomolecules place on the catalogue of cell designs? For instance, the number of carbon atoms n_c in the hydrocarbon chain of a lipid may be influenced by the following considerations:

- if the chains are too short ($n_c < 10$), the lipid concentration needed to produce a bilayer is high, and the cell wastes its production capacity by manufacturing unused lipids
- if the chains are too long ($n_c > 20$), the melting point of the lipids may be sufficiently high that double bonds must be added to the chain to keep the membrane fluid.

Thus, the lipid composition of the membrane reflects the cell's operating environment and the efficient use of materials. Any membrane with thickness in the 2–6 nm range would appear to be mechanically viable: the area compression modulus of a bilayer displays little dependence on d_{bl}. Even the observed variation of the bending modulus κ_b with bilayer thickness d_{bl} (namely $K_A d_{bl}^2$) can be accommodated in most cells.

Were cells much larger than microns in diameters, say by a factor of 100, it is likely that a new filament type (perhaps a thicker filament or a bundle of microtubules) would be needed to provide compressive elements that could assist cell division. Were cells much smaller than microns, say by a factor of 20, some structural elements such as microtubules or thick cell walls would be superfluous; for example, the bilayer of a cell 100 nm in diameter (more than a factor of two below today's smallest mycoplasma) could withstand an osmotic pressure of a few atmospheres. A smaller inventory of structural components means that less information need be coded in the cell's DNA, which reduces the DNA packaging requirements of the cell. Viewed as a random chain, the DNA of most procaryotic cells is somewhat larger than the dimension of the cell, but its effective volume can be reduced by a condensation mechanism or by supercoiling: if the DNA is twisted about its helical axis, pairs of coiled coils will radiate from the main axis, much like the tangles formed on a twisted telephone cord. However, the DNA of eucaryotic cells may be so long as to require additional packaging help, such as the histones which function as molecular barrels 11 nm in diameter, around which DNA may be wrapped like a garden hose. Given the length of non-coding regions on eucaryotic DNA, it appears more efficient for the cell to devise clever packaging strategies than to clean up its genetic ticker-tape.

10.4 Problems

Biological applications

10.1 Suppose that the tubulin (heterodimer) concentration [M] in a hypothetical cell of radius 10 μm is 1 μM.

(i)   What total length of microtubule could be made from this amount of protein if each heterodimer is 8 nm long?

(ii) If all filaments radiate from the center of the cell to its boundary, what is the average membrane area per microtubule.

(iii) At a force of 5 pN per filament, what would be the total pressure exerted on the cell membrane (quote your answer in J/m^3 and atmospheres).

10.2 Find the ratio of the flexural rigidities of the mast and rigging of an old sailboat. Use the following values for Young's modulus Y and diameter of the components:

• mast: $Y = 10^{10}$ J/m^3, diameter $= 30$ cm
• ropes: $Y = 2 \times 10^9$ J/m^3, diameter $= 5$ cm.

Compare this ratio to that for microtubules and actin, using data from Table 2.2.

10.3 Plot the minimal cell wall thickness d for a spherical cell to sustain a pressure difference of one atmosphere according to the bound in Section 10.2. Take the cell radius R to lie between 0.05 and 1.0 μm and assume the wall material to be peptidoglycan with $K_V = 3 \times 10^7$ J/m^3. Does $d < 1$ nm have a physical meaning? Given a rupture tension of 10^{-2} J/m^2, show the range in R for which a lipid bilayer alone can support this pressure.

10.4 Approximate the shape of the E. coli bacterium in Fig. 1.1(c), to be a cylinder 2 μm long and 1.0 μm in diameter. Equate the volume of this cylinder to that of a sphere to determine an equivalent radius of gyration R_g (you will have to determine R_g for a solid sphere). If the DNA of this bacterium obeyed ideal scaling, what must be its persistence length in order to have the same R_g? How does this compare with the measured persistence length of DNA? (*Note: the contour length of* E. coli *DNA is given in the text.*)

10.5 Suppose that a spherical cell were so small that the separation of its chromatids during cell division could be driven by actin filaments, rather than microtubules. What would be the radius of the cell such that an actin filament spanning the cell would not buckle when subject to a force of 5 pN. (*Note: choose a value for the actin persistence length in the middle of the range quoted in Table 2.2.*)

Formal development and extensions

10.6 A massless rod of length L lies horizontally with one end free and one end held so that it can neither translate nor rotate. A force F is applied in the upward direction to the free end. For gentle bends, show that the displacement $z(x)$ of the rod at any horizontal position x is given by

$$z(x) = (F/Y\mathcal{I}) \cdot (Lx^2/2 - x^3/6),$$

where Y and $\mathcal{I}$ are the beam's Young's modulus and moment of inertia of the cross-section. What is the displacement of the rod at $x = L$? (*Hint: find the analogue of Eq. (10.8) for a force perpendicular to the filament; after Feynmann et al., 1964.*)

10.7 A longitudinal compressive force F is applied to the ends of a cylinder of length L and radius R, as in the following diagram.

If F is less than $F_{\text{buckle}} = \pi^2 Y \mathcal{I} / L^2$ (see Section 10.1), the rod has a non-zero strain in the z-direction, where the axis of symmetry lies along the z-axis. Find u_{zz} in terms of R and L just at the buckling point $F = F_{\text{buckle}}$.

10.8 Suppose that a set of filaments are lined up in a spherical cell of radius R as shown: all filaments are uniformly distributed and point along the z-axis.

$2R \updownarrow$

Show that the average filament length is $4R/3$.

Appendix A

Animal cells and tissues

Focusing on the mechanical operation of a cell, this text tends to emphasize aspects of structure and organization common to all cells. However, cells differ in many details according to their function, evolution and environment, such that within a multicellular organism such as ourselves, there may be 100 or 200 hundred different cell types. Because frequent reference is made to these cell types in the text, we provide in this appendix a brief survey of animal cells and their organization in tissues. Readers seeking more than this cursory introduction to cell biology should consult one of many excellent textbooks, such as Alberts *et al.* (1994), Goodsell (1993), or Prescott *et al.* (1996).

Tissues

Cells act cooperatively in a multicellular system, and are hierarchically organized into tissues, organs and organ systems. Tissues contain the cells themselves, plus other material such as the extracellular matrix secreted by the cells. Several different tissues may function together as an organ, usually with one tissue acting as the "skin" of the organ and providing containment for other tissues in the interior. Lastly, organs themselves may form part of an organ system such as the respiratory system. Animal tissues are categorized as epithelial, connective, muscle or nervous tissue, whose relationship within an organ (in this case the intestinal wall) is illustrated in Fig. A.1. The intestine is bounded on the inside and outside by epithelial cell sheets (or epithelia), wherein epithelial cells are joined tightly to each other in order to restrict the passage of material out of the intestinal cavity. Connective tissue, which lies inside the boundaries provided by the epithelia, consists of fibroblastic cells that secrete an extracellular matrix which bears most of the stress in connective tissues. Lastly, several layers of smooth muscle tissue lie on the exterior of the connective tissue, both circling around the intestine (circular fibers) and running along the length of the intestine (longitudinal fibers).

Fig. A.1. Schematic cross section of the intestine, seen transverse to its cylindrical axis, showing three types of tissue: epithelial, connective and smooth muscle tissue, including both circular and longitudinal fibers (after Alberts *et al.*, 1994; reproduced with permission; © 1994 by Routledge Inc., part of The Taylor and Francis Group).

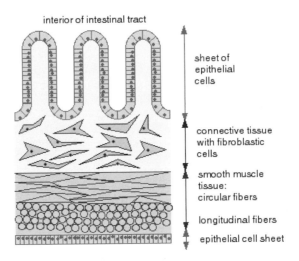

interior of intestinal tract

sheet of epithelial cells

connective tissue with fibroblastic cells

smooth muscle tissue: circular fibers

longitudinal fibers

epithelial cell sheet

Even a given cell type may exhibit several different forms, depending on the specific role that it plays. We now describe in more detail the four principal animal tissues.

Epithelial cells

The passage of material across the boundary of an organ must be selectively controlled, a task performed by epithelial cells. Figure A.2 displays one subtype of epithelial cell, namely an absorptive cell which possesses microvili to increase the effective area of the cell. The surfaces of the cell, denoted by their location as apical, lateral or basal in Fig. A.2, are attached by a variety of junctions both to other cells and to the basal lamina which underlies the epithelial cell sheet and is composed of the protein collagen. Linking the cells near their apical surface are tight junctions, which are not so much globs of glue as they are many strands of junction proteins which encircle the cell, such that a barrier of several strands faces any molecule attempting to slip by the cell along its lateral surface.

The mechanical stress on epithelial cells is borne largely by the cytoskeleton and anchoring junctions, of which there are several different types. Filaments of the protein actin are attached at adherens junctions, while thicker intermediate filaments are attached at desmosomes and hemidesmosomes. Bundles of actin filaments form an adhesion belt that encircles the cell; in Fig. A.2, the belt lies just below the ring of tight junctions. Adhesion belts on neighboring cells are aligned with each other, and are thought to be linked across the lateral surfaces by cell–cell adhesion molecules called cadherins. Epithelial cells also are linked to

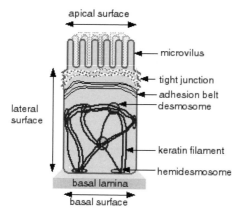

apical surface

microvilus

tight junction

adhesion belt

desmosome

lateral
surface

keratin filament

hemidesmosome

basal lamina

basal surface

Fig. A.2. Principal mechanical components of an absorptive epithelial cell. The apical surface faces into the intestine, while the basal surface is attached to the basal lamina. The view is quasi-perspective and the cell's organelles have been omitted for clarity. Typical cell width is about 25 μm (after Alberts *et al.*, 1994; reproduced with permission; © 1994 by Routledge Inc., part of The Taylor and Francis Group).

the basal lamina through cell–matrix adherens junctions. Not shown in the figure are gap junctions, which are regions in which the plasma membranes of adjacent cells lie close to each other, separated by a distance of about 3 nm. Gap junctions act as pores linking the cells and allowing small molecules to pass between them.

Fibroblasts and connective tissue

There may be several different cell types present in connective tissue, including nerve cells, capillaries lined with endothelial cells, and also macrophages. Fibroblastic cells, which are largely responsible for secreting the extracellular matrix, are also present. Lastly, connective tissue is usually criss-crossed by fibers such as collagen, secreted by the fibroblasts and forming a network whose density is reflected in the strength of the tissue. Aside from the usual organelles, fibroblasts contain stress fibers, which are bundles of the proteins actin and myosin, as illustrated in Fig. A.3. One end of a stress fiber terminates in a focal contact (or adhesion plaque) in the plasma membrane, by means of which a fibroblast can attach itself to a substrate. The other end of the fiber may be attached to a network of intermediate filaments surrounding the nucleus, or to a second focal contact. As their name implies, stress fibers

Fig. A.3. Schematic representation of a fibroblast attached to a substrate in culture, illustrating the network of actin stress fibers and microtubules; the focal contacts in this cell would be near the edges where the stress fibers terminate. Organelles are omitted for clarity. Grown in culture, a fibroblast may have a length in the 50–100 μm range when attached to a substrate (after Alberts *et al.*, 1994; reproduced with permission; © 1994 by Routledge Inc., part of The Taylor and Francis Group).

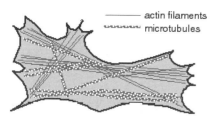

actin filaments
microtubules

can generate and bear a tensile stress, permitting the fibroblast to pull on the extracellular matrix of the connective tissue. Individually, stress fibers are not permanent but form when the cell is subject to tension. As in most cells, the cytoskeleton of fibroblasts also contains a network of microtubules.

Muscle cells

There are four structurally distinct forms of mammalian muscle cells, three of which make up most of the human body's muscle tissue. The first two categories of muscle are skeletal and cardiac, which together are known as striated muscle because of their visual appearance. What we often think of as muscle tissue, such as biceps etc., is skeletal muscle, which makes up most of the voluntary muscle mass of the human body. In contrast to the wide distribution of skeletal muscle, the involuntary cardiac muscle is found almost exclusively in the heart. A third category is smooth muscle, which is also involuntary and is present in the walls of blood vessels and the intestine, to name two examples. Myoepithelial cells, which are contractile cells present in some epithelial tissues, form a fourth category of muscle cells.

Skeletal muscle fibers are highly elongated, multinuclear cells: in humans, a skeletal muscle fiber may range up to 100 μm in diameter (although 50 μm is more typical) and up to half a meter long, a length-to-width ratio of 10^4! As displayed in Fig. A.4, many cylindrical myofibrils about 1–2 μm in diameter are packed together in a single skeletal muscle fiber. The multinuclear nature of the cells arises when mononuclear myoblasts fuse together early in the development of a vertebrate; their nuclei lie near the plasma membrane and do not reproduce once the skeletal muscle cell is formed. Individual multinuclear cells are organized into bundles and surrounded by connective tissue. Having a less cylindrical structure than skeletal muscle cells, the mononuclear smooth muscle cells are nevertheless highly elongated: the typical diameter of a smooth muscle cell in its relaxed state is 5–6 μm, while its length is 0.2 mm.

Fig. A.4. Schematic drawing of a multinuclear skeletal muscle cell, or muscle fiber, the interior of which is packed with myofibrils often running the length of the cell. Only a very short section of the cell is displayed. The envelope encasing the myofibrils is made to appear transparent (after Alberts *et al.*, 1994; reproduced with permission; © 1994 by Routledge Inc., part of The Taylor and Francis Group).

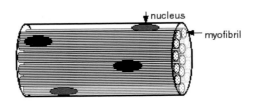

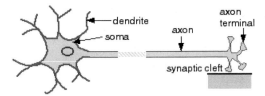

Fig. A.5. Schematic drawing of a neuron. The soma is typically 10–80 μm in diameter, while the axon and dendrites have diameters of a few microns. The axon is not drawn to scale, as it may be more than 10 000 times as long as it is wide.

Nerve cells

The human nervous system is composed of perhaps 10^{11} nerve cells or neurons, and although their visual appearance covers a range of shapes, neurons share common structural features, as illustrated in Fig. A.5. The central part of the cell is called the soma, a region with a diameter of 10–80 μm which contains the nucleus and is the main site of protein synthesis. Radiating from the soma are fine dendrites that collect information from sensory cells and other neurons, before passing it on to the soma in the form of an electrical impulse travelling along the plasma membrane. The highly branched structure of dendrites allows them to receive up to 10^5 inputs from other cells. The collective information-processing activities of the dendrites and soma may result in an impulse being sent along an extension of the cell body called the axon. Typically a few microns in diameter, the length of an axon can range from a few millimeters in the brain, up to a meter for large motor neurons; further, there are examples of neurons with more than one axon. At its remote end, an axon divides into branches, each with an axon terminal from which chemical neurotransmitters are released into the synaptic cleft between the axon and an adjacent cell. Because the neurotransmitters and other cell components must be delivered to the axon terminals from their distant production site in the soma, the axon contains microtubules to provide highways for the transport of chemically laden vesicles.

Blood cells

The final group of cells that we mention are the red and white blood cells of the human circulatory system. By far the most numerous cells in human blood, red blood cells (or erythrocytes) have the task of carrying oxygen and carbon dioxide from and to the lungs. A variety of white blood cells (or leukocytes) play a prime role in fighting infections, although they are outnumbered by red cells by about 500 to 1. Small incomplete cell fragments called platelets help repair damaged blood vessels, and are present in the blood at about one platelet for every 12–14 erythrocytes. Leukocytes can slip by the endothelial cells lining

Fig. A.6. Biconcave shape of the flaccid red cell, with a mean diameter of about 8 μm. The enlargement illustrates the membrane-associated cytoskeleton.

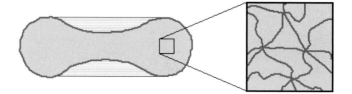

a capillary and, after passing through the basal lamina, can move into connective tissue. Within a day or two of entering such tissue, monocytes differentiate into macrophages, which can attack and remove foreign cells and debris.

With a biconcave shape approximately 8 μm in mean diameter, erythrocytes have a remarkably simple internal structure. The nucleus and other organelles that are present in red cells during their development are expelled before and shortly after the cells are released into the circulatory system, leaving the mature cells with no internal structural components other than the membrane-associated cytoskeleton. As represented in Fig. A.6, this cytoskeleton is a loose, two-dimensional network composed principally of tetrameric chains of the protein spectrin, with each tetramer having one or perhaps more attachment sites to the plasma membrane.

Appendix B

The cell's molecular building blocks

The purpose of this book is to explore the mesoscale mechanics of the cell: its structure and behavior on length scales of tens of nanometers to microns. This expedition cannot be undertaken without a biochemical lexicon, in part because the mechanical properties of some structures are a direct reflection of their molecular composition, and in part because the text draws constantly from the language of biochemistry. In this appendix, we describe in more detail the nomenclature and composition of several classes of compounds referred to in the main text, including:

membrane components: fatty acids and phospholipids

biopolymers: sugars, amino acids and proteins

the genetic blueprint: DNA and RNA.

For a more thorough treatment of biomolecular building blocks, the reader is directed to books such as Alberts *et al.* (1994), Gennis (1989), and Lehninger *et al.* (1993).

Fatty acids and phospholipids

One of the principal constituents of cellular membranes is the family of dual-chain lipids formed from *fatty acids*, which are carboxylic acids of the form RCOOH, where R represents a long hydrocarbon chain. In cells, fatty acids usually are not found in their free state, but rather are components of covalently bonded molecules such as phospholipids or triglycerides. A major component of cell membranes, phospholipids contain two fatty acids linked to a glycerol backbone. The remaining OH group in the glycerol is replaced by a phosphate group PO_4 which, in turn, is linked to yet another group, usually referred to as the polar head group:

Table B.1. *Some fatty acids commonly found in membrane lipids*

Acid name	Total number of carbon atoms	Number of double bonds	Position of double bond
lauric	12	0	
myristic	14	0	
palmitic	16	0	
palmitoleic	16	1	9-*cis*
stearic	18	0	
oleic	18	1	9-*cis*
linoleic	18	2	9-*cis*, 12-*cis*
arachidonic	20	4	9-*cis*, 8-*cis*, 11-*cis*, 14-*cis*

Notes:
Column 4 displays the C–C bond numbers on which any double bonds occur (see Fig. B.1).

The polar head groups of phospholipids may be chosen from a variety of organic compounds, including

Because the phosphate group has a negative charge, serine and glycerol phospholipids are negative while choline and ethanolamine phospholipids are neutral. The naming convention for a phospholipid mirrors, in part, its fatty acid composition. For example, a phospholipid containing two myristic acids (14 carbons each) and an ethanolamine group is called dimyristoyl phosphatidylethanolamine. Common abbreviations (italics provided for ease of pronounciation) include:

phosphatidyl*choline* PC phosphatidyl*ethanolamine* PE
phosphatidyl*glycerol* PG phosphatidyl*serine* PS.

Samples of the fatty acids found in the cell's membranes are given in Table B.1, including the position of any $C=C$ double bonds. Hydro-

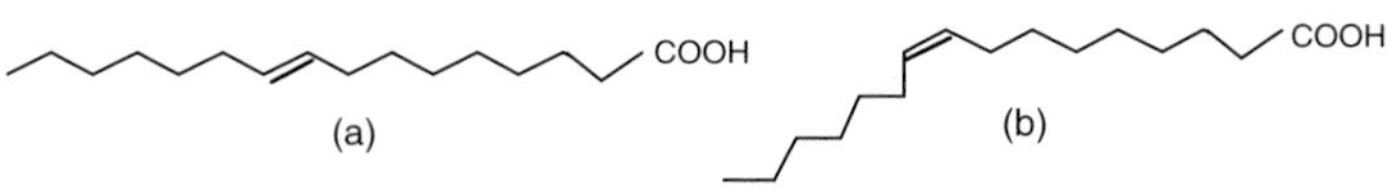

Fig. B.1. Geometry of *cis* and *trans* double bonds for palmitoleic acid. The double bond occuring at bond position 9 is *trans* in part (a) and *cis* in part (b).

carbon chains containing double bonds adopt one of the *cis* and *trans* forms displayed in Fig. B.1. Not all phospholipids are based solely on fatty acids. Sphingomyelin has the following structure

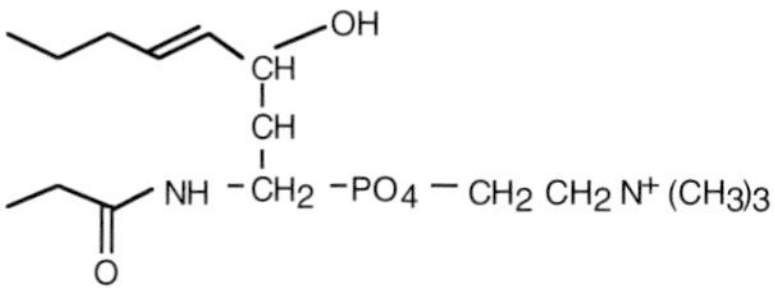

Further, there are other lipids present in cellular membranes which are not phospholipids, such as cholesterol and glycolipids.

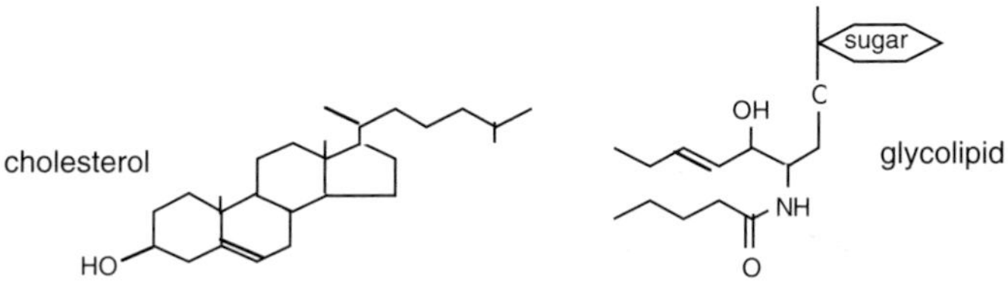

Cholesterol is a member of the steroid family and is frequently found in the membranes of eucaryotic cells. Glycolipids are similar to phospholipids in that they have a pair of hydrocarbon chains, but the phosphate group is replaced by a sugar residue (based on a ring of five carbons and one oxygen).

Sugars

Sugar molecules play an obviously important role in metabolism, and also are one component of DNA and of the peptidoglycan network of the bacterial cell wall. A single sugar molecule has the chemical formula $(CH_2O)_n$, the most biologically important sugars having $n = 5$ or 6. Examples of the conformations available for the glucose molecule ($n = 6$) are shown in Fig. B.2. Part (a) shows the molecule as a linear chain,

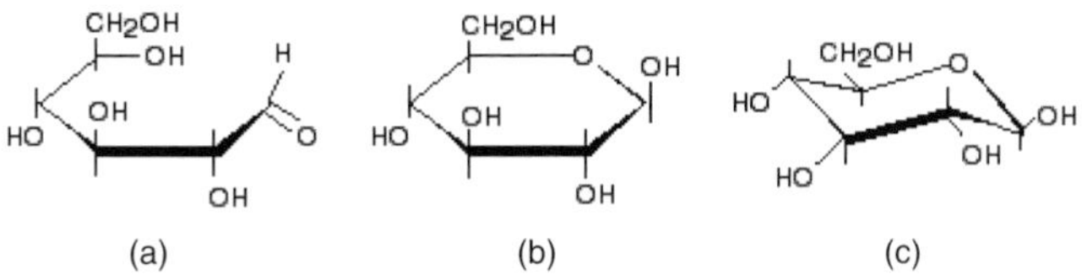

Fig. B.2. The sugar glucose can be a linear chain (a) or a six-member ring (b). Although often drawn as a planar ring, in fact glucose adopts configuration (c).

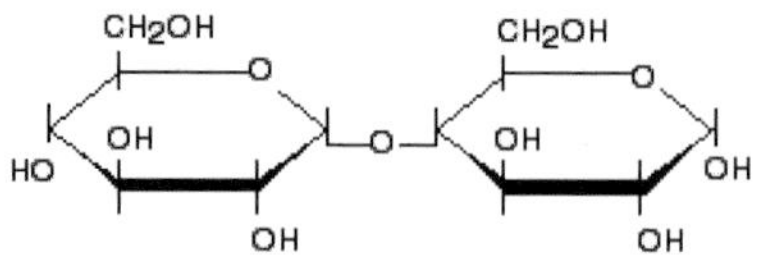

Fig. B.3. Two glucose molecules can combine to form maltose, a disaccharide, liberating H_2O as a product.

illustrating that there are several chemically inequivalent functional groups. Five of the oxygens are part of -OH groups while the sixth is double-bonded as an aldehyde. The double-bonded oxygen can be placed at one of several different positions on the chain, each corresponding to an inequivalent, yet related, molecule. The chain can be closed into a ring using one of the oxygens in a hydroxyl group (not the oxygen in the aldehyde group (RCOH)) as illustrated in Fig. B.2(b). As drawn, the ring in part (b) appears planar, in spite of the lack of in-plane double bonds such as are present in planar compounds like benzene. In fact, the actual configuration of a glucose ring is the bent form in part (c), just as it is in the single-bonded ring of cyclohexane. A single sugar molecule in isolation is referred to as a *monosaccharide*. But sugar molecules can polymerize through reactions in which two alcohol groups (one on each ring) combine to give a single bond between rings, liberating H_2O as a product. Two glucose molecules may combine to form the disaccharide maltose, as shown in Fig. B.3, or longer polysaccharides.

Amino acids and proteins

A principal component of the cytoskeleton and present in most membranes, proteins are linear chains of amino acids, a family of organic compounds containing an amino group ($-NH_3^+$) and a carboxyl group ($-COO^-$). Only 20 specific amino acids appear in protein construction, none of which has a high molecular mass, as can be seen from Fig. B.4. Amino acids can join together to form chains through an amide linkage ($-OC-N-$) referred to as a peptide bond. The reaction liberates H_2O, and has the general form:

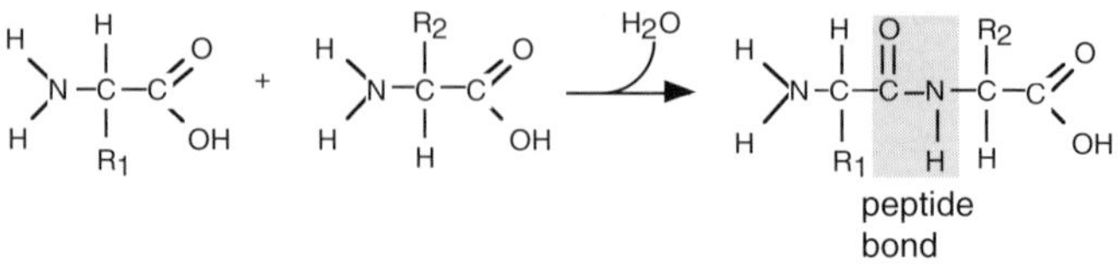

A protein chain of high molecular mass will be built up as this reaction occurs repeatedly; for example, the two inequivalent spectrin proteins in the human erythrocyte have molecular masses of $\sim$220 000 and $\sim$230 000 Da. Amino acids appear in a protein with varying relative abundance, and some, such as tryptophan, are uncommon. In a large

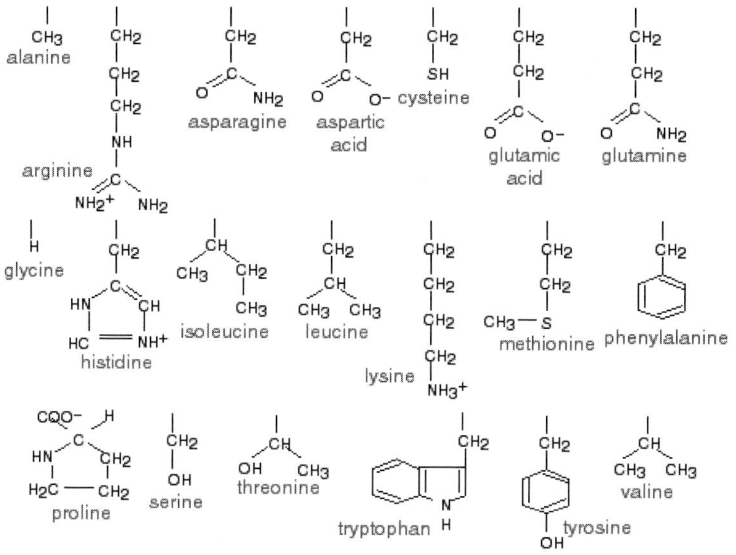

Fig. B.4. Summary of the 20 amino acids that appear in proteins. All but one of the acids (proline) have the form $NH_3^+{-}RCH{-}COO^-$ at pH 7.

[Figure: structures of the 20 amino acids — alanine, arginine, asparagine, aspartic acid, cysteine, glutamic acid, glutamine, glycine, histidine, isoleucine, leucine, lysine, methionine, phenylalanine, proline, serine, threonine, tryptophan, tyrosine, valine]

protein, the average molecular mass of the amino acids is 115 Da. This means that one spectrin monomer in the human red blood cell having a molecular mass of 230 000 Da would be composed of approximately 2000 amino acids.

Nucleotides and DNA

Ribonucleic acid, or RNA, directs protein synthesis in the cell, whereas **deoxy**ribonuclei acid, or DNA, is the carrier of genetic information. The elementary chemical units of DNA and RNA are nucleotides, which have a more complex structure than the amino-acid building blocks of proteins. Firstly, nucleotides are themselves composed of subunits, namely a sugar, an organic base and a phosphate group.

nucleotide:

The two different sugars found in nucleotides, ribose in RNA and deoxyribose in DNA, are five-membered rings differing from each other by only one oxygen atom.

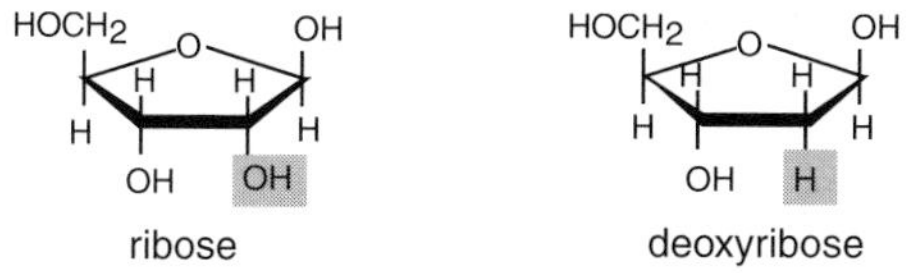

ribose   deoxyribose

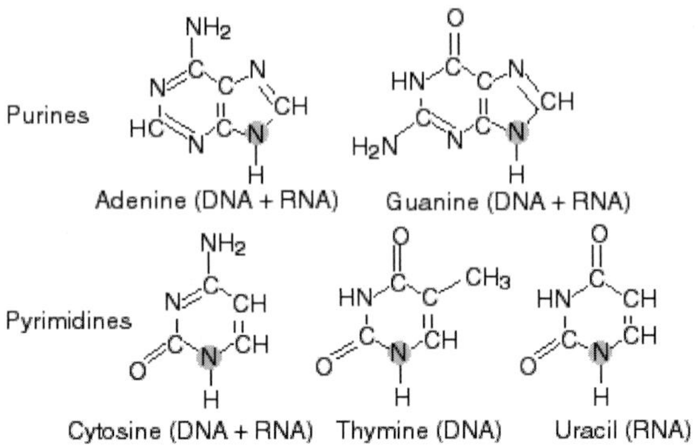

Fig. B.5. The five organic bases found in DNA and RNA. The shaded region is the reactive site of the base that joins onto the sugar of the nucleic acid.

Purines

Adenine (DNA + RNA)   Guanine (DNA + RNA)

Pyrimidines

Cytosine (DNA + RNA)   Thymine (DNA)   Uracil (RNA)

In either molecule, the OH groups are potential reaction sites for addition of a base. Five organic bases are found in nucleotides, and one can see from Fig. B.5 that they fall into two chemically similar groups – purines and pyrimidines. Only four of the five bases are present in a given DNA or RNA molecule, and the one "missing" base is different for each:

> RNA: adenine, guanine, cytosine, uracil
>
> DNA: adenine, guanine, cytosine, thymine.

The reaction of a sugar with a base releases water (an $-OH$ from the sugar plus an H from the base) and produces a sugar–base combination called a *nucleoside*. Addition of a phosphate to a nucleoside releases water and produces a *nucleotide*, two of which are shown in the chain of Fig. B.6(a). The nucleotides themselves can polymerize to form DNA and RNA, through a linkage between a sugar from one nucleotide and a phosphate from another, as schematically illustrated in Fig. B.6(a). In the double-stranded helix of DNA, the bases lie in the interior of the helix, and hold the helix together through hydrogen bonding between

Fig. B.6. (a) Schematic representation of the sugar/phosphate linkages in a single strand of DNA. (b) Hydrogen bonding between bases on different strands contributes to the interactions that stabilize the double helix structure of DNA.

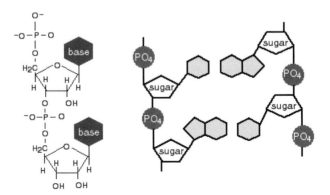

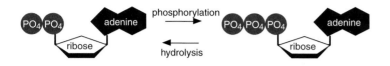

Fig. B.7. The most common denominations of the cell's energy currency are ADP (adenosine diphosphate) and ATP (adenosine triphosphate).

base-pairs. As illustrated in Fig. B.6(b), each matching base pair on the opposing strands consists of one purine and one pyrimidine: adanine/thymine and guanine/cytosine.

ADP and ATP

Chemical processes in cells involve changes in energy. In almost all examples, the cell captures or delivers energy by chemical means, minimizing the energy loss through heat or light. The energy economy of the cell has two components: a currency for local energy exchange, and a mechanism for long-term storage. The concept of an energy currency has parallels with a monetary currency: the amount of currency in circulation need not be nearly as large as the total economy, and long-term storage of economic output does not involve hoarding the currency itself. Long-term energy storage can be accomplished with triglycerides, whose structure is mentioned at the beginning of this appendix. However, the currency itself is based upon a sugar–base–phosphate nucleotide whose elementary components are the same as RNA, a very efficient use of materials. The sugar of the currency is ribose of RNA, and the most common base is adenine (or, much less frequently, guanine). The currency differs from RNA by the presence of more than one phosphate group, as illustrated in Fig. B.7. The lower energy state of the pair has two phosphate groups (adenosine diphosphate, or ADP), while the higher energy state has three (adenosine triphosphate, or ATP). The energy-consuming process of adding a phosphate is called phosphorylation, while the release of a phosphate is hydrolysis. Depending on conditions, the energy change for a single phosphate group is 11–13 kcal/mol, corresponding to 8×10^{-20} J $= 20k_{\mathrm{B}}T$ per reaction.

Appendix C

Elementary statistical mechanics

Soft materials, such as the filaments and membranes of the cell, may be subject to strong thermal fluctuations and sample a diverse configuration space. Each of these accessible configurations, of varying shape and energy, contribute to the ensemble average of an observable such as the mean length of a polymer chain. Statistical mechanics provides a formalism for describing the thermal fluctuations of a system; in this appendix, we review several of its cornerstones – entropy, the Boltzmann factor and the partition function – following a compact approach laid out in Reif (1965). As applications, the thermal fluctuations of a Hooke's law spring and the entropy of an ideal gas are calculated.

Temperature and entropy

Any configuration available to a system, be it a polymer or a gas in a container, can be characterized by its energy, although more than one configuration may have the same energy. At zero temperature, the system seeks out the state of lowest energy, but at non-zero temperature the system samples many configurations as it exchanges energy with its environment. To determine the magnitude of these energy fluctuations, let's examine two systems, L and S, in thermal contact only with each other, as in Fig. C.1. System S is the subject of our observations, while L is the reservoir; L has far more energy than S. Because L and S exchange energy only with each other, their total energy, E_{TOT}, is a constant:

$$E_{\text{TOT}} = E_L + E_S.$$ (C.1)

We now specify E_{TOT} to be a particular value E_o, say $E_{\text{TOT}} = E_o = 10^6$ J. We are interested in a particular state r of the small system with energy E_r, such that

$$E_S = E_r \text{ and } E_L = E_o - E_r.$$ (C.2)

Fig. C.1. Two subsystems (small and LARGE) in thermal contact can exchange energy; their energy is dominated by the large system.

Just to make sure that the notation is clear: E_S, E_L and E_{TOT} are general parameters, while E_r and E_o represent a specific choice of energies.

We define the number of states of the large system having energy E_L to be

$$\Omega_L(E_L) = [\text{number of states of large system with } E_L]. \tag{C.3}$$

The number of states of the large system $\Omega_L(E_L)$ varies, perhaps even rapidly, with E_L, but its logarithm varies more slowly, and can be expanded in a series around the specific value $E_L = E_o$:

$$\ln \Omega_L(E_o - E_r) = \ln \Omega_L(E_o) - [\partial \ln \Omega_L / \partial E_L]_o \, E_r + \dots \tag{C.4}$$

where the derivative of $\ln \Omega_L$ with respect to E_L is evaluated at $E_L = E_o$. The minus sign in front of the derivative arises because the energy E_L of the large system decreases by E_r when the energy of the small system increases by E_r. Higher order terms can be neglected since $E_r \ll E_o$. The derivative $[\partial \ln \Omega_L / \partial E_L]_o$ characterizes the large system around $E_L = E_o$, and does not depend on E_r; hence, it is notationally convenient to replace this derivative with a single symbol

$$[\partial \ln \Omega_L / \partial E_L]_o \equiv \beta. \tag{C.5}$$

It can be shown that β^{-1}, bearing units of energy, has the properties of a temperature; the physical temperature scale is set through $T = 1/k_B \beta$, where k_B is Boltzmann's constant ($k_B = 1.38 \times 10^{-23}$ J/K).

Returning now to Eq. (C.4), the derivative on the right-hand side can be replaced by β

$$\ln \Omega_L(E_o - E_r) = \ln \Omega_L(E_o) - \beta E_r + \dots,$$
$$\Rightarrow \Omega_L(E_o - E_r) = \Omega_L(E_o) \exp(-\beta E_r). \tag{C.6}$$

The number of configurations available to a system affects its behavior at finite temperature. For instance, the molecules of a gas are more likely to be found scattered throughout the volume of a container than collected together in one of its corners, all other things being equal. Mathematically, we say that a system at fixed volume minimizes its free energy $E - TS$, rather than just its energy E, where S denotes entropy; an even more general expression for the free energy is needed if the volume or the number of particles is not fixed. Entropy increases logarithmically with the number of states Ω accessible to the system,

$$S = k_B \ln \Omega(\langle E \rangle), \tag{C.7}$$

where Ω is a function of the mean energy $\langle E \rangle$ of the system (see Section 6.6 of Reif, 1965). Because TS enters the free energy with a minus sign, the free energy of a system falls as its entropy rises, a process that can occur spontaneously.

Boltzmann factor

We have selected a particular state r of the small system with energy E_r. The probability $\mathcal{P}_r$ of the small system being in this state is proportional to the number of states of the *large* system having the appropriate value of E_L; that is, $\mathcal{P}_r \propto \Omega_L(E_o - E_r)$ or

$$\mathcal{P}_r \equiv A\,\Omega_L(E_o - E_r), \tag{C.8}$$

where the proportionality constant A is a characteristic of the *large* system. The value of A can be determined through the condition that the small system must always occupy an available state, although it may occupy different states as time passes:

$$\sum_r \mathcal{P}_r = 1, \tag{C.9}$$

where the sum is over all of the states r available to the small system. We replace $\Omega_L(E_o - E_r)$ in Eq. (C.8) by its functional dependence on E_r in Eq. (C.6) to obtain

$$\mathcal{P}_r = A\,\Omega_L(E_o - E_r) = [A\,\Omega_L(E_o)]\exp(-\beta E_r). \tag{C.10}$$

This is the Boltzmann factor, and it shows that the probability of the small system being in a specific state r with energy E_r is a function of the energy of the state r and of the temperature of the large system with which it is in thermal contact. The two factors in the square braces of Eq. (C.10) are both constants, and can be rolled into one as

$$\mathcal{P}_r = Z^{-1}\exp(-\beta E_r) \tag{C.11}$$

with Z determined from

$$Z = \sum_r \exp(-\beta E_r), \tag{C.12}$$

so that $\mathcal{P}_r$ is properly normalized to unity. The quantity Z involves a sum over the states available to the small system, and is defined as the partition function. Note that if several configurations have the same E_r, then *each configuration* has the same probability of occurence $\mathcal{P}(E_r)$. Equation (C.11) tells us that if T is small, or β is correspondingly large, then the system is much more likely to be found with E_r near zero than with E_r much greater than zero (for which $\exp(-\beta E_r)$ is tiny), all other things being equal.

The probability also can be related to the entropy by manipulation of Eq. (C.11). Multiplying the logarithm of Eq. (C.11) by $\mathcal{P}_r$ yields $-\mathcal{P}_r \ln \mathcal{P}_r = \beta\,\mathcal{P}_r E_r + \mathcal{P}_r \ln Z$, or

$$-\sum_r \mathcal{P}_r \ln \mathcal{P}_r = \beta\langle E\rangle + \ln Z, \tag{C.13}$$

after summing over all states r and defining $\langle E \rangle = \Sigma_r \mathcal{P}_r E_r$. Now, it can be established from thermodynamics that

$$S = k_B(\beta\langle E \rangle + \ln Z), \tag{C.14}$$

(see Section 6.6 of Reif, 1965, for example). Thus, we expect

$$S = -k_B \Sigma_r \mathcal{P}_r \ln \mathcal{P}_r, \tag{C.15}$$

by comparing Eqs. (C.13) and (C.14).

Example: harmonic oscillator

As an application of the Boltzmann factor, we consider the one-dimensional motion of a particle in the quadratic potential $V(x)$ characteristic of Hooke's law for springs,

$$V(x) = k_{sp}x^2/2, \tag{C.16}$$

where k_{sp} is the spring constant and x is the displacement from equilibrium. For a given value of x, Eq. (C.16) is the corresponding potential energy, and the Boltzmann factor $\exp(-\beta k_{sp}x^2/2)$ provides the likelihood that the particle can be found with that energy.

Now, x is a continuous variable, so one talks of the probability $\mathcal{P}(x)dx$ of the particle having a displacement between x and $x + dx$. The continuous versions of Eqs. (C.11) and (C.12) become

$$\mathcal{P}(x)dx = dx \cdot \exp(-\alpha x^2) / \int \exp(-\alpha x^2)\, dx, \tag{C.17}$$

where the combination $\beta k_{sp}/2$ has been replaced with a single constant α

$$\alpha \equiv \beta k_{sp}/2. \tag{C.18}$$

Eq. (C.17) can be used to evaluate the fluctuations in x about $x = 0$:

$$\langle x^2 \rangle = \int x^2\, \mathcal{P}(x)\, dx$$

$$= \alpha^{-1} \int z^2 \exp(-z^2)\, dz / \int \exp(-z^2)\, dz$$

$$= \alpha^{-1} (\sqrt{\pi}/2)/\sqrt{\pi} = 1/2\alpha, \tag{C.19}$$

whence

$$\langle x^2 \rangle = 1/\beta k_{sp} = k_B T/k_{sp}. \tag{C.20}$$

Eq. (C.20) shows that $\langle x^2 \rangle$ increases with temperature T, as expected

- high temperature $\Rightarrow$ large fluctuations in x
- low temperature $\Rightarrow$ small fluctuations in x
- zero temperature $\Rightarrow$ no fluctuations in x (ground state at $x = 0$).

Partition function

The partition function introduced in Eq. (C.12) allows many thermody-namic quantities of interest to be written in a compact form. As an example, we determine the ensemble average of the system's energy, $\langle E \rangle$, which is the sum of the energy of each state according to the probabil-ity of its occurence: $\langle E \rangle = \sum_r E_r \mathcal{P}_r$, or

$$\langle E \rangle = Z^{-1} \sum_r E_r \exp(-\beta E_r). \tag{C.21}$$

Each term in the sum, $E_r \exp(-\beta E_r)$, can be written as the (negative of the) derivative of $\exp(-\beta E_r)$ with respect to β, so that

$$\langle E \rangle = - Z^{-1} \sum_r (\partial/\partial\beta)\exp(-\beta E_r)$$
$$= - Z^{-1} (\partial/\partial\beta) \sum_r \exp(-\beta E_r) = - Z^{-1} (\partial Z/\partial\beta), \tag{C.22}$$

or, equivalently,

$$\langle E \rangle = -\partial\ln Z/\partial\beta. \tag{C.23}$$

Similarly compact forms occur for other observables.

The sum-over-states in Eq. (C.12) involves a discrete set of states r. What happens in a classical system where there are many particles, each described by a position vector $\mathbf{r}$ and a momentum vector $\mathbf{p}$? In most systems, the potential energy of a particle depends upon its position, and the kinetic energy depends upon its momentum. If these variables are continuous, the sum over r must be replaced by a continuous inte-gral over $\mathbf{r}_i$ and $\mathbf{p}_i$ for each of the N particles in the system, as in

$$\exp[-\beta E(\mathbf{r}_1, \mathbf{r}_2, \ldots \mathbf{r}_N, \mathbf{p}_1, \mathbf{p}_2, \ldots \mathbf{p}_N)] \, d\mathbf{r}_1 \, d\mathbf{r}_2 \ldots d\mathbf{r}_N \, d\mathbf{p}_1 \, d\mathbf{p}_2 \ldots d\mathbf{p}_N.$$

As it stands, this expression has dimensions, whereas the partition function should be dimensionless. The complete expression for the partition function must therefore include a normalization factor reflecting the density of states: the number of states per unit volume of phase space (each point in *phase space* represents both a position and a momentum of a particle). We parametrize the phase space volume occupied by one state as h^3, without proving that the parameter h is actually Planck's constant, so that the normalized partition function becomes

$$Z = (N!)^{-1} h^{-3N} \int \int \ldots \int \exp[-\beta E(\mathbf{r}_1, \mathbf{r}_2, \ldots \mathbf{r}_N, \mathbf{p}_1, \mathbf{p}_2, \ldots \mathbf{p}_N)] \, d\mathbf{r}_1$$
$$d\mathbf{r}_2 \ldots d\mathbf{r}_N \, d\mathbf{p}_1 \, d\mathbf{p}_2 \ldots d\mathbf{p}_N. \tag{C.24}$$

The factor of $N!$ in Eq. (C.24) arises from the fact that, at the molecu-lar level, the particles are indistinguishable, and the $N!$ possible permu-tations implied by the integral do not correspond to physically distinguishable states (see Section 7.3 of Reif, 1965, for further details).

Example: entropy of an ideal gas

For most situations of interest in this text, the potential energy depends only on the positions of the particles, not their momenta. In other words, the conventional problem that one has to solve is described by a total energy, E, with

$$E = (p_1^2/2m + p_2^2/2m + \ldots + p_N^2/2m) + V(\mathbf{r}_1, \mathbf{r}_2, \ldots \mathbf{r}_N), \tag{C.25}$$

where m is the mass of an individual particle (taken to be identical) and $V(\mathbf{r}_1, \mathbf{r}_2, \ldots \mathbf{r}_N)$ is the total potential energy. Since the exponential of a sum is equal to the product of individual exponentials, the partition function then can be written as a product of $3N$ identical momentum integrals Z_p along with a $3N$-dimensional spatial integral Z_c:

$$Z = (N!)^{-1} Z_p^{3N} Z_c, \tag{C.26}$$

where

$$Z_p = h^{-1} \int \exp(-\beta p^2/2m) \, dp \tag{C.27}$$

$$Z_c = \int \int \ldots \int \exp[-\beta V(\mathbf{r}_1, \mathbf{r}_2, \ldots \mathbf{r}_N)] \, d\mathbf{r}_1 \, d\mathbf{r}_2 \ldots d\mathbf{r}_N. \tag{C.28}$$

The $3N$ identical momentum integrals arise from the fact that the momentum of each particle is independent and has three Cartesian components $(p^2 = p_x^2 + p_y^2 + p_z^2)$.

One can obtain Z_p using the expression $\int \exp(-x^2)dx = \sqrt{\pi}$, where the integration extends from $-\infty$ to $+\infty$. Changing variables in Eq. (C.27) to make the integral dimensionless, we find

$$Z_p = h^{-1} (2\pi m/\beta)^{1/2}. \tag{C.29}$$

For an ideal gas, all interactions between particles vanish, $V(\mathbf{r}_1, \mathbf{r}_2, \ldots \mathbf{r}_N) = 0$, so that the coordinate-space partition function Z_c just involves N integrals over the volume available to each particle:

$$Z_c = \int_v d\mathbf{r}_1 \int_v d\mathbf{r}_2 \ldots \int_v d\mathbf{r}_N = V^N. \tag{C.30}$$

Thus, the complete partition function is

$$Z = (N!)^{-1} V^N h^{-3N} (2\pi m/\beta)^{3N/2} \text{ (ideal gas).} \tag{C.31}$$

The quantities that we wish to calculate involve the logarithm of Z,

$$\ln Z = N [\ln V - 3\ln h - 3/2 \ln\beta + 3/2 \ln(2\pi m)] - \ln N! \tag{C.32}$$

This expression can be simplified by using Sterling's approximation for the logarithm of the factorial at large N,

$$\ln N! = N \ln N - N, \tag{C.33}$$

which is a familiar tool in statistical mechanics, and leads to

$$\ln Z = N\,[\ln(V/N) - 3/2\,\ln\beta + 3/2\,\ln(2\pi m/h^2) + 1]. \qquad (C.34)$$

Equation (C.34) can be substituted into Eq. (C.23) to obtain the average energy

$$\langle E \rangle = -\partial \ln Z / \partial \beta = 3/2\ N/\beta = 3/2\ Nk_{\mathrm{B}}T, \qquad (C.35)$$

which tells us that the average kinetic energy per particle is $3/2\ k_{\mathrm{B}}T$. The entropy of the gas can be found by returning to Eq. (C.14), $S = k_{\mathrm{B}}(\ln Z + \beta\langle E\rangle)$. Substituting Eq. (C.34) and $\beta\langle E\rangle = 3/2\ N$ from Eq. (C.35), we obtain

$$S = k_{\mathrm{B}}N[\ln(V/N) + 3/2\,\ln T + 3/2\,\ln(2\pi mk_{\mathrm{B}}/h^2) + 5/2], \qquad (C.36)$$

for the entropy of an ideal gas.

Appendix D

Elasticity

Our analysis of the elastic properties of biological materials draws heavily on concepts from continuum and statistical mechanics. In this appendix, we provide a limited review of several results from the theory of elasticity that are used routinely in the text, including:

- the description of deformations in terms of stress and strain
- the representation of the free energy density using elastic moduli and the strain tensor
- the relation between elastic moduli and fluctuations of an observable.

There is considerable variation in the notation and analytic approaches adopted by the spectrum of disciplines that make use of continuum mechanics; the notational convention adopted here is close to that of Landau and Lifshitz (1986).

Deformations and the strain tensor

Consider the deformation of a two-dimensional square as shown in Fig. D.1. A stress is applied to the square to change its shape into a rectangle, such that a mark on the square moves from its original position $\mathbf{x}$ to a new position $\mathbf{x}'$, according to a coordinate system that does *not* follow the object during deformation. The difference in positions is illustrated by Fig. D.1(b), in which the shaded region indicates the undeformed shape of the object and the wire outline is the deformed

Fig. D.1. Deformation of a square. The object before (a) and after (b) deformation is shown as the wire outline; for reference, the shaded region of part (b) also indicates the shape and labels of the undeformed object. The position of a specific element of the object shifts from $\mathbf{x}$ to $\mathbf{x}'$ under strain.

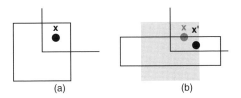

(a)   (b)

Fig. D.2. The displacement vector **u** describing a deformation varies with location on the object. Part (a) shows how the object deforms, while part (b) shows **u** as a function of location.

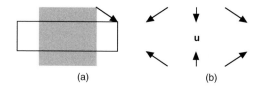

shape. The difference in the position vectors **x** and **x′** defines a displacement vector **u**

$$\mathbf{u} \equiv \mathbf{x'} - \mathbf{x}, \tag{D.1}$$

which, in general, is not uniform over the surface of the object: a constant displacement **u** of all positions **x** is simply a translation of the object. Rather, the direction and magnitude of **u** varies locally with **x**, as shown in Fig. D.2, and this variation provides a description of how the elements of the object move with respect to each other during the deformation.

Referring to Fig. D.3(a), consider two locations $\mathbf{x}^a$ and $\mathbf{x}^b$ which are separated by a vector d**x**,

$$\mathbf{dx} = \mathbf{x}^b - \mathbf{x}^a, \tag{D.2}$$

before a deformation. Denoting by $\mathbf{u}^a$ and $\mathbf{u}^b$ the change in the positions of the marks at $\mathbf{x}^a$ and $\mathbf{x}^b$ arising from the deformation, the separation $\mathbf{dx'} = \mathbf{x}^{\prime b} - \mathbf{x}^{\prime a}$ is given by

$$\mathbf{dx'} = \mathbf{dx} + \mathbf{u}^b - \mathbf{u}^a. \tag{D.3}$$

When points a and b are close to each other, Eq. (D.3) can be rewritten as

$$dx_i' = dx_i + \sum_j (\partial u_i / \partial x_j) dx_j, \tag{D.4}$$

where the components of each vector are indicated by the subscripts i and j. The distance between nearby points can be evaluated using Eq. (D.4). Before the deformation, the squared distance between neighboring points is just $d\ell^2$, where

$$d\ell^2 = \sum_i dx_i^2, \tag{D.5}$$

Fig. D.3. Change in vector d**x** to d**x′** during a shape-preserving deformation, where d**x** is the displacement between locations a and b. Parts (a) and (b) show the object before and after the deformation, respectively. The initial displacement vector d**x** is superimposed on part (b) to show the changes more clearly.

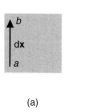

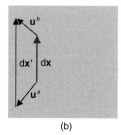

which changes to

$$d\ell'^2 = \sum_i dx_i'^2 = \sum_i [dx_i + \sum_j (\partial u_i/\partial x_j) dx_j]^2$$
$$= d\ell^2 + 2 \sum_{i,j} (\partial u_i/\partial x_j) \, dx_i \, dx_j$$
$$+ \sum_{i,j,k} (\partial u_i/\partial x_j)(\partial u_i/\partial x_k) \, dx_j \, dx_k, \qquad \text{(D.6)}$$

after the deformation. Rearranging the summation indices, Eq. (D.6) can be written as

$$d\ell'^2 = d\ell^2 + 2 \sum_{i,j} u_{ij} \, dx_i \, dx_j, \qquad \text{(D.7)}$$

where

$$u_{ij} \equiv 1/2 \, [\partial u_i/\partial x_j + \partial u_j/\partial x_i + \sum_k (\partial u_k/\partial x_i)(\partial u_k/\partial x_j)]. \qquad \text{(D.8)}$$

The quantity u_{ij} is called the *strain tensor*, which is clearly symmetric in indices i and j.

The utility of the strain tensor can be demonstrated by diagonalizing it and evaluating some geometrical quantities at small strain. Although it may not be possible to diagonalize the strain tensor globally (i.e., at all **x** simultaneously), it can be diagonalized locally to give three elements U^α, where $\alpha = 1, 2, 3$. The diagonal elements U^α have a direct geometrical interpretation, as can be seen by modifying Eq. (D.7) for the length squared to read

$$d\ell'^2 = \sum_{i,j} (\delta_{ij} + 2u_{ij}) \, dx_i \, dx_j, \qquad \text{(D.9)}$$

where δ_{ij} is a 3×3 matrix with diagonal elements $\delta_{ii} = 1$ and off-diagonal terms $\delta_{ij} = 0$. Equation (D.9) can be written in the diagonal representation ($u_{\alpha\alpha} = U^\alpha$, all other $u_{\alpha\beta} = 0$) as

$$d\ell'^2 = \sum_\alpha (1 + 2U^\alpha)(dx^\alpha)^2, \qquad \text{(D.10)}$$

in which terms such as $dx^\alpha \, dx^\beta$ with $\alpha \neq \beta$ vanish. The dx^α refers to the change in length along the direction associated with U^α. Equation (D.10) implies that during the deformation, an infinitesimal vector dx^α in the α direction changes to

$$dx^{\alpha'} = (1 + 2U^\alpha)^{1/2} \, dx^\alpha, \qquad \text{(D.11)}$$

demonstrating the relationship between U^α and the fractional change in length.

Small deformations

For many situations of interest, the deformations are small, so that the terms $\partial u_i/\partial x_j$ are small as well. Then the strain tensor in Eq. (D.8) can be approximated by

$$u_{ij} = 1/2 \, (\partial u_i/\partial x_j + \partial u_j/\partial x_i), \qquad \text{(D.12)}$$

Fig. D.4. Pairs of forces applied to a cube, seen in cross section. In (a), the forces are applied to the surfaces facing the x-axis, while in (b), the forces are applied parallel to surfaces facing the y-axis.

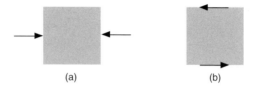

(a)   (b)

and the displacements in the α direction of Eq. (D.11) become

$$dx'^{\alpha} = (1 + U^{\alpha})\, dx^{\alpha}. \tag{D.13}$$

Finally, the volume $dV = \Pi_i dx_i$ around a small neighborhood changes to

$$dV' = dV\, \Pi_{\alpha}(1 + U^{\alpha}) = dV(1 + \mathrm{tr}\, U), \tag{D.14}$$

where Π_i is the product symbol ($\Pi_i a_i = a_1 \cdot a_2 \cdot a_3$) and $\mathrm{tr}\, U$ is the trace of the vector U, or $\mathrm{tr}\, U = \Sigma_{\alpha} U^{\alpha}$. But the trace of U is the same as the trace of the undiagonalized u_{ij}, so that the fractional change in volume can be written

$$(dV' - dV)/dV = \mathrm{tr}\, u, \tag{D.15}$$

which is a useful result for evaluating volume changes during compression.

Forces and the stress tensor

An object deforms in response to external forces, which may have arbitrary orientations with respect to the object. For instance, in Fig. D.4, a cube is subject to pairs of forces in the x direction: in part (a), the forces are applied perpendicular to the surfaces normal to the x-axis, resulting in a compression, while in part (b), the forces are applied parallel to the surfaces facing the y-axis, resulting in a shear. Now, a surface can be represented by a vector $\mathbf{a}$ whose magnitude is equal to the surface area and whose direction is parallel to the surface normal $\mathbf{n}$, related to $\mathbf{a}$ by

$$\mathbf{n} = \mathbf{a}/|\mathbf{a}|. \tag{D.16}$$

To account for the relative directions of the force $\mathbf{F}$ and the surface vector $\mathbf{a}$, we introduce a two-component object called the stress tensor σ_{ij} such that the components F_i are given in terms of σ_{ij} and the area element a_j via

$$F_i = \Sigma_j \sigma_{ij} a_j, \tag{D.17}$$

where σ_{ij} has units of force per unit area (or energy density) in three dimensions. By considering moments of forces (products of the force and the distance from a rotational axis) it can be shown that

$$\sigma_{ij} = \sigma_{ji}, \tag{D.18}$$

i.e., the stress tensor is symmetric (see Landau and Lifshitz, 1986).

As a simple example of the stress tensor, consider an object under hydrostatic pressure P, in which the force has a constant magnitude throughout the medium. The force $\mathbf{F}$ on a surface is in the opposite direction to the vector $\mathbf{a}$ describing the surface:

$$F_i = -P\, a_i = -P \sum_j \delta_{ij}\, a_j. \tag{D.19}$$

By comparing Eqs. (D.17) and (D.19), the form of the stress tensor for hydrostatic pressure must be

$$\sigma_{ij} = -P\, \delta_{ij}. \tag{D.20}$$

In general, the diagonal elements of the stress tensor correspond to compressions and the off-diagonal elements to shear. The fact that the off-diagonal elements vanish for hydrostatic pressure indicates that the system is not subject to shear in this example.

Hooke's law and elastic moduli

A spring is an example of a Hooke's law material in which the stress (the restoring force) is proportional to the strain (the displacement from equilibrium). In general, Hooke's law involves stress and strain tensors, so that the simple spring constant must be replaced by a more complex proportionality constant. There is a choice of conventions for defining the generalized spring constants: one can write

$$u_{ij} = \sum_{k,l} S_{ijkl}\,\sigma_{kl}, \tag{D.21}$$

where the constants S_{ijkl} are called the elastic compliance constants (or elastic constants), or one can choose

$$\sigma_{ij} = \sum_{k,l} C_{ijkl}\,u_{kl}, \tag{D.22}$$

where the constants C_{ijkl} are called the elastic stiffness constants (or elastic moduli). In this text, we use the elastic moduli C_{ijkl} which, in three dimensions, have units of [force]/[area] or [energy]/[volume], since the strain tensor is dimensionless. The elastic parameters must have four indices, meaning that there are, in principle, $3^4 = 81$ components of C_{ijkl} or S_{ijkl} in three dimensions, although symmetry considerations dramatically reduce this number.

We now describe in some detail the elastic parameters of an isotropic material, meaning that its characteristics are independent of direction. The first step is to determine the minimum number of elastic moduli needed to describe the material. In Chapters 3 and 4, we apply symmetry considerations relentlessly to C_{ijkl} to find relationships among its elements, ultimately reducing the number of independent components to two for an isotropic material. An alternative approach that we use here is to determine the lowest order contributions to an expansion of the free energy density $\mathcal{F}$ in powers of the strain tensor and then to identify the elastic moduli with specific deformation modes. Just as the potential energy for small deformations of a spring is quadratic in its displacement from equilibrium, the smallest deformations of a continuous material have an energy that is quadratic in the strain tensor. To describe a scalar quantity like the energy, the quadratic combinations of u also must be scalars, of which there are only two possibilities: the squared sum of the diagonal elements, $(\mathrm{tr}u)^2$, and the sum of the elements squared, $\Sigma_{i,j} u_{ij}^2$. Thus, for small strains, the free energy density can be written as

$$\mathcal{F} = \mathcal{F}_o + (\lambda/2)(\mathrm{tr}u)^2 + \mu \, \Sigma_{i,j} u_{ij}^2. \tag{D.23}$$

The constant $\mathcal{F}_o$ represents the free energy density of the unperturbed system and can be safely omitted from the equations that follow. The constants λ and μ are called the Lamé coefficients and have units of energy density, just like C_{ijkl}.

The Lamé coefficients can be related to the elastic moduli by rewriting the strain tensor as the sum of a pure shear and a hydrostatic compression through the simple rearrangement

$$u_{ij} = (u_{ij} - \delta_{ij} \, \mathrm{tr}u/3) + \delta_{ij} \, \mathrm{tr}u/3. \tag{D.24}$$

If the deformation is a pure shear (no change in volume), then only the first term on the right-hand side survives. In contrast, for a hydrostatic compression, the trace of the first term vanishes and the second term measures the change in volume associated with the deformation, according to Eq. (D.15). Thus, the first and second terms correspond to shear and compression modes, respectively. The factor of 3 appearing in both terms of Eq. (D.24) is dimension-dependent and arises from the number of elements in the trace; for two-dimensional systems, 3 is replaced by 2.

The free energy density also can be decomposed into independent deformation modes by substituting Eq. (D.24) into Eq. (D.23):

$$\mathcal{F} = (\lambda/2)(\mathrm{tr}u)^2 + \mu \, \Sigma_{i,j} [(u_{ij} - \delta_{ij} \, \mathrm{tr}u/3) + \delta_{ij} \, \mathrm{tr} \, u/3]^2$$
$$= \mu \, \Sigma_{i,j} (u_{ij} - \delta_{ij} \, \mathrm{tr}u/3)^2 + 1/2 \, (\lambda + 2\mu/3) \, (\mathrm{tr}u)^2, \tag{D.25}$$

where the first term is a pure shear and the second is a hydrostatic compression. Just as the elastic parameter k_{sp} multiplies the square of the displacement in the potential energy of a Hookean spring, so too here: the energy density is proportional to the relevant elastic modulus times the square of the corresponding strain. Hence, the shear modulus is simply the Lamé coefficient μ, while the compression modulus K_V must be

$$K_V = \lambda + 2\mu/3. \tag{D.26}$$

It is clear from the quadratic form of Eq. (D.25) that both K_V and μ must be positive, or else the system could spontaneously deform by a shear or compression mode. However, the Lamé coefficient λ is not constrained to be positive and a very limited number of systems are known in which λ is negative.

As they are directly related to specific deformation modes, the pair K_V and μ provide a useful parametrization of the two independent elastic moduli expected for isotropic materials. The stress–strain relations can be expressed in terms of K_V and μ by substituting Eq. (D.25) into the fundamental expression

$$\sigma_{ij} = \partial \mathcal{F} / \partial u_{ij}, \tag{D.27}$$

(a result proven in standard texts on continuum mechanics, see, for example, Landau and Lifshitz, 1986) to quickly obtain

$$\sigma_{ij} = \delta_{ij} K_V \, \text{tr} u + 2\mu (u_{ij} - \delta_{ij} \, \text{tr} u \, / \, 3), \tag{D.28}$$

or its inverse

$$u_{ij} = \delta_{ij} \, \text{tr} \sigma / 9 K_V + (\sigma_{ij} - \delta_{ij} \, \text{tr} \sigma / 3) / 2\mu. \tag{D.29}$$

Under isotropic pressure, for example, $\sigma_{ij} = - P \, \delta_{ij}$ from Eq. (D.20), so that the trace of Eq. (D.28) yields $P = - K_V \, \text{tr} u$, where $\text{tr} u$ is the relative change in volume. In two dimensions, this expression becomes $\tau = K_A(u_{xx} + u_{yy})$ for an isotropic tension τ.

Another representation for the elastic characteristics of isotropic materials can be found by analyzing the deformation of a uniformly thin rod by a tensile force applied perpendicular to its ends. The coordinate system is chosen such that the rod lies along the z-axis, and the forces are applied along the z-axis as well. With no x or y components to the force, all components of σ_{ij} vanish except σ_{zz}. Under such a simple stress tensor, Eq. (D.29) can be solved easily for the components of the strain tensor:

$$u_{xx} = u_{yy} = \sigma_{zz} \, (1/3K_V - 1/2\mu)/3 \tag{D.30a}$$

$$u_{zz} = \sigma_{zz} \, (1/3K_V + 1/\mu)/3. \tag{D.30b}$$

Note the sign convention: a tensile force corresponds to $\sigma_{zz} > 0$ [opposite sign to compression in Eq. (D.20)] so that $u_{zz} > 0$ in Eq. (D.30b) and the rod stretches in the direction of the applied force. If $K_V \sim 3\mu$, as it is for many materials, both u_{xx} and u_{yy} are less than zero, and the rod shrinks transversely. The Poisson ratio σ_p is a measure of how much a material *contracts* in the x- or y-direction when it is stretched in the z-direction:

$$\sigma_p = -u_{xx}/u_{zz}. \tag{D.31}$$

Substituting Eq. (D.30) from the analysis of the thin rod gives

$$\sigma_\mathrm{p} = (3K_\mathrm{V} - 2\mu)/(6K_\mathrm{V} + 2\mu) \quad \text{(three dimensions)}. \tag{D.32}$$

The expression for the Poisson ratio depends upon the dimensionality of the object, and for objects in two dimensions, σ_p is

$$\sigma_\mathrm{p} = (K_\mathrm{A} - \mu)/(K_\mathrm{A} + \mu) \quad \text{(two dimensions)}, \tag{D.33}$$

where K_A and μ are two-dimensional elastic moduli.

Lastly, we mention Young's modulus Y, defined by the stress–strain relation

$$u_{zz} = \sigma_{zz}/Y. \tag{D.34}$$

This commonly measured elastic quantity can be expressed in terms of the fundamental elastic constants K_V and μ by substituting Eq. (D.30) for the strain tensor into Eq. (D.34)

$$Y = 9K_\mathrm{V}\mu/(3K_\mathrm{V} + \mu). \tag{D.35}$$

Many materials display $\sigma_\mathrm{p} = 1/3$, which corresponds to $K_\mathrm{V} = (8/3)\mu = Y$.

Fluctuations

The strain tensor describes a deformation from one specific configuration to another. However, systems at finite temperature exchange energy with their surroundings, allowing their shapes and volumes to fluctuate, as illustrated in Fig. D.5. Thus, the deformation of an object under stress at non-zero temperature reflects changes in the average positions of its elements, owing to thermal fluctuations. The magnitude of the shape fluctuations is inversely proportional to the elastic moduli of the object, as can be seen by considering fluctuations in volume. The volume compression modulus (at constant temperature) K_V is related to the change in volume V with respect to pressure P via

$$K_\mathrm{V}^{-1} = -V^{-1}(\partial V/\partial P)_\mathrm{T}. \tag{D.36}$$

That is, the smaller the change in volume for a given change in pressure (i.e., $\partial V/\partial P$ small), the larger the compression modulus. In terms of fluctuations, this means that a system with a large K_V undergoes only

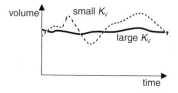

Fig. D.5. Volume fluctuations as a function of time for different compression moduli.

small fluctuations in volume if held at a fixed pressure, as illustrated in Fig. D.5.

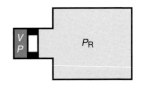

We obtain a relation between the compression modulus and volume fluctuations following Reif (1965). Consider a small system at pressure P in thermal and mechanical contact with a large reservoir at pressure P_R, as displayed in Fig. D.6. Being in thermal contact, both systems have the same temperature T, a result which we do not bother to prove. The small system has a volume V, which it is free to exchange with the large system by moving a piston connecting the systems. As the piston moves, the pressure of the small system may change, but the pressure of the reservoir does not. We define a Gibbs free energy for the small system, G_S, in terms of the temperature and pressure of the reservoir,

$$G_S = E + P_R V - TS, \tag{D.37}$$

Fig. D.6. A small system with volume V at pressure P in contact with a large reservoir at fixed pressure P_R. The piston is able to slide and conduct heat, allowing the small system to exchange energy and volume with the reservoir. Both systems have the same temperature.

where G_S is a function of V at fixed T.

For a given pressure and temperature of the reservoir, the volume of the small system fluctuates about a value V_o that minimizes G_S at a value G_{min}. That is, writing the free energy as a function of the volume, $G_S(V)$, then

$$G_{min} = G_S(V_o). \tag{D.38}$$

For small fluctuations around V_o, the free energy can be expanded as a power series in the change in volume ΔV,

$$G_S(V) = G_{min} + (\partial G_S / \partial V)_T \, \Delta V + 1/2 \, (\partial^2 G_S / \partial V^2)_T \, (\Delta V)^2, \tag{D.39}$$

where

$$\Delta V = V - V_o, \tag{D.40}$$

and the notation $(\ldots)_T$ indicates that the derivative is taken at constant temperature. To find G_{min}, we must evaluate derivatives of G_S with respect to V, obtaining

$$(\partial G_S / \partial V)_T = (\partial E / \partial V)_T + P_R - T(\partial S / \partial V)_T, \tag{D.41}$$

from Eq. (D.37). However, the fundamental relation of thermodynamics

$$T \, dS = dE + P \, dV, \tag{D.42}$$

applied to the small system gives

$$T \, (\partial S / \partial V)_T = (\partial E / \partial V)_T + P, \tag{D.43}$$

so that the first derivative of G_S becomes

$$(\partial G_S / \partial V)_T = P_R - P, \tag{D.44}$$

when Eq. (D.43) is substituted into Eq. (D.41).

For G_S to be a minimum at V_o, the first derivative of G_S must vanish, and hence $P_R = P$, according to Eq. (D.44). Just as importantly, Eq. (D.44) can be differentiated to yield an expression for the second derivative of G_S,

$$(\partial^2 G_S/\partial V^2)_T = -(\partial P/\partial V)_T. \tag{D.45}$$

The first and second derivatives of G_S now can be substituted into Eq. (D.39) to give

$$\Delta G_S = G_S - G_{min} = -1/2 \, (\partial P/\partial V)_T \, (\Delta V)^2, \tag{D.46}$$

or, in terms of the compression modulus from Eq. (D.36)

$$\Delta G_S = + K_V \, (\Delta V)^2/2V_o. \tag{D.47}$$

Knowing the change in free energy with volume allows us to determine the relative probability for the system to have a particular volume; as discussed in Appendix C on statistical mechanics, the relative probability $\mathcal{P}(V)$ for the system to have a volume V is

$$\mathcal{P}(V) \sim \exp(-\beta \Delta G_S) \sim \exp\{-\beta K_V \, (\Delta V)^2/2V_o\}. \tag{D.48}$$

Having a Gaussian form, the probability distribution can be integrated easily to give the dispersion in the volume. The steps are the same as those in evaluating the mean square displacement of the harmonic oscillator in Appendix C, and yield

$$\langle(\Delta V)^2\rangle = V_o/\beta K_V, \tag{D.49}$$

or in the more familiar form

$$1/\beta K_V = \langle(\Delta V)^2\rangle/V_o. \tag{D.50}$$

As expected from Fig. D.5, this expression shows that the fluctuations in V rise as the compression modulus falls.

Glossary

α-actinin   rod-shaped actin-binding protein which links parallel actin filaments into bundles.

actin   protein of mass 42 kDa; forms a polymeric filament about 8 nm in diameter; mechanical element of the cytoskeleton, ubiquitous from yeasts to humans; major component of muscle tissue.

actin-binding protein   type of protein capable of binding to actin, often to form a network or bundle.

ADP   (adenosine diphosphate; see ATP).

amino acid   organic molecule containing both carboxyl and amino groups attached to the same carbon atom (see Appendix B for a summary of the 20 amino acids commonly found in proteins).

amphiphile   molecule with both polar and non-polar regions; simultaneously attractive to immiscible fluids such as oil and water.

anisotropic   possessing properties that are direction-dependent.

ankyrin   protein that links β-spectrin filaments of the red cell cytoskeleton to the protein band 3 embedded in the plasma membrane.

archaebacteria   a subkingdom of bacteria capable of thriving in environments of extreme acidity or temperature.

arc length (s)   distance along a curve following its contour.

ATP   (adenosine triphosphate) along with GTP, a fundamental unit of energy currency of the cell, releasing energy upon hydrolysis to ADP.

avidin   tetrameric protein of mass 68 kDa capable of forming adhesive bonds with ligands on adjacent membranes (similar to **streptavidin**).

axisymmetric   unchanged by rotation about an axis (e.g. a cylinder).

axon   long, string-like section of a nerve cell extending from the soma (which contains the cell's nucleus).

bacteria   single-celled micro-organisms without a nucleus, classified into the subkingdoms of archaebacteria and eubacteria.

basal lamina   thin mat of interconnected proteins adjacent to an epithelial sheet, separating it from connective tissue (see Appendix A).

base-pair   complementary organic bases linking two strands of DNA by means of hydrogen bonds.

bending rigidity (κ_b)   elastic constant characterizing a membrane's resistance to a change in its mean curvature; commonly has a value of 10^{-19} J for biomembranes.

bilayer   quasi-two-dimensional structure composed of two monolayers, each an extended array just a single molecule in thickness.

biotin   a vitamin of molecular mass 0.244 kDa; binds tightly (but not covalently) to the protein avidin.

Bjerrum length (ℓ_B)   temperature-dependent length scale for a medium of permittivity ε containing ions of charge q; $\ell_B \equiv q^2 / 4\pi\varepsilon k_B T$.

Boltzmann constant (k_B)   fundamental physical constant equal to 1.38 $\times 10^{-23}$ J/K; classically, $k_B T / 2$ is the mean thermal energy of each independent mode of motion.

Boltzmann factor [$\exp(-\Delta E / k_B T)$]   relative probability for a system at temperature T to occupy a specific state having energy ΔE with respect to a reference state (derived in Appendix C).

branched polymer   polymeric structure whose chains have multiple branch points.

cadherin   adhesion protein present in desmosomes of animal cells.

cell wall   stress-bearing fabric enclosing the plasma membranes of most plant cells and bacteria, but not animal cells; composition varies according to cell type.

cellulose   principal component of the plant cell wall, consisting of bundles of linear polysaccharides.

centriole   cylindrical structure present (usually in pairs) in the centrosome of a cell.

centrosome   diffuse region of the cell near the nucleus, from which radiate microtubule components of the cell's cytoskeleton.

chloroplast   organelle found in plant and algae cells; site of photosynthesis.

cholesterol   a type of sterol; often present in cell membranes.

clathrin   a three-armed protein that assembles into a two-dimensional coat on a local region of a membrane during the formation of a vesicle in endocytosis.

collagen fibrous protein found in connective tissue; the most abundant protein in animals.

compression deformation mode characterized by a change in area or volume, depending on the dimensionality of the system; magnitude of deformation is determined by a compression modulus K_A or K_V in two or three dimensions, respectively.

connective tissue heterogeneous tissue in animal cells providing mechanical support, among its other roles.

contour length (L_c) total length of a filament as measured along its contour.

critical micelle concentration (CMC) in aqueous solutions of amphiphiles, the minimum concentration for the formation of condensed phases such as micelles.

curvature (C) measure of the rate of change with arc length of a unit tangent vector to a line or normal vector to a surface; at any location on a surface, there are two **principal curvatures**, C_1 and C_2, which may be combined as the **mean curvature** ($C_1 + C_2$) or **Gaussian curvature** ($C_1 C_2$).

cytoplasm interior contents of the cell, including the cytosol and most organelles but excluding the nucleus.

cytoskeleton filamentous network permeating the cell's interior; composed of microtubules, intermediate filaments, actin and sometimes spectrin, depending on the cell.

cytosol heterogeneous fluid component of the cell's cytoplasm, excluding organelles.

dalton (Da) elementary unit of atomic mass, defined as 1/12 of the mass of the ^{12}C-atom; equal to 1.66054×10^{-27} kg.

Debye length (ℓ_D) length scale characterizing the exponential decay of the electrostatic potential in an electrolyte; for a neutral electrolyte of monovalent ions, $\ell_D^{-2} \equiv 8\pi\ell_B\rho_s$, where ℓ_B is the Bjerrum length and ρ_s is the number density of the ions.

dendrites branched structures emanating from the main cell body of a neuron; collect signals from adjacent cells and propagate them to the cell body.

desmosome local junction between adjacent cells (as in epithelia), mechanically linking their cytoskeletons.

diagonal matrix mathematically, a matrix m_{ij} with vanishing off-diagonal elements ($i \neq j$).

diffusion constant (D) proportionality constant relating the mean square displacement $\langle r^2 \rangle$ of a randomly moving object to the elapsed time t of its movement; $\langle r^2 \rangle = 2dDt$, where d is the embedding dimension.

dimer   a composite of two molecular subunits, which, in the case of **heterodimers**, need not be identical.

DLVO theory   theory of colloids based on electrostatic and van der Waals interactions; formulated by Derjaguin, Landau, Verway, and Overbeek.

DNA (deoxyribonucleic acid)   genetic material of the cell consisting of a linear chain whose elemental unit is a troika of a sugar derivative (deoxyribose), an organic base, and a phosphate group; pairs of chains intertwine as a double helix.

echinocyte   a conformation of the red blood cell adopted under unusual conditions and characterized by the appearance of bumps on its surface.

edge energy   energy associated with the free boundary of a surface; in its simplest form, equal to $\lambda\Gamma$, where Γ is the length of the boundary and λ is the material-specific edge tension (energy per unit length).

elastic moduli (C_{ijkl}) general set of elastic constants relating stress to strain; for Hooke's law materials, $\sigma_{ij} = \Sigma_{k,l} C_{ijkl} u_{kl}$, where σ_{ij} and u_{kl} are the stress and strain tensors, respectively.

electrolyte   a solution or molten state that conducts electricity owing to the presence of mobile ions.

electrostatic potential (ψ) for a charged object, $\psi = V/q$, where V is the potential energy of the object and q is its charge.

endocytosis   process in which material enters a cell via the formation of vesicles by invagination of the plasma membrane.

endoplasmic reticulum   an organelle with an extensive membrane area, attached to the nuclear envelope; location of protein synthesis.

end-to-end displacement (r_{ee}) displacement vector between the ends of a linear polymer.

enthalpy (H) in thermodynamics, the sum of the internal energy E of a system and the work of deformation under pressure P ($H = E + PA$ or $H = E + PV$ in two or three dimensions, respectively).

entropy (S) in statistical mechanics, a measure of the number of configurations Ω available to a system; $S = k_B \ln \Omega$, where k_B is Boltzmann's constant (see Gibbs free energy).

epithelium   a sheet-like tissue, such as the skin or intestinal wall, isolating the body from its environment.

equilateral   having sides of equal length, as in equilateral triangle.

erythrocyte   blood cell containing hemoglobin, responsible for oxygen transport in the circulatory system.

eucaryote   cell with a distinct nucleus containing most, but not all, of its DNA; all cells are eucaryotic except bacteria.

exocytosis   process by which material is expelled from the cell by the fusion of vesicles with its plasma membrane.

extracellular matrix   stress-bearing fabric secreted by cells, composed of proteins and polysaccharides.

fatty acid   organic molecule consisting of a single hydrocarbon chain terminating at a carboxyl group ($-COOH$).

fibroblast   cell commonly found in connective tissue; function includes secretion of extracellular matrix.

filamin   as a V-shaped dimer, a protein capable of cross-linking actin filaments at wide angles.

fimbrin   short protein that links parallel actin filaments into dense bundles.

flagellum   long, whip-like structure extending from a cell; undulations of the flagellum may generate cell locomotion.

flexural rigidity (κ_f)   material parameter characterizing the bending resistance of a filament; $\kappa_f = \pi Y R^4/4$ for a solid cylindrical rod of radius R and Young's modulus Y.

freely jointed chain   model for a polymer in which linked segments may pivot about their junctions without restriction.

freely-rotating chain   model for a polymer in which the polar angle between neighboring straight segments is fixed, although the segments are free to rotate azimuthally (like an alkane chain).

Gaussian distribution   in statistics, the normalized probability distribution for an observable x given by $(2\pi\sigma^2)^{-1/2} \exp(-[x-\mu]^2/2\sigma^2)$, where σ^2 is the variance and μ is the mean.

Gaussian rigidity (κ_G)   elastic constant characterizing a membrane's resistance to a change in its Gaussian curvature.

GDP   (guanosine diphosphate; see GTP).

Gibbs free energy (G)   in thermodynamics, the free energy of a system at fixed temperature T and pressure P; in three dimensions, $G = E + PV - TS$, where E, V, and S are the internal energy, volume and entropy of the system, respectively.

glucose   a specific sugar molecule (one of many with the composition $C_6H_{12}O_6$) used in cell metabolism.

glycocalyx   sugar-based coat surrounding most eucaryotic cells.

Golgi apparatus   organelle with extensive folded membranes, whose function includes protein sorting.

Gram-negative bacterium   bacterium whose thin cell wall is sandwiched between the inner (plasma) membrane and an outer membrane.

Gram-positive bacterium   bacterium with a single membrane encapsulated by a thick cell wall.

GTP (guanosine triphosphate)   one element of the cell's energy currency, releasing energy during hydrolysis to GDP.

Hooke's law   behavior of materials for which stress is linearly proportional to strain; the force $\mathbf{F}$ experienced by a Hookean spring obeys $\mathbf{F} = -k_{sp}\mathbf{x}$, where k_{sp} is an elastic parameter and $\mathbf{x}$ is the displacement from equilibrium.

hydrocarbon   molecule, or section thereof, containing exclusively hydrogen and carbon atoms.

hydrolysis   chemical reaction involving the breakage of covalent bonds by the addition of water.

hydrophilic   literally, water-loving; in a molecule, usually an electrically polar region with an attraction to water.

hydrophobic   literally, water-avoiding; in a molecule, generally a non-polar region.

ideal gas   gas of N point-like constituents obeying the ideal gas law $PV = Nk_{B}T$, where P, V, and T are pressure, volume, and temperature, respectively.

ideal scaling   particular relationship between the effective size of a polymer and its contour length L_c; the end-to-end displacement $\mathbf{r}_{ee}$ of ideal polymers obeys $\langle \mathbf{r}_{ee}^2 \rangle \propto L_c$.

intermediate filament   one of the common filaments of the cytoskeleton; a bundle of intertwined and cross-linked proteins about 10 nm in diameter.

isosceles   a triangle with two sides of equal length.

isotropic   describing a medium or material whose physical properties are independent of direction.

keratin   one of the fibrous proteins capable of forming intermediate filaments.

kinesin   a type of molecular motor that (generally) moves towards the plus end of a microtubule.

lamellipodium   thin, sheet-like extension of a cell spread on a substrate; generally rich in actin filaments and part of the cell locomotion apparatus.

lamin   one of the proteins capable of forming intermediate filaments; present in the nuclear lamina.

lipids   broad class of organic molecules that are insoluble in water; phospholipids and cholesterol are the principal lipid constituents of cellular membranes.

ligand   a molecule capable of binding to a specific receptor molecule.

lysis   mechanical failure of the plasma membrane under tension.

macrophage   specialized white blood cell responsible for removing bacteria and cell debris from the body.

macroscopic   associated with large length scales, such as a millimeter or more, visible to the unaided eye.

matrix   mathematically, a two-dimensional array of elements m_{ij}.

mean value   mathematically, for a set of N elements x_i, the mean value is $(1/N) \sum_i x_i$.

membrane   in the cell, a sheet-like structure that forms the boundaries of a cell and its organelles; derives its two-dimensional structure from a lipid bilayer, which may also contain cholesterol and embedded proteins.

mesophase   phase of matter with properties intermediate between solids and isotropic liquids; examples include some phases of liquid crystals.

mesoscopic   having a length scale larger than molecular sizes yet smaller than macroscopic dimensions.

micelle   an aggregate of amphiphilic molecules in a fluid, having roughly spherical or cylindrical shapes.

microtubule   polymeric protein filament composed of tubulin monomers, present in the cytoskeleton, as well as flagella and cilia, with a hollow structure 25 nm in diameter.

minus end   slow-growing or unstable end of a polymeric protein filament such as actin or a microtubule.

mitochondrion   one of two organelles responsible for producing ATP; a few nanometers in length and bounded by a double membrane.

moment of inertia of cross section ($\mathcal{I}$)   geometrical quantity influencing the energy needed to bend a filament; equal to $\pi R^4/4$ for a cylinder of radius R.

monolayer   a quasi-two-dimensional array just one atom or molecule thick.

monomer   chemical building block capable of associating into polymeric structures such as filaments.

motor protein   a class of proteins (dynein, kinesin, and myosin) capable of movement along a cytoskeletal filament, often for the purpose of vesicle transport or cell contraction.

muscle tissue   individual or composite cells capable of producing movement through contraction; muscle cells are classified as skeletal, smooth, cardiac or myoepithelial (see Appendix A).

mycoplasma   class of small procaryotic cells lacking a cell wall and having diameters of 0.8 μm or less.

myosin   a family of motor proteins that move along actin filaments.

nematic   a mesophase characterized by long-range orientational order without positional order.

neuron   a cell of the nervous system capable of receiving and transmitting electrical signals; principal components of a neuron are dendrites, soma and axon.

normal vector (n)   unit vector locally perpendicular to a surface or line.

nuclear lamina   connected mat of intermediate filaments attached to the inner surface of the inner nuclear membrane.

nucleotide   a subunit of DNA and RNA consisting of a sugar ring to which are separately attached an organic base and phosphate group; polymerization occurs between the sugar and phosphate of successive nucleotides.

nucleus   unique to eucaryotic cells, an organelle containing most, but not all, of a cell's DNA.

oblate   axially symmetric shape like a pancake (short along the axis of rotation).

organelle   membrane-bound compartment of a eucaryotic cell; examples include the nucleus, mitochondria etc.

organic acid   an organic compound containing the carboxyl group ($-COOH$).

organic base   in chemistry, an organic compound which is often basic by virtue of accepting a proton from aquaeous solution, leaving OH^-.

peptide bond   a covalent bond formed by the linkage of carboxyl ($-COOH$) and amino ($-NH_2$) groups, eliminating H_2O.

peptidoglycan   material of the bacterial cell wall, composed of linear polysaccharides cross-linked by polypeptide chains.

percolation   in physics, a concentration threshold above which a continuous path traverses a network or lattice.

persistence length (ξ_p)   a measure of the distance scale over which tangent vectors to a curve, or normal vectors to a surface, become uncorrelated.

phantom membrane   a mathematical surface which is permitted to intersect itself.

phosphate   the chemical group PO_4.

phosphorylation   the covalent addition of a phosphate group to a molecule.

pipette   in chemistry, a glass tube used for transferring liquids; laboratory study of cells may involve micropipettes with a tapered end just a micron in diameter.

plaquette   elementary geometrical unit of a surface.

plasma membrane   semi-permeable membrane surrounding a cell and controlling the passage of molecular material to and from its interior.

plus end   rapidly growing end of a polymeric filament such as actin or a microtubule.

Poisson-Boltzmann equation   an equation describing the electrostatic potential experienced by a charge at non-zero temperature (see Section 8.2).

Poisson ratio   for an object under uniaxial stress, the ratio of the fractional contraction in the transverse direction compared to the fractional extension in the direction of the applied stress.

polar head-group   polarized terminal group attached to the phosphate group of a phospholipid.

polymer   a structure formed through the association of monomeric subunits; bonding need not be covalent, and the structure may be dynamic, continually adding and releasing monomers.

polysaccharide   linear or branched polymer of linked sugar molecules.

procaryote   a mechanically simple cell or aggregate without a nucleus; bacteria are procaryotic cells.

prolate   axially symmetric shape like a cigar (long along the axis of rotation).

protein   linear polymer of amino acids, covalently linked by peptide bonds.

purine   a specific family of organic bases with a molecular structure of two rings (five and six members each) sharing a common bond; present in DNA and RNA.

pyrimidine   a specific family of organic bases in DNA and RNA comprising a single six-member ring.

radius of curvature   reciprocal of the curvature of a line or surface.

radius of gyration (R_g)   for a collection of N objects with positions $\mathbf{r}_i$ with respect to the center-of-mass, $R_g^2 = \Sigma_i \, r_i^2 / N$.

receptor protein   a protein that binds to specific ligands, often as part of a signaling mechanism.

ribose   specific sugar with a five-member ring, present in RNA; a related compound, deoxyribose, is present in DNA.

ribosome   complex cellular body constructed from protein and ribosomal RNA; site of protein synthesis.

RNA (ribonucleic acid)   molecular chain of linked nucleotides where ribose is the sugar unit; plays a variety of roles in protein synthesis.

root mean square   for a set of N elements x_i, the root mean square is $(\Sigma_i x_i^2 / N)^{1/2}$.

self-avoiding walk   an otherwise random trajectory which does not cross itself.

self-avoiding membrane   a mathematical surface which is not permitted to intersect itself.

shear   a force applied tangentially to a surface; the **shear modulus** (μ) of isotropic materials is the shear stress divided by the angle of deformation (in radians).

shear storage and loss moduli (G' and G'')   frequency-dependent mechanical parameters of viscoelastic materials; in the limit where the deformation frequency ω vanishes, $G' \rightarrow \mu$, and $G'' \rightarrow \eta/\omega$, where η is the viscosity.

spectrin   two closely related proteins (α- and β-spectrin) of mass about 220 kDa; as $\alpha\beta$-heterodimers, form a filament with a contour length of 200 nm; principal component of the human erythrocyte cytoskeleton.

strain   deformation of an object in response to a stress; a dimensionless quantity, the **strain tensor** (u_{ij}) is given by ($\partial u_i/\partial x_j + \partial u_j/\partial x_i$)/2 for small displacements **u**, where **x** is the coordinate system.

stomatocyte   cup-shaped conformation of a red blood cell.

stress   in three dimensions, a force per unit area; the **stress tensor** (σ_{ij}) is defined in Appendix D.

sugar   a family of compounds with the chemical formula $(CH_2O)_n$; for some values of n, the molecule may adopt either a linear chain or ring configuration.

surface tension (γ)   energy required to increase a surface by a unit area.

surfactant   a chemical compound that reduces the surface tension at an interface of a liquid (from surface active agent).

synapse   junction between a cell and the axon of a nerve cell, across which a signal may by carried chemically or electrically.

tangent vector (**t**)   unit vector locally tangent to a line or surface.

tensile strength   stress required to cause mechanical failure of a material.

thermal ratchet   model for molecular motion in which thermally activated movement is biased in one direction.

tissue   an extended collection of cells and associated materials acting cooperatively with other tissues as part of an organ (see Appendix A for classification of tissues).

trace (tr)   mathematically, the sum of the diagonal elements of a matrix; tr$m = \Sigma m_{ii}$.

treadmilling   dynamic state of a polymer fillament occuring when the addition rate of monomers at one end equals the release rate at the other.

tubulin   monomeric component of microtubules; the proteins α- and β-tubulin form a heterodimer 8 nm in length.

vacuole   enclosed space within plant or animal cells often used for storage; may occupy a sizeable fraction of the volume of a plant cell.

van der Waals forces   a class of attractive forces arising from the interaction of permanent and instantaneous electric dipole moments; corresponding potential energy falls like $1/r^6$ for large separations r.

variance   in statistics, for a set of N measured values x_i,
$$\text{Var}(x_1 \ldots x_N) = (N-1)^{-1} \Sigma_i (x_i - X)^2,$$ where the mean value is $X = \Sigma_i x_i / N$.

vesicle   small membrane-bounded sphere used for transport in the cytoplasm.

vimentin   a protein of mass 54 kDa capable of forming intermediate filaments.

virus   parasitic life-form consisting of a core of tightly packed DNA or RNA protected by a proteinaceous capsid; can replicate inside a host cell but is otherwise inert.

viscosity (η)   a measure of the resistance of a fluid to deformation under stress.

Young equation   relationship between the contact angle of a droplet on a substrate and the interfacial energies of the materials.

Young's modulus (Y)   an elastic property of materials; ratio of stress to strain for a uniaxial deformation.

References

Abraham, F.F. and Kardar, M. (1991). Folding and unbinding transitions in tethered membranes. *Science*, **252**, 419–422.

Abraham, F.F. and Nelson, D.R. (1990). Diffraction from polymerized membranes. *Science*, **249**, 393–397.

Abraham, V.C., Krishnamurthi, V., Taylor, D.L. and Lanni, F. (1999). The actin-based nanomachine at the leading edge of migrating cells. *Biophys. J.*, **77**, 1721–1732.

Abramowitz, M. and Stegun, I.A. (1970). *Handbook of Mathematical Functions*. New York, NY: Dover.

Aebi, U., Cohn, J., Buhle, L. and Gerace, L. (1986). The nuclear lamina is a meshwork of intermediate-type filaments. *Nature*, **323**, 560–564.

Ajdari, A. and Prost, J. (1992). Mouvement induit par un potential périodique de basse symétrie: diélectrophorèse pulsée. *C. R. Acad. Sci. Paris Ser. II.*, **315**, 1635–1639.

Albersdörfer, A., Feder, T. and Sackmann, E. (1997). Adhesion-induced domain formation by interplay of long-range repulsion and short-range attraction force: a model membrane study. *Biophys. J.*, **73**, 245–257.

Alberts, B., Bray, D., Lewis, J., Raff, M., Roberts, K. and Watson, J.D. (1994). *Molecular Biology of the Cell*, 3rd edn. New York, NY: Garland.

Amos, L.A. and Amos, W.B. (1991). *Molecules of the Cytoskeleton*. London: MacMillan.

Arnoldi, M., Fritz, M., Bäuerlein, E., Radmacher, M., Sackmann, E. and Boulbitch, A. 2000. Bacterial turgor pressure can be measured by atomic force microscopy. *Phys. Rev. E*, **62**, 1034–1044.

Aronovitz, J.A. and Lubensky, T.C. (1988). Fluctuations of solid membranes. *Phys. Rev. Lett.*, **60**, 2634–2637.

Ashkin, A., Schütze, K., Dziedzic, J.M., Euterneuer, U. and Schliwa, M. (1990). Force generation of organelle transport measured *in vivo* by an infrared laser trap. *Nature*, **348**, 346–348.

Astumian, R.D. and Bier, M. (1994). Fluctuation driven ratchets: Molecular motors. *Phys. Rev. Lett.*, **72**, 1766–1769.

Bailey, S.M., Chiruvolu, S., Israelachvili, J.N. and Zasadzinski, J.A.N. (1990). Measurements of forces involved in vesicle adhesion using freeze-fracture electron microscopy. *Langmuir*, **6**, 1326–1329.

Bärmann, M., Käs, J., Kurzmeier, H. and Sackmann, E. (1992). A new cell model: actin networks encaged by giant vesicles. In *The Structure and Conformation of Amphiphilic Membranes*, eds. R. Lipowsky, D. Richter, and K. Kremer, p. 137–143. Springer-Verlag, Berlin.

Bar-Ziv, R. and Moses, E. (1994). Instability and "pearling" states produced in tubular membranes by competition of curvature and tension. *Phys. Rev. Lett.*, **73**, 1392–1395.

Bar-Ziv, R., Moses, E. and Nelson, P. (1998). Dynamic excitations in membranes induced by optical tweezers. *Biophys. J.*, **75**, 294–320.

Baumgärtner, A. (1982). Statistics of self-avoiding ring polymers. *J. Chem. Phys.*, **76**, 4275–4282.

Baumgärtner, A. (1991). Does a polymerized membrane crumple? *J. Phys. I France*, **1**, 1549–1556.

Baumgärtner, A. and Ho, J.-S. (1990). Crumpling of fluid vesicles. *Phys. Rev. A*, **41**, 5747–5750.

Bausch, A.R., Möller, W. and Sackmann, E. (1999). Measurement of local viscoelasticity and forces in living cells by magnetic tweezers. *Biophys. J.*, **76**, 573–579.

Bausch, A.R., Ziemann, F., Boulbitch A.A., Jacobson, K. and Sackmann, E. (1998). Local measurements of viscoelastic parameters of adherent cell surfaces by magnetic bead microrheology. *Biophys. J.*, **75**, 2038–2049.

Bayley, P.M., Sharma, K.K. and Martin, S.R. (1994). Microtubule dynamics *in vitro*. In *Microtubules*, eds. J.S. Hyams and C.W. Lloyd, pp. 111–137. New York, NY: Wiley-Liss.

Beck, W.S., ed. (1991). *Hematology*, 5th edn. Boston, MA: MIT Press.

Bell, G.I. (1978). Models for the specific adhesion of cells to cells. *Science*, **200**, 618–627.

Berg, H.C. (1983). *Random Walks in Biology*. Princeton, NJ: Princeton University Press.

Berg, H.C. (1995). Torque generation by the flagellar rotary motor. *Biophys. J.*, **68**, 163s–167s.

Berg, H.C. (2000). Motile behavior of bacteria. *Physics Today*, **53**, 24–29.

Berg, H.C. and Turner, L. (1993). Torque generated by the flagellar motor of *Escherichia coli*. *Biophys. J.*, **65**, 2201–2216.

Berg, O.G. and von Hippel, P.H. (1985). Diffusion-controlled macromolecular interactions. *Ann. Rev. Biophys. Biophys. Chem.*, **14**, 131–160.

Bergen, L.G. and Borisy, G.G. (1980). Head-to-tail polymerization of microtubules *in vitro*. *J. Cell Biol.*, **84**, 141–150.

Berndl, K., Käs, J., Lipowsky, R., Sackmann, E. and Seifert, U. (1990). Shape transformations of giant vesicles: Extreme sensitivity to bilayer asymmetry. *Europhys. Lett.*, **13**, 659–664.

Berry, R.M. (1993). Torque and switching in the bacterial flagellar motor: an electrostatic model. *Biophys. J.*, **64**, 961–973.

Berry, R.M. and Berg, H.C. (1999). Torque generated by the flagellar motor of *Escherichia coli* while driven backward. *Biophys. J.*, **76**, 580–587.

Bessis, M. (1973). *Living Blood Cells and Their Ultrastructure*. New York, NY: Springer-Verlag.

Bier, M. and Astumian, R.D. (1993). Matching a diffusive and a kinetic approach for escape over a fluctuating barrier. *Phys. Rev. Lett.*, **71**, 1649–1652.

Bloom, M., Evans, E. and Mouritsen, O.G. (1991). Physical properties of the fluid lipid-bilayer component of cell membranes: a perspective. *Quart. Rev. Biophys.*, **24**, 293–397.

Bloomfield, V.A. (1991). Condensation of DNA by multivalent ions: considerations on mechanism. *Biophys. J.*, **31**, 1471–1481.

Boal, D.H. (1991). Phases of two-dimensional vesicles under pressure. *Phys. Rev.*, **A43**, 6771–6777.

Boal, D.H. (1994). Computer simulation of a model network for the erythrocyte cytoskeleton. *Biophys. J.*, **67**, 521–529.

Boal, D.H. and Rao, M. (1992a). Scaling behavior of fluid membranes in three dimensions. *Phys. Rev. A*, **45**, R6947–R6950.

Boal, D.H. and Rao, M. (1992b). Topology changes in fluid membranes. *Phys. Rev. A*, **46**, 3037–3045.

Boal, D.H., Seifert, U. and Shillcock, J.C. (1993). Negative Poisson ratio in two-dimensional networks under tension. *Phys. Rev. E*, **48**, 4274–4283.

Bonder, E.M., Fishkind, D.J. and Mooseker, M.S. (1983). Direct measurement of critical concentrations and assembly rate constants at the two ends of an actin filament. *Cell*, **34**, 491–501.

Braun, V., Gnirke, U., Henning, U. and Rehn, K. (1973). Model for the structure of the shape-maintaining layer of *Escherichia coli* cell envelope. *J. Bacteriol.*, **114**, 1264–1270.

Brochard, F. and Lennon, J.F. (1975). Frequency spectrum of the flicker phenomenon in erythrocytes. *J. Phys. Paris*, **36**, 1035–1047.

Brotschi, E.A., Hartwig, J.H. and Stossel, T.P. (1978). The gelation of actin by actin-binding protein. *J. Biol. Chem.*, **253**, 8988–8993.

Bruinsma, R., Goulian, M. and Pincus, P. (1994). Self-assembly of membrane junctions. *Biophys. J.*, **67**, 746–750.

Bull, B.S. (1977). The resting shape of normal red blood cells is determined by bending and ___? *Blood Cells*, **3**, 321–323.

Burge, R.E., Fowler, A.G. and Reaveley, D.A. (1977). Structure of the peptidoglycan of bacterial cell walls. I. *J. Mol. Biol.*, **117**, 927–953.

Bustamante, C., Marko, J.F., Siggia, E.D. and Smith, S. (1994). Entropic elasticity of λ-phage DNA. *Science*, **265**, 1599–1600.

Byers, T. and Branton, D. (1985). Visualizations of the protein associations in the erythrocyte membrane skeleton. *Proc. Natl. Acad. Sci. USA*, **82**, 6153–6157.

Canham, P.B. (1970). The minimum energy of bending as a possible explanation of the biconcave shape of the human red blood cell. *J. Theor. Biol.*, **26**, 61–81.

Cao, L., Fishkind, D.J. and Wang, Y.-L. (1993). Localization and dynamics of nonfilamentous actin in cultured cells. *J. Cell Biol.*, **123**, 173–181.

Carlier, M.-F. and Pantaloni, D. (1981). Kinetic analysis of guanosine 5′-triphosphate hydrolysis associated with tubulin polymerization. *Biochemistry*, **20**, 1918–1924.

Cardini, G., Bareman, J.P. and Klein, M.L. (1988). Characterization of a Langmuir–Blodgett monolayer using molecular dynamics calculations. *Chem. Phys. Lett.*, **145**, 493–498.

Cevc, G. and Marsh, D. (1987). *Phospholipid Bilayers. Physical Principles and Models*. New York, NY: Wiley-Interscience.

Chandrasekhar, S. (1943). *Rev. Mod. Phys.*, **15**, 1.

Chen, Z. and Rand, R.P. (1997). The influence of cholesterol on phospholipid membrane curvature and bending elasticity. *Biophys. J.*, **73**, 267–276.

Chernomordik, L.V. and Chizmadzhev, Y.A. (1989). Electrical breakdown of lipid bilayer membranes: phenomenology and mechanism. In *Electroporation and Electrofusion in Cell Biology*, eds. E. Neumann, A.E. Sowers and C.A. Jordan, pp. 83–95. New York, NY: Plenum.

Cook-Röder, J. and Lipowsky, R. (1992). Adhesion and unbinding for bunches of fluid membranes. *Europhys. Lett.*, **18**, 433–438.

Coppin, C.M. and Leavis, P.C. (1992). Quantitation of liquid-crystalline ordering in F-actin solutions. *Biophys. J.*, **63**, 794–807.

Cortese, J.D., Schwab, B. III, Freiden, C. and Elson, E.L. (1989). Actin polymerization induces a shape change in actin-containing vesicles. *Proc. Natl. Acad. Sci. USA*, **86**, 5773–5777.

Creighton, T.E. (1993). *Proteins: Structures and Molecular Properties*, 2nd edn. New York, NY: Freeman.

Crick, F.H.C. and Hughes, A.F.W. (1950). The physical properties of cytoplasm. A study by means of magnetic particle method part I. Experimental. *Exp. Cell Res.*, **1**, 37–80.

Cros, S., Garnier, C., Axelos, M.A.V., Imberty, A. and Pérez, S. (1996). Solution conformations of pectin polysaccharides: determination of chain characteristics by small angle neutron scattering, viscometry, and molecular modeling. *Biopolymers*, **39**, 339–352.

Crowther, R.A., Finch, J.T. and Pearse, B.M.F. (1976). On the structure of coated vesicles. *J. Mol. Biol.*, **103**, 785–798.

Culver, H.B. (1992). *The Book of Old Ships*. New York, NY: Dover.

Dabiri, G.A., Sanger, J.M., Portnoy, D.A. and Southwick, F.S. (1990). *Listeria monocytogenes* moves rapidly through the host-cell cytoplasm by inducing directional actin assembly. *Proc. Natl. Acad. Sci. USA*, **87**, 6068–6072.

Dammer, U., Popescu, O., Wagner, P., Anselmetti, D., Güntherodt, H.-J. and Misevic, G.N. (1995). Binding strength between cell adhesion proteoglycans measured by atomic force microscopy. *Science*, **267**, 1173–1175.

David, F. (1989). Geometry and field theory of random surfaces and membranes. In *Statistical Mechanics of Membranes and Surfaces*, eds. D. Nelson, T. Piran and S. Weinberg, pp. 157–223. Singapore: World Scientific.

Deam, R.T. and Edwards, S.F. (1976). *Phil. Trans. R. Soc. Lond.*, A**280**, 317–353.

de Duve, C.R. (1995). *Vital Dust*. New York, NY: Basic Books.

de Gennes, P.-G. (1979). *Scaling Concepts in Polymer Physics*. Ithaca, NY: Cornell University Press.

de Gennes, P.G. and Prost, J. (1993). *The Physics of Liquid Crystals*, 2nd edn. Oxford: Oxford University Press.

de Gennes, P.G. and Taupin, C. (1982). Microemulsions and the flexibility of oil/water interfaces. *J. Phys. Chem.*, **86**, 2294–2304.

Delsemme, A. (1998). *Our Cosmic Origins*. Cambridge: Cambridge University Press.

Dembo, M. (1989). Mechanics and control of the cytoskeleton in *Amoeba proteus*. *Biophys. J.*, **55**, 1053–1080.

Dembo, M. and Harlow, F. (1986). Cell motion, contractile networks, and the physics of interpenetrating reactive flow. *Biophys. J.*, **50**, 109–121.

Dembo, M., Torrey, D.C., Saxman, K. and Hammer, D. (1988). The reaction-limited kinetics of membrane-to-surface adhesion and detachment. *Proc. R. Soc. Lond., B***234**, 55–83.

Derjaguin, B.V. (1949). *Theory of Stability of Colloids and Thin Films*, trans. R.K. Johnston. New York, NY: Consultants Bureau.

Derjaguin, B.V., and Landau, L. (1941). *Acta Physicochim. URSS*, **14**, 633–662.

Derrida, B. and Stauffer, D. (1985). Corrections to scaling and phenomenological renormalization for 2-dimensional percolation and lattice animal problems. *J. Phys. (Paris)*, **46**, 1623–1630.

Deryagin, B.V. and Gutop, Yu.V. (1962). Theory of the breakdown (rupture) of free films. *Kolloidnyi Zh.*, **24**, 431–437.

Deuling, H.J. and Helfrich, W. (1976). The curvature elasticity of fluid membranes: a catalogue of vesicle shapes. *J. Physique*, **37**, 1335–1345.

Discher, D.E., Boal, D.H. and Boey, S.K. (1997). Phase transitions and anisotropic responses of planar triangular nets under large deformation. *Phys. Rev. E*, **55**, 4762–4772.

Discher, D.E., Boal, D.H. and Boey, S.K. (1998). Simulations of the erythrocyte cytoskeleton at large deformation. II. Micropipette aspiration. *Biophys. J.*, **75**, 1584–1597.

Discher, D.E., Mohandas, N. and Evans, E.A. (1994). Molecular maps of red cell deformation: hidden elasticity and *in situ* connectivity. *Science*, **266**, 1032–1035.

Döbereiner, H.-G., Evans, E., Kraus, M., Seifert, U. and Wortis, M. (1997). Mapping vesicle shapes into the phase diagram: A comparison of experiment and theory. *Phys. Rev. E*, **55**, 4458–4474.

Döbereiner, H.-G., Selchow, O. and Lipowsky, R. (1999). Spontaneous curvature of fluid vesicles induced by trans-bilayer sugar asymmetry. *Eur. Biophys. J.*, **28**, 174–178.

Doering, C.R. and Gadoua, J.C. (1992). Resonant activation over a fluctuating barrier. *Phys. Rev. Lett.*, **69**, 2318–2321.

Dogterom, M. and Leibler, S. (1993). Physical aspects of the growth and regulation of microtubule structures. *Phys. Rev. Lett.*, **70**, 1347–1350.

Doi, M. and Edwards, S.F. (1986). *The Theory of Polymer Dynamics*. Oxford: Oxford University Press.

Dustin, P. (1978). *Microtubules*. Berlin: Springer.

Duplantier, B. (1990). Exact fractal area of two-dimensional vesicles. *Phys. Rev. Lett.*, **64**, 493.

Duwe, H.P., Kaes, J. and Sackmann, E. (1990). Bending elastic moduli of lipid bilayers: modulation by solutes. *J. Phys. France*, **51**, 945–962.

Egberts, E., and Berendsen, H.J.C. (1988). Molecular dynamics simulation of a smectic liquid crystal with atomic detail. *J. Chem. Phys.*, **89**, 3718–3732.

Elbaum, M., Fygenson, D.K. and Libchaber, A. (1996). Buckling microtubules in vesicles. *Phys. Rev. Lett.*, **76**, 4078–4081.

Elson, E.L. (1988). Cellular mechanics as an indicator of cytoskeletal structure and function. *Ann. Rev. Biophys. Chem.*, **17**, 397–430.

Elston, T.C. and Oster, G. (1997). Protein turbines I: The bacterial flagellar motor. *Biophys. J.*, **73**, 703–721.

Eppenga, R. and Frenkel, D. (1984). Monte Carlo study of the isotropic and nematic phases of infinitely thin hard platelets. *Molec. Phys.*, **52**, 1303–1334.

Euteneuer, U. and Schliwa, M. (1984). Persistent, directional motility of cells and cytoplasmic fragments in the absence of microtubules. *Nature*, **310**, 58–61.

Evans, E.A. (1973a). A new material concept for the red cell membrane. *Biophys. J.*, **13**, 926–940.

Evans, E.A. (1973b). New membrane concept applied to the analysis of fluid shear- and micropipette-deformed red blood cells. *Biophys. J.*, **13**, 941–954.

Evans, E.A. (1974). Bending resistance and chemically induced moments in membrane bilayers. *Biophys. J.*, **14**, 923–931.

Evans, E.A. (1983). Bending elastic modulus of red blood cell membrane derived from buckling instability in micropipet aspiration tests. *Biophys. J.*, **43**, 27–30.

Evans, E. (1993). New physical concepts for cell amoeboid motion. *Biophys. J.*, **64**, 1306–1322.

Evans, E., Klingenberg, D.J., Rawicz, W. and Szoka, F. (1996). Interactions between polymer-grafted membranes in concentrated solutions of free polymer. *Langmuir*, **12**, 3031–3037.

Evans, E. and Metcalfe, M. (1984). Free energy potential for aggregation of giant, neutral lipid bilayer vesicles by van der Waals attraction. *Biophys. J.*, **46**, 423–426.

Evans, E. and Needham, D. (1987). Physical properties of surfactant bilayer membranes: thermal transitions, elasticity, rigidity, cohesion and colloidal interactions. *J. Phys. Chem.*, **91**, 4219–4228.

Evans, E. and Needham, D. (1988). Attraction between lipid bilayer membranes in concentrated solutions of nonadsorbing polymers: comparison of mean-field theory with measurements of adhesion energy. *Macromolecules*, **21**, 1822–1831.

Evans, E.A. and Parsegian, V.A. (1986). Thermal-mechanical fluctuations enhance repulsion between bimolecular layers. *Proc. Natl. Acad. Sci. USA*, **83**, 7132–7136.

Evans, E. and Rawicz, W. (1990). Entropy-driven tension and bending elasticity in condensed-fluid membranes. *Phys. Rev. Lett.*, **64**, 2094–2097.

Evans, E. and Ritchie, K. (1997). Dynamic strength of molecular adhesion bonds. *Biophys. J.*, **72**, 1541–1555.

Evans, E. and Skalak, R. (1980). *Mechanics and Thermodynamics of Biomembranes*. Boca Raton, FL: CRC Press.

Evans, E.A. and Waugh, R. (1977a). Osmotic correction to elastic area compressibility measurements on red cell membrane. *Biophys. J.*, **20**, 307–313.

Evans, E.A. and Waugh, R. (1977b). Mechano-chemistry of closed vesicular membranes. *J. Coll. Interfac. Sci.*, **60**, 286–298.

Evans, E.A. and Waugh, R.E. (1980). Mechano-chemical study of red cell membrane structure *in situ*. In *Erythrocyte Mechanics and Blood Flow*. New York, NY: Liss.

Everaers, R. and Kremer, K. (1996). Topological interactions in model polymer networks. *Phys. Rev. E*, **53**, R37–R40.

Falcioni, M., Bowick, M.J., Guitter, E. and Thorleifsson, G. (1997). The Poisson ratio of crystalline surfaces. *Europhys. Lett.*, **38**, 67–72.

Falvo, M.R., Washburn, S., Superfine, R., Finch, M., Brooks, F.P. Jr., Chi, V. and Taylor, R.M. II (1997). Manipulation of individual viruses: friction and mechanical properties. *Biophys. J.*, **72**, 1396–1403.

Faucon, J.F., Mitov, M.D., Méléard, P., Bivas, I. and Bothorel, P. (1989). Bending elasticity and thermal fluctuations of lipid membranes. Theoretical analysis and experimental requirements. *J. Phys. France*, **50**, 2389–2414.

Felgner, H., Frank, R. and Schliwa, M. (1996). Flexural rigidity of microtubules measured with the use of optical tweezers. *J. Cell Sci.*, **109**, 509–516.

Feng, S., and Sen, P.N. (1984). Percolation on elastic networks: new exponent and thresholds. *Phys. Rev. Lett.*, **52**, 216–219.

Feng, S., Thorpe, M.F. and Garboczi, E. (1985). Effective-medium theory of percolation on central-force elastic networks. *Phys. Rev. B*, **31**, 276–280.

Ferris, J.P., Hill, A.R. Jr., Liu, R. and Orgel, L.E. (1996). Synthesis of long pre-biotic oligomers on mineral surfaces. *Nature*, **381**, 59–61.

Ferry, J.D. (1980). *Viscoelastic Properties of Polymers*, 3rd edn. New York, NY: Wiley.

Feynmann, R.P., Leighton, R.B. and Sands, M. (1964). *The Feynmann Lectures on Physics*. vol. II. Reading, MA: Addison-Wesley.

Finer, J.T., Simmons, R.M. and Spudich, J.A. (1994). Single myosin molecule mechanics: piconewton forces and nanometre steps. *Nature*, **368**, 113–119.

Florin, E.-L., Moy, V.T. and Gaub, H.E. (1994). Adhesion forces between individual ligand–receptor pairs. *Science*, **264**, 415–417.

Flory, P.J. (1953). *Principles of Polymer Chemistry*. Ithaca, NY: Cornell University Press.

Flory, P.J. (1969). *Statistical Mechanics of Chain Molecules*. New York, NY: Wiley.

Flory, P.J. (1976). Statistical thermodynamics of random networks. *Proc. R. Soc. Lond.*, **A351**, 351–380.

Flügge, W. (1973). *Stresses in Shells*, 2nd edn. New York, NY: Springer-Verlag.

Forscher, P. and Smith, S.J. (1988). Actions of cytochalasins on the organization of actin filaments and microtubules in a neuronal growth cone. *J. Cell Biol.*, **107**, 1505–1516.

Foty, R.A., Forgacs, G., Pfleger, C.M. and Steinberg, M.S. (1994). Liquid properties of embryonic tissues: measurement of interfacial tensions. *Phys. Rev. Lett.*, **72**, 2298–2301.

Fourcade, B., Miao, L., Rao, M., Wortis, M. and Zia, R.K.P. (1994). Scaling analysis of narrow necks in curvature models of fluid lipid-bilayer vesicles. *Phys. Rev. E*, **49**, 5276–5286.

Francis, N.R., Irikura, V.M., Yamaguchi, S., DeRosier, D.J. and Macnab, R.M. (1992). Localization of the *Salmonella typhimurium* flagella switch protein FliG to the cytoplasmic M-ring face of the basal body. *Proc. Natl. Acad. Sci. USA*, **89**, 6304–6308.

Fromhertz, P. (1983). Lipid-vesicle structure: size control by edge-active agents. *Chem. Phys. Lett.*, **94**, 259–266.

Fromhertz, P., Röcker, C. and Rüppel, D. (1986). From discoid micelles to spherical vesicles: The concept of edge activity. *Faraday Discuss. Chem. Soc.*, **81**, 39–48.

Fung, Y.C. (1994). *A First Course in Continuum Mechanics*. Englewood Cliffs, NJ: Prentice-Hall.

Furukawa, R., Kundra, R. and Fechheimer, M. (1993). Formation of liquid crystals from actin filaments. *Biochem.*, **32**, 12 346–12 352.

Gelbart, W.M., Bruinsma, R.F., Pincus, P.A. and Parsegian, V.A. (2000). DNA-inspired electrostatics. *Physics Today*, **53**(9), 38–44.

Gennis, R.B. (1989). *Biomembranes: Molecular Structure and Function*. New York, NY: Springer-Verlag.

Girouard, M. (1985). *Cities and People: A Social and Architectural History*, p. 68. New Haven, CT: Yale University Press.

Gittes, F. and MacKintosh, F.C. (1998). Dynamic shear modulus of a semiflexible polymer network. *Phys. Rev. E*, **58**, R1241–R1244.

Gittes, F. Mickey, B., Nettleton, J. and Howard, J. (1993). Flexural rigidity of microtubules and actin filaments measured from thermal fluctuations in shape. *J. Cell Biol.*, **120**, 923–934.

Glaus, U. (1988). Monte Carlo study of self-avoiding surfaces. *J. Stat. Phys.*, **50**, 1141–1166.

Goetz, A. (1970). *Introduction to Differential Geometry*. Reading, MA: Addison-Wesley.

Goetz, R. and Lipowsky, R. (1998). Computer simulations of bilayer membranes: Self-assembly and interfacial tension. *J. Chem. Phys.*, **108**, 7397–7407.

Goetz, R., Gompper, G. and Lipowsky, R. (1999). Mobility and elasticity of self-assembled membranes. *Phys. Rev. Lett.*, **82**, 221–224.

Goldberg, M.B. and Theriot, J.A. (1995). *Shigella flexneri* surface protein IcsA is sufficient to direct actin-based motility. *Proc. Natl. Acad. Sci. USA.*, **92**, 6572–6576.

Goldstein, R.E., Nelson, P., Powers, T. and Seifert, U. (1996). Front propagation in the pearling instability of tubular vesicles. *J. Phys. II*, **6**, 767–796.

Gompper, G. and Kroll, D.M. (1989). Steric interactions in multi-membrane systems: a Monte Carlo study. *Europhys. Lett.*, **9**, 59–64.

Gompper, G. and Kroll, D.M. (1991). Fluctuations of a polymerized membrane between walls. *J. Phys. France*, **1**, 1411–1432.

Gompper, G. and Kroll, D.M. (1995). Phase diagram and scaling behavior of fluid vesicles. *Phys. Rev. E*, **51**, 514–525.

Gompper, G. and Kroll, D.M. (1997). Freezing flexible vesicles. *Phys. Rev. Lett.*, **78**, 2859–2862.

Gompper, G. and Schick, M. (1994). Self-assembling amphiphilic systems. In *Phase Transitions and Critical Phenomena*, Vol. 16, eds. C. Domb and J.L. Lebowitz. London Academic Press.

Goodell, E.W. (1985). Recycling of murein by *Escherichia coli. J. Bacteriol.*, **163**, 305–310.

Goodsell, D.S. (1993). *The Machinery of Life*. New York, NY: Springer.

Goulian, M., Lei, N., Miller, J. and Sinha, S.K. (1992). Structure factor for randomly oriented self-affine membranes. *Phys. Rev. A*, **46**, R6170–R6173.

Gradshteyn, I.S. and Ryzhik, I.M. (1980). *Table of Integrals, Series and Products*, 4th edn. New York, NY: Academic Press.

Green, N.M. (1975). Avidin. *Adv. Prot. Chem.*, **29**, 85–133.

Grest, G.S. and Murat, M. (1993). Structure of grafted polymeric brushes in solvents of varying quality: a molecular dynamics study. *Macromolecules* **26**, 3108–3117.

Griffith, L.M. and Pollard, T.D. (1982). The interaction of actin filaments with microtubules and microtubule-associated proteins. *J. Biol. Chem.*, **257**, 9143–9151.

Grønbech-Jensen, N., Mashl, R.J., Bruinsma, R.F. and Gelbart, W.M. (1997). Counterion-induced attraction between rigid polyelectrolytes. *Phys. Rev. Lett.*, **78**, 2477–2480.

Gross, D.J. (1984). The size of random surfaces. *Phys. Lett.*, **138B**, 185–190.

Guitter, E., David, F., Leibler, S. and Peliti, L. (1989). Thermodynamical behavior of polymerized membranes. *J. Phys. France*, **50**, 1787–1819.

Guitter, E. and Palmeri, J. (1992). Tethered membranes with long-range interactions. *Phys. Rev. A*, **45**, 734–744.

Ha, B.-Y. and Liu, A.J. (1997). Counterion-mediated attraction between two like-charged rods. *Phys. Rev. Lett.*, **79**, 1289–1292.

Harbich, W., Servuss, R.M. and Helfrich, W. (1976). Optical studies of lecithin-membrane melting. *Phys. Lett.*, **57A**, 294–296.

Harley, R., James, D., Miller, A. and White, J.W. (1977). Phonons and the elastic moduli of collagen and muscle. *Nature*, **267**, 285–287.

Helfrich, W. (1973). Elastic properties of lipid bilayers: Theory and possible experiments. *Z. Naturforsch.*, **28c**, 693–703.

Helfrich, W. (1974a). The size of bilayer vesicles generated by sonication. *Phys. Lett.*, **50A**, 115–116.

Helfrich, W. (1974b). Blocked lipid exchanges in bilayers and its possible influence on the shape of vesicles. *Z. Naturforsch.*, **29c**, 510–515.

Helfrich, W. (1975). Out-of-plane fluctuations of lipid bilayers. *Z. Naturforsch.*, **30c**, 841–842.

Helfrich, W. (1978). Steric interaction of fluid membranes in multi-layer systems. *Z. Naturforsch.*, **33a**, 305–315.

Helfrich, W. (1986). Size distributions of vesicles: the role of the effective rigidity of membranes. *J. Phys. France*, **47**, 321–329.

Helfrich, W. and Servuss, R.-M. (1984). Undulations, steric interaction and cohesion of fluid membranes. *Nuovo Cimento*, **3D**, 137–151.

Heller, H., Schaefer, M. and Schulten, K. (1993). Molecular dynamics simulation of a bilayer of 200 lipids in the gel and in the liquid crystal phase. *J. Phys. Chem.*, **97**, 8343–8360.

Hénon, S., Lenormand, G., Richert, A. and Gallet, F. (1999). A new determination of the shear modulus of the human erythrocyte membrane using optical tweezers. *Biophys. J.*, **76**, 1145–1151.

Hesketh, J.E. and Prym, I.F., eds. (1996). *The Cytoskeleton.* Greenwich, CT: JAI Press.

Heuser, J. (1980). Three-dimensional visualization of coated vesicle formation in fibroblasts. *J. Cell Biol.*, **84**, 560–583.

Heuser, J.E. (1983). Procedure for freeze-drying molecules adsorbed to mica flakes. *J. Cell Biol.*, **169**, 155–195.

Hill, T.L. (1981). Microfilament or microtubule assembly or disassembly against a force. *Proc. Natl. Acad. Sci. USA*, **78**, 5613–5617.

Hinner, B., Tempel, M., Sackmann, E., Kroy, K. and Frey, E. (1998). Entanglement, elasticity, and viscous relaxation of actin solutions. *Phys. Rev. Lett.*, **81**, 2614–2617.

Hirokawa, N. (1982). Cross-linker system between neurofilaments, microtubules and membranous organelles in frog axons revealed by the quick-freeze, deep etching method. *J. Cell Biol.*, **94**, 129–142.

Hochmuth, R.M. and Needham, D. (1990). The viscoelasticity of neutrophils and their transit times through small pores. *Biorheol.*, **27**, 817–828.

Holley, M.C. and Ashmore, J.F. (1990). Spectrin, actin and the structure of the cortical lattice in mammalian cochlear outer hair cells. *J. Cell Sci.*, **96**, 283–291.

Höltje, J.-V. (1993). "Three-for-one" – a simple growth mechanism that guarantees a precise copy of the thin rod-shaped murein sacculus of *Escherichia coli*. In *Bacterial Growth and Lysis*, eds. M.A. de Pedro, V.-J. Höltje and W. Löffelhardt. New York, NY: Plenum.

Höltje, J.-V. (1998). Growth of stress-bearing and shape-maintaining murein sacculus of *Escherichia coli. Microbiol. Mol. Biol. Rev.*, **62**, 181–203.

Horio, T. and Hotani, H. (1986). Visualization of the dynamic instability of individual microtubules by dark-field microscopy. *Nature*, **321**, 605–607.

Hotani, H. and Horio, T. (1988). Dynamics of microtubules visualized by darkfield microscopy: treadmilling and dynamic instability. *Cell Motil. Cytoskeleton*, **10**, 229–236.

Howard, J. (1995). The mechanics of force generation by kinesin. *Biophys. J.*, **68**, 245s-255s.

Huang, C. and Mason, J.T. (1978). Geometrical packing constraints in egg phosphatidylcholine vesicles. *Proc. Natl. Acad. Sci. USA*, **75**, 308–310.

Hunt, A.J., Gittes, F. and Howard, J. (1994). The force extended by a single kinesin molecule against a viscous load. *Biophys. J.*, **67**, 766–781.

Hwang, W.C. and Waugh, R.E. (1997). Energy of dissociation of lipid bilayer from the membrane skeleton of red blood cells. *Biophys. J.*, **72**, 2669–2678.

Hynes, R.O. (1992). Integrins: versatility, modulation and signalling in cell adhesion. *Cell*, **69**, 11–25.

Ingber, D.E. (1997). Tensegrity: the architectural basis of cellular mechanotransduction. *Ann. Rev. Physiol.*, **59**, 575–599.

Isambert, H. and Maggs, A.C. (1996). Dynamics and rheology of actin solutions. *Macromolecules*, **29**, 1036–1040.

Isambert, H., Venier, P., Maggs, A.C., Fattoum, A., Kassab, R., Pantaloni, D. and Carlier, M.-F. (1995). Flexibility of actin filaments derived from thermal fluctuations: Effect of bound nucleotide, phalloidin, and muscle regulatory proteins. *J. Biol. Chem.*, **270**, 11 437–11 444.

Israelachvili, J.N. (1991). *Intermolecular and Surface Forces*, 2nd edn. London: Academic Press.

Israelachvili, J.N., Mitchell, D.J. and Ninham, B.W. (1976). Theory of self-assembly of hydrocarbon amphiphiles into micelles and bilayers. *J. Chem. Soc. Faraday Trans.*, II **72**, 1525–1568.

Iwazawa, J., Imae, Y. and Kobayasi, S. (1993). Study of the torque of the bacterial flagellar motor using a rotating electric field. *Biophys. J.*, **64**, 925–933.

Jakosky, B. (1998). *The Search for Life on Other Planets.* Cambridge: Cambridge University Press.

James, H.M. and Guth, E. (1943). *J. Chem. Phys.*, **11**, 470.

Janke, W. and Kleinert, H. (1987). Fluctuation pressure of a stack of membranes. *Phys. Rev. Lett.*, **58**, 144–147.

Janmey, P.A., Hvidt, S., Lamb, J. and Stossel, T.P. (1990). Resemblance of actin-binding protein/actin gels to covalently crosslinked networks. *Nature*, **345**, 89–92.

Janmey, P.A., Euteneuer, U., Traub, P. and Schliwa, M. (1991). Viscoelastic properties of vimentin compared with other filamentous biopolymer networks. *J. Cell Biol.*, **113**, 155–159.

Janmey, P.A., Hvidt, S., Käs, J., Lerche, D., Maggs, A., Sackmann, E., Schliwa, M. and Stossel, T.P. (1994). The mechanical properties of actin gels. Elastic modulus and filament motions. *J. Biol. Chem.*, **269**, 32 503–32 513.

Jeppesen, C. and Ipsen, J.H. (1993). Scaling properties of self-avoiding surfaces with free topology. *Europhys. Lett.*, **22**, 713–716.

Jülicher, F. (1996). The morphology of vesicles of higher topological genus: conformal degeneracy and conformal modes. *J. Phys. II France*, **6**, 1797–1824.

Kantor, Y. (1989). Properties of tethered surfaces. In *Statistical Mechanics of Membranes and Surfaces*, eds. D. Nelson, T. Piran and S. Weinberg, pp. 115–136. Singapore: World Scientific.

Kantor, Y., Kardar, M. and Nelson, D.R. (1986). Statistical mechanics of tethered surfaces. *Phys. Rev. Lett.*, **57**, 791–794.

Kantor, Y. and Nelson, D.R. (1987). Crumpling transition in polymerized membranes. *Phys. Rev. Lett.*, **58**, 2774–2777.

Käs, J. and Sackmann, E. (1991). Shape transitions and shape stability of giant phospholipid vesicles in pure water induced by area-to-volume changes. *Biophys. J.*, **60**, 825–844.

Käs, J., Strey, H., Tang, J.X., Finger, D., Ezzell, R., Sackmann, E. and Janmey, P.A. (1996). F-actin, a model polymer for semiflexible chains in dilute, semidilute and liquid crystalline solutions. *Biophys. J.*, **70**, 609–625.

Kauzmann, W. (1966). *Kinetic Theory of Gases*. New York, NY: Benjamin.

Khan, S. and Macnab, R.M. (1980). The steady-state counter-clockwise/clockwise ratio of bacterial flagellar motors is regulated by protonmotive force. *J. Mol. Biol.*, **138**, 563–597.

Kirkwood, J.G. and Auer, P.L. (1951). The visco-elastic properties of solutions of rod-like macromolecules. *J. Chem. Phys.*, **19**, 281–283.

Kittel, C. (1966). *Introduction to Solid State Physics*. 3rd edn. New York, NY: Wiley.

Koch, A.L. (1990). Growth and form of the bacterial cell wall. *Amer. Scientist*, **78**, 327–341.

Koch, A.L. (1993). Stresses on the surface stress theory. In *Bacterial Growth and Lysis*, eds. M.A. de Pedro, V.-J. Höltje and W. Löffelhardt. New York, NY: Plenum.

Koch, A.L. and Woeste, S. (1992). Elasticity of the sacculus of *Escherichia coli*. *J. Bacteriol.*, **174**, 4811–4819.

Koenig, B.W., Strey, H.H. and Gawirsch, K. (1997). Membrane lateral compressibility determined by NMR and X-ray diffraction: Effect of acyl chain unsaturation. *Biophys. J.*, **73**, 1954–1966.

Komura, S. and Baumgärtner, A. (1991). Tethered vesicles at constant pressure: Monte Carlo study and scaling analysis. *Phys. Rev. A*, **44**, 3511–3518.

Korn, E.D., Carlier, M.-F. and Pantaloni, D. (1987). Actin polymerization. *Science*, **238**, 638–644.

Kreyszig, E. (1959). *Differential Geometry*. Toronto, ON: University of Toronto Press.

Kroll, D.M. and Gompper, G. (1992). The conformation of fluid membranes: Monte Carlo simulations. *Science*, **255**, 968–971.

Kroy, K. and Frey, E. (1996). Force-extension relation and plateau modulus for worm-like chains. *Phys. Rev. Lett.*, **77**, 306–309.

Kubori, T., Shimamoto, N., Yamaguchi, S., Namba, K. and Aizawa, S.-I. (1992). Morphological pathway of flagellar assembly in *Salmonella typhimurium*. *J. Mol. Biol.*, **266**, 433–446.

Kuhn, W. and Grün, F. (1942). *Kolloid Z.*, **101**, 248.

Kumar, P.B.S., Gompper, G. and Lipowsky, R. (1999). Modulated phases in multicomponent fluid membranes. *Phys. Rev. E*, **60**, 4610–4618.

Kuo, S. and Sheetz, M.P. (1993). Force of single kinesin molecules measured by optical tweezers. *Science*, **260**, 232–234.

Kurz, J.C. and Williams, R.C. Jr. (1995). Microtubule-associated proteins and the flexibility of microtubules. *Biochemistry*, **34**, 13 374–13 380.

Kwok, R. and Evans, E. (1981). Thermoelasticity of large lecithin bilayer vesicles. *Biophys. J.*, **35**, 637–652.

Lai, P.-Y. and Binder, K. (1992). Structure and dynamics of polymer brushes near the Θ-point: a Monte Carlo simulation. *J. Chem. Phys.*, **97**, 586–595.

Laidler, K.J. (1987). *Chemical Kinetics*, 3rd edn. New York, NY: Harper and Row.

Lal, A.A., Korn, E.D. and Brenner, S.L. (1984). Rate constants for actin polymerization in ATP determined using cross-linked actin trimers as nuclei. *J. Biol. Chem.*, **259**, 8794–8800.

Lammert, P. and Discher, D.E. (1998). Tethered networks in two dimensions: A low temperature view. *Phys. Rev. E*, **57**, 4368–4374.

Landau, L.D. and Lifshitz, E.M. (1986). *Theory of Elasticity*. Oxford: Pergamon Press.

Landau, L.D. and Lifshitz, E.M. (1987). *Fluid Mechanics*, 2nd edn. Oxford: Pergamon Press.

Lange, Y., Hadesman, R.A. and Steck, T.L. (1982). Role of the reticulum in the stability and shape of the isolated human erythrocyte membrane. *J. Cell Biol.*, **92**, 714–721.

Lauffenburger, D.A. and Horwitz, A.F. (1996). Cell migration: a physically integrated molecular process. *Cell*, **84**, 359–369.

Lee, J., Ishihara, A., Theriot, J.A. and Jacobson, K. (1993). Principles of locomotion for simple-shaped cells. *Nature*, **362**, 167–171.

Le Doussal, P. and Radzihovsky, L. (1992). Self-consistent theory of polymerized membranes. *Phys. Rev. Lett.*, **69**, 1209–1212.

Lehninger, A.L., Nelson, D.L. and Cox, M.M. (1993). *Principles of Biochemistry*. New York, NY: Worth.

Leibler, S. (1989). Equilibrium statistical mechanics of fluctuating films and membranes. In *Statistical Mechanics of Membranes and Surfaces*, eds. D. Nelson, T. Piran and S. Weinberg, pp. 45–103. Singapore: World Scientific.

Leibler, S. and Maggs, A.C. (1989). Entropic interactions between polymerized membranes. *Phys. Rev. Lett.*, **63**, 406–409.

Leibler, S., Singh, R.R.P. and Fisher, M.E. (1987). Thermodynamic behavior of two-dimensional vesicles. *Phys. Rev. Lett.* **59**, 1989–1992.

LeNeveu, D.M., Rand, R.P. and Parsegian, V.A. (1975). Measurement of forces between lecithin bilayers. *Nature*, **259**, 601–603.

Levinson, E.A. (1991). Monte Carlo studies of crumpling for Sierpinski gaskets. *Phys. Rev. A*, **43**, 5233–5239.

Lewis, B.A. and Engelman, D.A. (1983). Lipid bilayer thickness varies linearly with acyl chain length in fluid phosphatidylcholine vesicles. *J. Mol. Biol.*, **166**, 211–217.

Lifshitz, E.M. (1956). Theory of molecular attractive forces between solids. *Soviet Physics – JETP*, **2**, 73–83.

Lin, C.-H. and Forscher, P. (1993). Cytoskeleton remodeling during growth cone-target interactions. *J. Cell Biol.*, **121**, 1369–1383.

Lipowsky, R. (1994). Generic interactions of flexible membranes. In *Structure and Dynamics of Membranes*, eds. R. Lipowsky and E. Sackmann. Amsterdam: Elsevier.

Lipowsky, R. (1996). Adhesion of membranes *via* anchored stickers. *Phys. Rev. Lett.*, **77**, 1652–1655.

Lipowsky, R., Döbereiner, H.-G., Hiergeist, C. and Indrani, V. (1998). Membrane curvature induced by polymers and colloids. *Physica*, **A249**, 536–543.

Lipowsky, R. and Girardet, M. (1990). Shape fluctuations of polymerized or solidlike membranes. *Phys. Rev. Lett.*, **65**, 2893–2896.

Lipowsky, R. and Leibler, S. (1986). Unbinding transitions of interacting membranes. *Phys. Rev. Lett.*, **56**, 2541–2544; **59**, 1983(E).

Lipowsky, R. and Seifert, U. (1991). Adhesion of membranes: a theoretical perspective. *Langmuir*, **7**, 1867–1873.

Lipowsky, R. and Zielinska, B. (1989). Binding and unbinding of lipid membranes: a Monte Carlo study. *Phys. Rev. Lett.*, **62**, 1572–1575.

Litster, J.D. (1975). Stability of lipid bilayers and red blood cell membranes. *Phys. Lett.*, **53A**, 193–194.

Liu, S.-C., Derick, L.H. and Palek, J. (1987). Visualization of the hexagonal lattice in the erythrocyte membrane skeleton. *J. Cell Biol.*, **104**, 527–536.

Liu, D. and Plischke, M. (1992). Monte Carlo studies of tethered membranes with attractive interactions. *Phys. Rev. A*, **45**, 7139–7144.

Lorrain, P., Corson, D.R. and Lorrain, F. (1988). *Electromagnetic Fields and Waves*, 3rd edn. New York, NY: Freeman.

MacKintosh, F.C., Käs, J. and Janmey, P.A. (1995). Elasticity of semiflexible biopolymer networks. *Phys. Rev. Lett.*, **75**, 4425–4428.

Maggs, A.C., Leibler, S., Fisher, M.E. and Camacho, C.J. (1990). The size of an inflated vesicle in two dimensions. *Phys. Rev. A*, **42**, 691–695.

Magnasco, M.O. (1993). Forced thermal ratchets. *Phys. Rev. Lett.*, **71**, 1477–1481.

Maniotis, A.J., Chen, C.S. and Ingber, D.E. (1997). Demonstration of mechanical connections between integrins, cytoskeletal filaments, and nucleoplasm that stabilize nuclear structure. *Proc. Natl. Acad. Sci.*, **94**, 849–854.

Manson, M.D., Tedesco, P.M. and Berg, H.C. (1980). Energetics of flagellar rotation in bacteria. *J. Mol. Biol.*, **138**, 541–561.

Marcelja, S. (1974). Chain ordering in liquid crystals II. Structure of bilayer membranes. *Biochim. Biophys. Acta*, **367**, 165–176.

Marko, J.F. and Siggia, E.D. (1995). Stretching DNA. *Macromolecules*, **28**, 8759–8770.

Marra, J. and Israelachvili, J.N. (1985). Direct measurement of forces between phosphatidylcholine and phosphatidylethanolamine bilayers in aqueous electrolyte solutions. *Biochemistry*, **24**, 4608–4618.

Marsh, D. (1990). *CRC Handbook of Lipid Bilayers*. Boca Raton, FL: CRC Press.

Marsh, D. (1996). Intrinsic curvature in normal and inverted lipid structures and in membranes. *Biophys. J.*, **70**, 2248–2255.

Marsh, D. (1997). Renormalization of the tension and area expansion modulus in fluid membranes. *Biophys. J.*, **73**, 865–869.

Maruyama, K., Kaibara, M. and Fukuda, E. (1974). Rheology of F-actin II: Effect of tropomyosin and troponin. *Biochim. Biophys. Acta*, **371**, 30–38.

Mashi, R.J. and Bruinsma, R. (1998). Spontaneous curvature theory of clathrin coated membranes. *Biophys. J.*, **74**, 2862–2875.

Maxwell, J.C. (1864). *Philos. Mag.*, **27**, 294.

McGough, A.M. and Josephs, R. (1990). On the structure of erythrocyte spectrin in partially expanded membrane skeletons. *Proc. Natl. Acad. Sci. USA*, **87**, 5208–5212.

McGrath, J.L., Tardy, Y., Dewey, C.F. Jr., Meister, J.J. and Hartwig, J.H. (1998). Simultaneous measurements of actin filament turnover, filament friction, and monomer diffusion in endothelial cells. *Biophys. J.*, **75**, 2070–2078.

McKay, D.S., Gibson, E.K. Jr., Thomas-Keprta, K.L., Vali, H., Romanek, C.S., Clemett, S.J., Chillier, X.D.F., Maechling, C.R. and Zare, R.N. (1996). Search for past life on Mars: possible relic biogenic activity in martian meteorite ALH84001. *Science*, **273**, 924–930.

McKenzie, D.S. (1976). Polymers and scaling. *Phys. Rep.*, **27**, 35–88.

McKeon, F.D., Kirschner, M.W. and Caput, D. (1986). Homologies in both primary and secondary structure between nuclear envelope and intermediate filament proteins. *Nature*, **319**, 463–468.

McKiernan, A.E., MacDonald, R.I., MacDonald, R.C. and Axelrod, D. (1997). Cytoskeleton protein binding kinetics at planar phospholipid membranes. *Biophys. J.*, **73**, 1987–1998.

Meier, T. and Fahrenholz, F. (1996). *A Laboratory Guide to Biotin-labelling in Biomolecule Analysis*. Basel: Birkhäuser Verlag.

Méléard, P., Gerbeaud, C., Pott, T., Fernandez-Puente, L., Bivas, I., Mitov, M.D., Dufourcq, J. and Bothorel, P. (1997). Bending elasticities of model membranes: influences of temperature and sterol content. *Biophys. J.*, **72**, 2616–2629.

Merkel, R., Nassoy, P., Leung, A., Ritchie, K. and Evans, E. (1999). Energy landscapes of receptor-ligand bonds explored with dynamic force spectroscopy. *Nature*, **397**, 50–53.

Miao, L., Fourcade, B., Rao, M., Wortis, M. and Zia, R.K.P. (1991). Equilibrium budding and vesiculation in the curvature model of fluid lipid vesicles. *Phys. Rev. A*, **43**, 6843–6856.

Miao, L., Seifert, U., Wortis, M. and Döbereiner, H.-G. (1994). Budding transitions of fluid-bilayer vesicles: the effect of area-difference elasticity. *Phys. Rev. E*, **49**, 5389–5407.

Michalet, X., Bensimon, D. and Fourcade, B. (1994). Fluctuating vesicles of non-spherical topology. *Phys. Rev. Lett.*, **72**, 168–171.

Miller, S.L. (1953). A production of amino acids under possible primitive Earth conditions. *Science*, **117**, 528–529.

Milner, S.T. (1991). Polymer brushes. *Science*, **251**, 905–914.

Milner, S.T. and Safran, S.A. (1987). Dynamical fluctuations of droplet microemulsions and vesicles. *Phys. Rev. A*, **36**, 4371–4379.

Mitaku, S., Ikegami, A. and Sakanishi, A. (1978). Ultrasonic studies of lipid bilayer. Phase transition in synthetic phosphatidylcholine liposomes. *Biophys. Chem.*, **8**, 295–304.

Mitchison, T.J., and Cramer, L.P. (1996). Actin-based cell motility and cell locomotion. *Cell*, **84**, 371–379.

Mitchison, T., and Kirschner, M. (1984). Dynamic instability of microtubule growth. *Nature*, **312**, 237–242.

Mitchison, T. and Salmon, E.D. (1992). Poleward kinetochore fiber movement occurs during both metaphase and anaphase-A in newt lung cell mitosis. *J. Cell Biol.*, **119**, 569–582.

Miyata, H. and Hotani, H. (1992). Morphological changes in liposomes caused by polymerization of encapsulated actin and spontaneous formation of actin bundles. *Proc. Natl. Acad. Sci. USA.*, **89**, 11 547–11 551.

Mohanty, J. and Ninham, B.W. (1976). *Dispersion Forces.* London: Academic Press.

Moligner, A. and Oster, G. (1996). Cell motility driven by actin polymerization. *Biophys. J.*, **71**, 3030–3045.

Moroz, D. and Nelson, P. (1997). Dynamically stabilized pores in bilayer membranes. *Biophys. J.*, **72**, 2211–2216.

Morse, D.C. (1998). Viscoelasticity of tightly entangled solutions of semiflexible polymers. *Phys. Rev. E*, **58**, R1237–R1240.

Mouritsen, O.G. (1984). *Computer Simulations of Phase Transitions and Critical Phenomena.* Berlin: Springer-Verlag.

Mutz, M. and Bensimon, D. (1991). Observation of toroidal vesicles. *Phys. Rev. A*, **43**, 4525–4527.

Mutz, M. and Helfrich, W. (1990). Bending rigidities of some biological model membranes as obtained from the Fourier analysis of contour sections. *J. Phys. France*, **51**, 991–1002.

Mutz, M. and Helfrich, W. (1989). Unbinding transition of a biological model membrane. *Phys. Rev. Lett.*, **62**, 2881–2884.

Nagle, J. and Tristram-Nagle, S. (2000). Structure of lipid bilayers. *Biochim. Biophys. Acta*, **1469**, 159–195.

Napper, D.H. (1983). *Polymeric Stabilization of Colloidal Dispersions.* London: Academic Press.

Needham, D. and Nunn, R.S. (1990). Elastic deformation and failure of lipid bilayer membranes containing cholesterol. *Biophys. J.*, **58**, 997–1009.

Nelson, D. (1989). Theory of the crumpling transition. In *Statistical Mechanics of Membranes and Surfaces*, eds. D. Nelson, T. Piran and S. Weinberg, pp. 137–155. Singapore: World Scientific.

Nelson, D.R. and Peliti, L. (1987). Fluctuations in membranes with crystalline and hexatic order. *J. Physique*, **48**, 1085–1092.

Nelson, P., Powers, T. and Seifert, U. (1995). Dynamical theory of the pearling instability in cylindrical vesicles. *Phys. Rev. Lett.*, **74**, 3384–3387.

Netz, R.R. and Lipowsky, R. (1993). Unbinding of symmetric and asymmetric stacks of membranes. *Phys. Rev. Lett.*, **71**, 3596–3599.

Netz, R.R. and Lipowsky, R. (1995). Stacks of fluid membranes under pressure and tension. *Europhys. Lett.*, **29**, 345–350.

Netz, R.R. and Schick, M. (1996). Pore formation and rupture in fluid bilayers. *Phys. Rev. E*, **53**, 3875–3885.

Nicklas, R.B. (1983). Measurements of the force produced by the mitotic spindle in anaphase. *J. Cell Biol.*, **97**, 542–548.

Nicklas, R.B. (1988). The forces that move chromosomes in mitosis. *Ann. Rev. Biophys. Biophys. Chem.*, **17**, 431–449.

Nielsen, M. (1999). Ph. D. thesis (McGill).

Nienhuis, B. (1982). Exact critical point and critical exponents of $O(n)$ models in two dimensions. *Phys. Rev. Lett.*, **49**, 1062–1065.

Ninham, B.W., Parsegian, V.A. and Weiss, G.H. (1970). On the macroscopic theory of temperature-dependent van der Waals forces. *J. Stat. Phys.*, **2**, 323–328.

Oakley, B.R. (1994). γ-Tubulin. In *Microtubules*, eds. J.S. Hyams and C.W. Lloyd, pp. 33–45. New York, NY: Wiley-Liss.

Oghalai, J.S., Patel, A.A., Nakagawa, T. and Brownell, W.F. (1998). Fluorescence-imaged microdeformation of the outer hair cell lateral wall. *J. Neuroscience*, **18**, 48–58.

Olbrich, K., Rawicz, W., Needham, D. and Evans, E. (2000). Water permeability and mechanical strength of polyunsaturated lipid bilayers. *Biophys. J.*, **79**, 321–327.

Onsager, L. (1949). The effects of shape on the interaction of colloidal particles. *Ann. N.Y. Acad. Sci. USA.*, **51**, 627–659.

Oosawa, F. and Asakura, S. (1975). *Thermodynamics of the Polymerization of Proteins.* New York, NY: Academic Press.

Osborn, M., Webster, R. and Weber, K. (1978). Individual microtubules viewed by immunofluorescence and electron microscopy in the same PtK2 cell. *J. Cell Biol.*, **77**, R27–R34.

Palmer, A., Mason, T.G., Xu, J., Kuo, S.C. and Wirtz, D. (1999). Diffusing wave spectroscopy microrheology of actin filament networks. *Biophys. J.*, **76**, 1063–1071.

Pantaloni, D., Hill, T.L., Carlier, M.F. and Korn, E.D. (1985). A model for actin polymerization and the kinetic effects of ATP hydrolysis. *Proc. Natl. Acad. Sci. USA.*, **82**, 7202–7211.

Parsegian, V.A. (1966). Theory of liquid-crystalline phase transitions in lipid + water systems. *Trans. Faraday Soc.*, **62**, 848–860.

Parsegian, V.A., Fuller, N. and Rand, R.P. (1979). Measured work of deformation and repulsion of lecithin bilayers. *Proc. Natl. Acad. Sci. USA.*, **76**, 2750–2754.

Pastor, R.W. (1994). Molecular dynamics and Monte Carlo simulations of lipid bilayers. *Curr. Opin. Struct. Biol.*, **4**, 486–492.

Peliti, L. and Leibler, S. (1985). Effects of thermal fluctuations on systems with small surface tension. *Phys. Rev. Lett.*, **54**, 1690–1693.

Perkins, T.T., Smith, D.E., Larson, R.G. and Chu, S. (1995). Stretching of a single tethered polymer in a uniform flow field. *Science*, **268**, 83–87.

Peskin, C.S., Odell, G.M. and Oster, G. (1993). Cellular motion and thermal fluctuations: the Brownian ratchet. *Biophys. J.*, **65**, 316–324.

Peterson, M.A., Strey, H. and Sackmann, E. (1992). Theoretical and phase contrast microscope eigenmode analysis of erythrocyte flicker: amplitudes. *J. Phys. II France*, **2**, 1273–1285.

Petrov, A.G. (1999). *The Lyotropic State of Matter: Molecular Physics and Living Matter Physics*. Amsterdam: Gordon and Breach.

Pink, D.A., Green, T.J. and Chapman, D. (1980). Raman scattering in bilayers of unsaturated phophatidylcholines. Experiment and theory. *Biochemistry*, **19**, 349–356.

Plischke, M. and Bergersen, B. (1994). *Equilibrium Statistical Physics*. 2nd edn. Singapore: World Scientific.

Plischke, M. and Boal, D.H. (1988). Absence of a crumpling transition in strongly self-avoiding tethered membranes. *Phys. Rev. A*, **38**, 4943–4945.

Plischke, M. and Fourcade, B. (1991). Monte Carlo simulation of bond-diluted tethered membranes, *Phys. Rev. A*, **43**, 2056–2058.

Plischke, M. and Joos, B. (1998). Entropic elasticity of diluted central force networks. *Phys. Rev. Lett.*, **80**, 4907–4910.

Podolski, J.L. and Steck, T.L. (1990). Length distribution of F-actin in *Dictyostelium discoideum*. *J. Biol. Chem.*, **265**, 1312–1318.

Pollard, T.D. (1986). Rate constants for the reactions of ATP- and ADP-actin with the ends of actin filaments. *J. Cell Biol.*, **103**, 2747–2754.

Powers, D.L. (1999). *Boundary Value Problems*, 4th edn. London: Academic Press.

Prescott, L.M., Harley, J.P. and Klein, D.A. (1996). *Microbiology*, 3rd edn. Dubuque, IA: Brown.

Prost, J., Chauwin, J.-F., Peliti, L. and Ajdari, A. (1994). Asymmetric pumping of particles. *Phys. Rev. Lett.*, **72**, 2652–2655.

Rädler, J., Feder, T.J., Strey, H.H. and Sackmann, E. (1995). Fluctuation analysis of tension-controlled undulation forces between giant vesicles and solid substrates. *Phys. Rev. E*, **51**, 4526–4536.

Rädler, J. and Sackmann, E. (1993). Imaging optical thickness and separation distances of phospholipid vesicles at solid surfaces. *J. Phys. II France*, **3**, 727–747.

Radmacher, M., Fritz, M., Kacher, C.M., Cleveland, J.P. and Hansma, P.K. (1996). Measuring the viscoelastic properties of human platelets with the atomic force microscope. *Biophys. J.*, **70**, 556–567.

Ragsdale, G.K., Phelps, J. and Luby-Phelps, K. (1997). Viscoelastic response of fibroblasts to tension transmitted through adherens junctions. *Biophys. J.*, **73**, 2798–2808.

Rand, R.P. and Parsegian, V.A. (1989). Hydration forces between phospholipid bilayers. *Biochim. Biophys. Acta*, **988**, 351–376.

Rawicz, W., Olbrich, K.C., McIntosh, T., Needham, D. and Evans, E. (2000). Effects of chain length and unsaturation on elasticity of lipid bilayers. *Biophys. J.*, **79**, 328–339.

Rayment, I., Holden, H.M., Whittaker, M., Yohn, S.B., Lorenz, M., Holmes, K.C. and Milligan, R.A. (1993b). Structure of the actin-myosin complex and its implications for muscle contraction. *Science*, **261**, 58–65.

Rayment, I., Rypniewski, W.R., Schmidt-Bäse, K., Smith, R., Tomchick, D.R., Benning, M.W., Winkelmann, D.A., Wesenberg, G. and Holden, H.M. (1993a). Three-dimensional structure of myosin subfragment-1: A molecular motor. *Science*, **261**, 50–58.

Reif, F. (1965). *Fundamentals of Statistical and Thermal Physics*. New York, NY: McGraw-Hill.

Rief, M., Pascual, J., Saraste, M. and Gaub, H.E. (1999). Single molecule force spectroscopy of spectrin repeats: Low unfolding forces in helix bundles. *J. Mol. Biol.*, **286**, 553–561.

Riveline, D., Wiggins, C.H., Goldstein, R.E. and Ott, A. (1997). Elastohydrodynamic study of actin filaments using fluorescence microscopy. *Phys. Rev. E*, **56**, R1330–R1333.

Rodionov, V.I., Nadezhdina, E. and Borisy, G.G. (1999). Centrosomal control of microtubule dynamics. *Proc. Natl. Acad. Sci. USA*, **96**, 115–120.

Ross, R.S. and Pincus, P. (1992). Bundles: end-grafted polymers in poor solvent. *Europhys. Lett.*, **19**, 79–84.

Sackmann, E. (1990). Molecular and globular structure and dynamics of membranes and lipid bilayers. *Can. J. Phys.*, **68**, 999–1012.

Saenger, W. (1984). *Principles of Nucleic Acid Structure*. New York, NY: Springer.

Safran, S.A. (1994). *Statistical Thermodynamics of Surfaces, Interfaces, and Membranes*. Reading, MA: Addison-Wesley.

Safran, S.A., Pincus, P. and Andelman, D. (1990). Theory of spontaneous vesicle formation in surfactant mixtures. *Science*, **248**, 354–356.

Sanger, J.M., Sanger, J.W. and Southwick, F.S. (1992). Host cell actin assembly is necessary and likely to provide the propulsive force for intracellular movement of *Listeria monocytogenes*. *Infect. Immun.*, **60**, 3609–3619.

Satcher, R.L. Jr. and Dewey, C.F. Jr. (1996). Theoretical estimates of mechanical properties of the endothelial cell cytoskeleton. *Biophys. J.*, **71**, 109–118.

Sato, M., Schwarz, W.H. and Pollard, T.D. (1987). Dependence of the mechanical properties of actin/α-actinin gels on deformation rate. *Nature*, **325**, 828–830.

Sawin, K.E. and Mitchison, T.J. (1991). Poleward microtubule flux in mitotic spindles assembled *in vitro*. *J. Cell Biol.*, **112**, 941–954.

Schmidt, C.F., Bärmann, M., Isenberg, G. and Sackmann, E. (1989). Chain dynamics, mesh size, and diffusive transport of polymerized actin. A quasielastic light scattering and microfluorescence study. *Macromolecules*, **22**, 3638–3649.

Schmidt, C.F., Svoboda, K., Lei, N., Petsche, I.B., Berman, L.E., Safinya, C.R. and Grest, G.S. (1993). Existence of a flat phase in red cell membrane skeletons. *Science*, **259**, 952–955.

Schneider, M.B., Jenkins, J.T. and Webb, W.W. (1984). Thermal fluctuations of large cylindrical phospholipid vesicles. *Biophys. J.*, **45**, 891–899.

Schnurr, B., Gittes, F., MacKintosh, F.C. and Schmidt, C.F. (1997). Determining microscopic viscoelasticity in flexible and semi-flexible polymer networks from thermal fluctuations. *Macromolecules*, **30**, 7781–7792.

Schopf, J.W. (1993). Microfossils of the early archean Apex chert: new evidence of the antiquity of life. *Science*, **260**, 640–646.

Schopf, J.W. (1999). *Cradle of Life*. Princeton, NJ: Princeton University Press.

Seifert, U. (1991). Vesicles of toroidal topology. *Phys. Rev. Lett.*, **66**, 2404–2407.

Seifert, U. (1993). Curvature-induced lateral phase segregation in two-component vesicles. *Phys. Rev. Lett.*, **70**, 1335–1338.

Seifert, U. (1995). Self-consistent theory of bound vesicles. *Phys. Rev. Lett.*, **74**, 5060–5063.

Seifert, U., Berndl, K. and Lipowsky, R. (1991). Shape transformations of vesicles: Phase diagram for spontaneous-curvature and bilayer-couple models. *Phys. Rev. A*, **44**, 1182–1202.

Seifert, U. and Lipowsky, R. (1990). Adhesion of vesicles. *Phys. Rev. A*, **42**, 4768–4771.

Selve, N. and Wegner, A. (1986). Rate of treadmilling of actin filaments *in vitro*. *J. Mol. Biol.*, **187**, 627–631.

Servuss, R.M., Harbich, W. and Helfrich, W. (1976). Measurement of the curvature-elastic modulus of egg lecithin bilayers. *Biochim. Biophys. Acta*, **436**, 900–903.

Servuss, R.M. and Helfrich, W. (1989). Mutual adhesion of lecithin membranes at ultralow tensions. *J. Phys. France.*, **50**, 809–827.

Shao, J.-Y. and Hochmuth, R.M. (1999). Mechanical anchoring strength of *L*-selectin, β_2 integrins and CD45 to neutrophil cytoskeleton and membrane. *Biophys. J.*, **77**, 587–596.

Sheetz, M.P. and Singer, S.J. (1974). Biological membranes as bilayer couples. A molecular mechanism of drug–erythrocyte interactions. *Proc. Natl. Acad. Sci. USA.*, **71**, 4457–4461.

Sheterline, P., Clayton, J. and Sparrow, J.C. (1998). *Actin*, 4th edn. Oxford: Oxford University Press.

Shillcock, J.C. and Boal, D.H. (1996). Entropy-driven instability and rupture of fluid membranes. *Biophys. J.*, **71**, 317–326.

Shillcock, J.C. and Seifert, U. (1998). Thermally induced proliferation of pores in a model fluid membrane. *Biophys. J.*, **74**, 1754–1766.

Shlyakhtenko, L.S., Gall, A.A., Weimer, J.J., Hawn, D.D. and Lyubchenko, Y.L. (1999). Atomic force microscopy imaging of DNA covalently immobilized on a functionalized mica substrate. *Biophys. J.*, **77**, 568–576.

Simson, R., Wallraff, E., Faix, J., Niewöhner, J., Gerisch, G. and Sackmann, E. (1998). Membrane bending modulus and adhesion energy of wild-type and mutant cells of *Dictyostelium* lacking talin or cortexillins. *Biophys. J.*, **74**, 514–522.

Sit, P.S., Spector, A.A., Lue, A.J.C., Popel, A.S. and Brownell, W.E. (1997). Micropipette aspiration on the outer hair cell wall. *Biophys. J.*, **72**, 2812–2819.

Sleep, J., Wilson, D., Simmons, R. and Gratzer, W. (1999). Elasticity of the red cell membrane and its relation to hemolytic disorders: an optical tweezers study. *Biophys. J.*, **77**, 3085–3095.

Small, J.V., Herzog, M. and Anderson, K. (1995). Actin filament organization in the fish keratocyte lamellipodium. *J. Cell Biol.*, **129**, 1275–1286.

Soga, K.G., Guo, H. and Zuckermann, M. (1995). Polymer brushes in a poor solvent. *Europhys. Lett.*, **29**, 531–536.

Southam, G., Firtel, M., Blackford, B.L., Jericho, M.H., Xu, W., Mulhern, P.J. and Beveridge, T.J. (1993). Transmission electron microscopy, scanning tunneling microscopy and atomic force microscopy of the cell envelope layers of the archeabacterium *Methanospirillum hungatei* GP1. *J. Bacteriol.*, **175**, 1946–1955.

Stauffer, D. and Aharony, A. (1992). *Introduction to Percolation Theory*, 2nd edn. London: Taylor and Francis.

Steck, T.L. (1989). Red cell shape. In *Cell Shape: Determinants, Regulation and Regulatory Role*, eds. W.D. Stein and F. Bronner, pp. 205–246. New York, NY: Academic Press.

Stigter, D. and Bustamante, C. (1998). Theory for the hydrodynamic and electrophoretic stretch of tethered B-DNA. *Biophys. J.*, **75**, 1197–1210.

Stokke, T. and Brant, D.A. (1990). The reliability of wormlike polysaccharide chain dimensions estimated from electron micrographs. *Biopolymers*, **30**, 1161–1181.

Stokke, B.T., Mikkelsen, A. and Elgsaeter, A. (1985a). Human erythrocyte spectrin dimer intrinsic viscosity: temperature dependence and implications for the molecular basis of the membrane free energy. *Biochim. Biophys. Acta*, **816**, 102–110.

Stokke, B.T., Mikkelsen, A. and Elgsaeter, A. (1985b). Some viscoelastic properties of human erythrocyte spectrin networks end-linked *in vitro*. *Biochim. Biophys. Acta*, **816**, 111–121.

Strick, T.R., Allemand, J.-F., Bensimon, D., Bensimon, A. and Croquette, V. (1996). The elasticity of a single supercoiled DNA molecule. *Science*, **271**, 1835–1837.

Suzuki, A., Maeda, T. and Ito, T. (1991). Formation of liquid crystalline phase of actin filament solutions and its dependence on filament length as studied by optical birefringence. *Biophys. J.*, **59**, 25–30.

Svetina, S., Brumen, M. and Zeks, B. (1985). *Stud. Biophys.*, **110**, 177.

Svetina, S. and Zeks, B. (1985). Bilayer couple as a possible mechanism of biological shape formation. *Biomed. Biochim. Acta*, **44**, 979–986.

Svoboda, K. and Block, S.M. (1994). Force and velocity measured for single kinesin molecules. *Cell*, **77**, 773–784.

Svoboda, K., Mitra, P.P. and Block, S.M. (1994). Fluctuation analysis of motor protein movement and single enzyme kinetics. *Proc. Natl. Acad. Sci. USA.*, **91**, 11 782–11 786.

Svoboda, K., Schmidt, C.F., Branton, D. and Block, S.M. (1992). Conformation and elasticity of the isolated red blood cell membrane skeleton. *Biophys. J.*, **63**, 784–793.

Svoboda, K., Schmidt, C.F., Schnapp, B.J. and Block, S.M. (1993). Direct observation of kinesin stepping by optical trapping interferometry. *Nature*, **365**, 721–727.

Swift, D.G., Posner, R.G. and Hammer, D.A. (1998). Kinetics of adhesion of IgE-sensitized rat basophilic leukemia cells to surface-immobilized antigen in Couette flow. *Biophys. J.*, **75**, 2597–2611.

Takeuchi, M., Miyamoto, H., Sako, Y., Komizu, H. and Kusumi, A. (1998). Structure of the erythrocyte membrane skeleton as observed by atomic force microscopy. *Biophys. J.*, **74**, 2171–2183.

Tanford, C. (1980). *The Hydrophobic Effect*, 2nd edn. New York, NY: Wiley.

Tang, J.X., Janmey, P.A., Stossel, T.P. and Ito, T. (1999). Thiol oxidation of actin produces dimers that enhance the elasticity of the F-actin network. *Biophys. J.*, **76**, 2208–2215.

Taupin, C., Dvolaitzky, M. and Sauterey, C. (1975). Osmotic pressure induced pores in phospholipid vesicles. *Biochemistry*, **14**, 4771–4775.

Taylor, W.H. and Hagerman, P.J. (1990). Application of the method of phage T4 DNA ligase-catalyzed ring-closure to the study of DNA structure II. NaCl-dependence of DNA flexibility and helical repeat. *J. Mol. Biol.*, **212**, 363–376.

Tempel, M., Isenberg, G. and Sackmann, E. (1996). Temperature-induced sol-gel transition and microgel formation in α-actinin cross-linked actin networks: A rheological study. *Phys. Rev. E*, **54**, 1802–1810.

Theriot, J.A. and Mitchison, T.J. (1991). Actin microfilament dynamics in locomoting cells. *Nature*, **352**, 126–131.

Theriot, J.A., Mitchison, T.J., Tilney, L.G. and Portnoy, D.A. (1992). The rate of actin-based motility of intracellular *Listeria monocytogenes* equals the rate of actin polymerization. *Nature*, **357**, 257–260.

Thorpe, M.F. (1986). Elastic properties of network glasses. *Ann. N.Y. Acad. Sci.*, **484**, 206–213.

Thwaites, J.J. (1993). Growth and control of the cell wall: a mechanical model for *Bacillus subtilis*. In *Bacterial Growth and Lysis*, eds. M.A. de Pedro, V.-J. Höltje and W. Löffelhardt. New York, NY: Plenum.

Thwaites, J.J. and Surana, U.C. (1991). Mechanical properties of *Bacillus subtilis* cell walls: effects of removing residual culture medium. *J. Bacteriol.*, **173**, 197–203.

Tolomeo, J.A., Steele, C.R. and Holley, M.C. (1996). Mechanical properties of the lateral cortex of mammalian auditory outer hair cells. *Biophys. J.*, **71**, 421–429.

Tran-Son-Tay, R. (1993). In *Physical Forces and the Mammalian Cell*, ed. J.A. Frangos. San Diego, CA: Academic Press.

Treloar, R.L.G. (1975). *The Physics of Rubber Elasticity*. Oxford: Oxford University Press.

Ungewickell, E. and Branton, D. (1981). Assembly units of clathrin coats. *Nature*, **289**, 420–422.

Ursitti, J.A. and Wade, J.B. (1993). Ultrastructure and immunocytochemistry of the isolated human erythrocyte membrane skeleton. *Cell Motil. Cytoskeleton*, **25**, 30–42.

van der Ploeg, P. and Berendsen, H.J.C. (1983). Molecular dynamics of a bilayer membrane. *Mol. Phys.*, **49**, 233–248.

Van de Ven, T.G.M. (1989). *Colloidal Dynamics*. London: Academic Press.

Vance, D.E. and Vance, J.E. (1996). *Biochemistry of Lipids, Lipoproteins and Membranes*. Amsterdam: Elsevier.

Veerman, J.A.C. and Frenkel, D. (1992). Phase behavior of disklike hard-core mesogens. *Phys. Rev. A*, **45**, 5632–5648.

Veigel, C., Bartoo, M.L., White, D.C.S., Sparrow, J.C. and Molley, J.E. (1998). The stiffness of rabbit skeletal actomyosin cross-bridges determined with an optical tweezers transducer. *Biophys. J.*, **75**, 1424–1438.

Venier, P., Maggs, A.C., Carlier, M.-F. and Pantaloni, D. (1994). Analysis of microtubule rigidity using hydrodynamic flow and thermal fluctuations. *J. Biol. Chem.*, **269**, 13 353–13 360.

Verde, F., Dogterom, M., Stelzer, E., Karsenti, E. and Leibler, S. (1992). Control of microtubule dynamics and length by cyclin A- and cyclin B-dependent kinases in *Xenopus* egg extracts. *J. Cell Biol.*, **118**, 1097–1108.

Vertogen, G. and de Jeu, W.H. (1988). *Thermotropic Liquid Crystals, Fundamentals*. Berlin: Springer-Verlag.

Verway, E.J.W. and Overbeek, J.Th.G. (1948). *Theory of Stability of Lyophobic Colloids*. Amsterdam: Elsevier.

Verwer, R.W.H., Nanninga, N., Keck, W. and Schwarz, U. (1978). Arrangement of glycan chains in the sacculus of *E. coli*. *J. Bacteriol.*, **136**, 723–729.

Vogel, S. (1998). *Cat's Paws and Catapults*. New York, NY: Norton.

Walker, R.A., O'Brien, E.T., Pryer, N.K., Soboeiro, M.F., Voter, W.A. and Erickson, H.P. (1988). Dynamic instability of individual microtubules analysed by video light microscopy: rate constants and transition frequencies. *J. Cell Biol.*, **107**, 1437–1448.

Waugh, R.E. and Agre, P. (1988). Reductions of erythrocyte membrane viscoelastic coefficients reflect spectrin deficiencies in hereditary spherocytosis. *J. Clin. Invest.*, **81**, 133–141.

Waugh, R.E. and Bauserman, R.G. (1995). Physical measurements of bilayer-skeletal separation forces. *Ann. Biomed. Eng.*, **23**, 308–321.

Waugh, R. and Evans, E.A. (1976). Viscoelastic properties of erythrocyte membranes of different vertebrate animals. *Microvasc. Res.*, **12**, 291–304.

Waugh, R. and Evans, E.A. (1979). Thermoelasticity of red blood cell membrane. *Biophys. J.*, **26**, 115–132.

Waugh, R.E. Song, J., Svetina, S. and Zeks, B. (1992). Local and nonlocal curvature elasticity in bilayer membranes by tether formation from lecithin vesicles. *Biophys. J.*, **61**, 974–982.

Weast, R.E., ed. (1970). *Handbook of Chemistry and Physics*, 50th edn. Boca Raton, FL: CRC Press.

Weber, G. (1975). Energetics of ligand binding to proteins. *Adv. Prot. Chem.*, **29**, 1–83.

Weber, P.C., Wendolski, J.J., Pantoliano, M.W. and Salemme, F.R. (1992). Crystallographic and thermodynamic comparison of natural and synthetic ligands bound to streptavidin. *J. Am. Chem. Soc.*, **114**, 3197–3200.

Wegner, A. (1976). Head to tail polymerization of actin. *J. Mol. Biol.*, **108**, 139–150.

Wegner, A. (1982). Treadmilling of actin at physiological salt concentrations. *J. Mol. Biol.*, **161**, 607–615.

Weikl, T.R., Netz, R.R. and Lipowsky, R. (2000). Unbinding transitions and phase separation of multicomponent membranes. *Phys. Rev. E*, **62**, R45–R48.

Weston, R.E. Jr. and Schwarz, H.A. (1972). *Chemical Kinetics*. Englewood Cliffs, NJ: Prentice-Hall.

Wientjes, F.B., Woldringh, C.L. and Nanninga, N. (1991). Amount of peptidoglycan in cell walls of gram-negative bacteria. *J. Bacteriol.*, **173**, 7684–7691.

Wilhelm, C., Winterhalter, M., Zimmermann, U. and Benz, R. (1993). Kinetics of pore size during irreversible electrical breakdown of lipid bilayer membranes. *Biophys. J.*, **64**, 121–128.

Wintz, W., Döbereiner, H.-G. and Seifert, U. (1996). Starfish vesicles. *Europhys. Lett.*, **33**, 403–408.

Wintz, W., Everaers, R. and Seifert, U. (1997). Mesh collapse in two-dimensional elastic networks under compression. *J. Phys. I France*, **7**, 1097–1111.

Wolfe, J., Dowgert, M.F. and Steponkus, P.L. (1985). Dynamics of membrane exchange of the plasma membrane and the lysis of isolated protoplasts during rapid expansion in area. *J. Membrane Biol.*, **86**, 127–138.

Wong, J.Y., Kuhl, T.L., Israelachvili, J.N., Mullah, N. and Zapilsky, S. (1997). Direct measurement of a tethered ligand-receptor interaction potential. *Science*, **275**, 820–822.

Xu, W., Mulhern, P.J., Blackford, B.L., Jericho, M.H., Firtel, M. and Beveridge, T.J. (1996). Modeling and measuring the elastic properties of an archeal surface, the sheath of *Methanospirillum hungatei*, and the implication for methane production. *J. Bacteriol.*, **178**, 3106–3112.

Xu, J., Wirtz, D. and Pollard, T.D. (1998a). Dynamic cross-linking by α-actinin determines the mechanical properties of actin filament networks. *J. Biol. Chem.*, **273**, 9570–9576.

Xu, J., Schwarz, W.H., Käs, J.A., Stossel, T.P., Janmey, P.A. and Pollard, T.D. (1998b). Mechanical properties of actin filament networks depend on preparation, polymerization conditions and storage of actin monomers. *Biophys. J.*, **74**, 2731–2740.

Yao, X., Jericho, M., Pink, D. and Beveridge, T. (1999). Thickness and elasticity of Gram-negative murein sacculi measured by atomic force microscopy. *J. Bacteriol.*, **181**, 6865–6875.

Yeung, A. and Evans, E. (1995). Unexpected dynamics in shape fluctuations of bilayer vesicles. *J. Phys. II France*, **5**, 1501–1523.

Yeung, C., Balazs, A.C. and Jasnow, D. (1993). Lateral instabilities in a grafted layer in a poor solvent. *Macromolecules*, **26**, 1914–1921.

Young, T. (1805). *Phil. Trans. R. Soc. Lond.*, **95**, 65.

Zhang, Z., Davis, H.T. and Kroll, D.M. (1996). Molecular dynamics simulations of tethered membranes with periodic boundary conditions. *Phys. Rev. E*, **53**, 1422–1429.

Zhelev, D.V. (1998). Material property characteristics for lipid bilayers containing lysolipid. *Biophys. J.*, **75**, 321–330.

Zhelev, D.V. and Needham, D. (1993). Tension-stabilized pores in giant vesicles: determination of pore size and pore line tension. *Biochim. Biophys. Acta*, **1147**, 89–104.

Zhou, Z. and Joós, B. (1997). Mechanisms of membrane rupture: From cracks to pores. *Phys. Rev. B*, **56**, 2997–3009.

Ziemann, F., Rädler, J. and Sackmann, E. (1994). Local measurements of viscoelastic moduli of entangled actin networks using an oscillating bead micro-rheometer. *Biophys. J.*, **66**, 2210–2216.

Zilker, A., Ziegler, M. and Sackmann, E. (1992). Spectral analysis of erythrocyte flickering in the 0.3–4 μm^{-1} regime by microinterferometry combined with fast image processing. *Phys. Rev. A*, **46**, 7998–8001.

Zipperle, G.F. Jr., Ezzell, J.W. Jr. and Doyle, R.J. (1984). Glucosamine substitution and muramidase susceptibility in *Bacillus anthracis*. *Can. J. Microbiol.*, **30**, 553–559.

Zuckermann, D. and Bruinsma, R. (1995). Statistical mechanics of membrane adhesion by reversible molecular bonds. *Phys. Rev. Lett.*, **74**, 3900–3903.

Index